RENEWALS 458-4574
DATE DUE

SIGNAL PROCESSING FOR REMOTE SENSING

SIGNAL PROCESSING FOR REMOTE SENSING

EDITED BY
C. H. CHEN

CRC Press
Taylor & Francis Group
Boca Raton London New York

CRC Press is an imprint of the
Taylor & Francis Group, an **informa** business

The material was previously published in *Signal and Image Processing for Remote Sensing* © Taylor and Francis 2006.

Library
University of Texas
at San Antonio

CRC Press
Taylor & Francis Group
6000 Broken Sound Parkway NW, Suite 300
Boca Raton, FL 33487-2742

© 2008 by Taylor & Francis Group, LLC
CRC Press is an imprint of Taylor & Francis Group, an Informa business

No claim to original U.S. Government works
Printed in the United States of America on acid-free paper
10 9 8 7 6 5 4 3 2 1

International Standard Book Number-13: 978-1-4200-6666-1 (Hardcover)

This book contains information obtained from authentic and highly regarded sources. Reprinted material is quoted with permission, and sources are indicated. A wide variety of references are listed. Reasonable efforts have been made to publish reliable data and information, but the author and the publisher cannot assume responsibility for the validity of all materials or for the consequences of their use.

Except as permitted under U.S. Copyright Law, no part of this book may be reprinted, reproduced, transmitted, or utilized in any form by any electronic, mechanical, or other means, now known or hereafter invented, including photocopying, microfilming, and recording, or in any information storage or retrieval system, without written permission from the publishers.

For permission to photocopy or use material electronically from this work, please access www.copyright.com (http://www.copyright.com/) or contact the Copyright Clearance Center, Inc. (CCC) 222 Rosewood Drive, Danvers, MA 01923, 978-750-8400. CCC is a not-for-profit organization that provides licenses and registration for a variety of users. For organizations that have been granted a photocopy license by the CCC, a separate system of payment has been arranged.

Trademark Notice: Product or corporate names may be trademarks or registered trademarks, and are used only for identification and explanation without intent to infringe.

Library of Congress Cataloging-in-Publication Data

Signal processing for remote sensing / [edited by] C.H. Chen.
    p. cm.
  Includes bibliographical references and index.
  ISBN-13: 978-1-4200-6666-1
  ISBN-10: 1-4200-6666-8
  1. Remote sensing--Data processing. 2. Signal processing. I. Chen, C. H. (Chi-hau), 1937- II. Title.

G70.4.S536 2008
621.36'7801154--dc22   2007030189

Visit the Taylor & Francis Web site at
http://www.taylorandfrancis.com

and the CRC Press Web site at
http://www.crcpress.com

Preface

Signal processing has been playing an increasingly important role in remote sensing, though most remote sensing literatures are concerned with remote sensing images. Many data received by remote sensors such as microwave and geophysical sensors, are signals or waveforms, which can be processed by analog and digital signal processing techniques.

This volume is a spin-off edition derived from *Signal and Image Processing for Remote Sensing*. It focuses on signal processing for remote sensing, and presents for the first time a comprehensive and up-to-date treatment of the subject. The progress in signal processing itself has been enormous in the last 30 years, but signal processing application in remote sensing has received more attention only in recent years. This volume covers important signal processing topics like principal component analysis, projected principal component analysis, Kalman adaptive filtering, prediction error filtering for interpolation, factor analysis, time series analysis, neural network classification, neural network parameter retrieval, blind source separation algorithm, independent component analysis, etc. The book presents for the first time the use of Huang–Hilbert transform in remote sensing data. As there are so many areas in remote sensing that can benefit from signal processing, we hope the book can help to attract more talents in signal processing to work on remote sensing problems that may involve environmental monitoring, resource management and planning, as well as energy exploration, and many others with the use of remotely sensed data.

Original Preface from *Signal and Image Processing for Remote Sensing*
Both signal processing and image processing have been playing increasingly important roles in remote sensing. While most data from satellites are in image forms and thus image processing has been used most often, signal processing can contribute significantly in extracting information from the remotely sensed waveforms or time series data. In contrast to other books in this field which deal almost exclusively with the image processing for remote sensing, this book provides a good balance between the roles of signal processing and image processing in remote sensing. The book covers mainly methodologies of signal processing and image processing in remote sensing. Emphasis is thus placed on the mathematical techniques which we believe will be less changed as compared to sensor, software and hardware technologies. Furthermore, the term "remote sensing" is not limited to the problems with data from satellite sensors. Other sensors which acquire data remotely are also considered. Thus another unique feature of the book is the coverage of a broader scope of the remote sensing information processing problems than any other book in the area.

The book is divided into two parts [now published as separate volumes under the following titles]. Part I [comprising the present volume], *Signal Processing for Remote Sensing*, has 12 chapters and Part II, *Image Processing for Remote Sensing*, has 16 chapters. The chapters are written by leaders in the field. We are very fortunate, for example, to have Dr. Norden Huang, inventor of the Huang–Hilbert transform, along with Dr. Steven Long, to write a chapter on the application of the transform to remote sensing problem, and Dr. Enders A. Robinson, who has made many major contributions to geophysical signal processing for over half a century, to write a chapter on the basic problem of constructing seismic images by ray tracing.

In Part I, following Chapter 1 by Drs. Long and Huang, and my short Chapter 2 on the roles of statistical pattern recognition and statistical signal processing in remote sensing, we start from a very low end of the electromagnetic spectrum. Chapter 3 considers the classification of infrasound at a frequency range of 0.001 Hz to 10 Hz by using a parallel bank neural network classifier and a 11-step feature selection process. The >90% correct classification rate is impressive for this kind of remote sensing data. Chapter 4 through Chapter 6 deal with seismic signal processing. Chapter 4 provides excellent physical insights on the steps for construction of digital seismic images. Even though the seismic image is an image, this chapter is placed in Part I as seismic signals start as waveforms. Chapter 5 considers the singular value decomposition of a matrix data set from scalar-sensors arrays, which is followed by independent component analysis (ICA) step to relax the unjustified orthogonality constraint for the propagation vectors by imposing a stronger constraint of fourth-order independence of the estimated waves. With an initial focus of the use of ICA in seismic data and inspired by Dr. Robinson's lecture on seismic deconvolution at the 4th International Symposium, 2002, on Computer Aided Seismic Analysis and Discrimination, Mr. Zhenhai Wang has examined approaches beyond ICA for improving seismic images. Chapter 6 is an effort to show that factor analysis, as an alternative to stacking, can play a useful role in removing some unwanted components in the data and thereby enhancing the subsurface structure as shown in the seismic images. Chapter 7 on Kalman filtering for improving detection of landmines using electromagnetic signals, which experience severe interference, is another remote sensing problem of higher interest in recent years. Chapter 8 is a representative time series analysis problem on using meteorological and remote sensing indices to monitor vegetation moisture dynamics. Chapter 9 actually deals with the image data for digital elevation model but is placed in Part I mainly because the prediction error (PE) filter is originated from the geophysical signal processing. The PE filter allows us to interpolate the missing parts of an image. The only chapter that deals with the sonar data is Chapter 10, which shows that a simple blind source separation algorithm based on the second-order statistics can be very effective to remove reverberations in active sonar data. Chapter 11 and Chapter 12 are excellent examples of using neural networks for retrieval of physical parameters from the remote sensing data. Chapter 12 further provides a link between signal and image processing as the principal component analysis and image sharpening tools employed are exactly what are needed in Part II.

With a focus on image processing of remote sensing images, Part II begins with Chapter 13 [Chapter 1 of the companion volume] that is concerned with the physics and mathematical algorithms for determining the ocean surface parameters from synthetic aperture radar (SAR) images. Mathematically Markov random field (MRF) is one of the most useful models for the rich contextual information in an image. Chapter 14 [now Chapter 2] provides a comprehensive treatment of MRF-based remote sensing image classification. Besides an overview of previous work, the chapter describes the methodological issues involved and presents results of the application of the technique to the classification of real (both single-date and multitemporal) remote sensing images. Although there are many studies on using an ensemble of classifiers to improve the overall classification performance, the random forest machine learning method for classification of hyperspectral and multisource data as presented in Chapter 15 [now Chapter 3] is an excellent example of using new statistical approaches for improved classification with the remote sensing data. Chapter 16 [now Chapter 4] presents another machine learning method, AdaBoost, to obtain robustness property in the classifier. The chapter further considers the relations among the contextual classifier, MRF-based methods, and spatial boosting. The following two chapters are concerned with different aspects of the change detection problem. Change detection is a uniquely important problem in remote sensing as the images acquired at different times over the same geographical area can be used in the areas of environmental monitoring, damage management, and so on. After discussing change detection methods for multitemporal SAR images, Chapter 17 [now Chapter 5] examines an adaptive scale–driven technique for change detection in medium resolution SAR data. Chapter 18 [now Chapter 6] evaluates the Wiener filter-based method, Mahalanobis distance, and subspace projection methods of change detection, with the change detection performance illustrated by receiver operating characteristics (ROC) curves. In recent years, ICA and related approaches have presented many new potentials in remote sensing information processing. A challenging task underlying many hyperspectral imagery applications is decomposing a mixed pixel into a collection of reflectance spectra, called endmember signatures, and the corresponding abundance fractions. Chapter 19 [now Chapter 7] presents a new method for unsupervised endmember extraction called vertex component analysis (VCA). The VCA algorithms presented have better or comparable performance as compared to two other techniques but require less computational complexity. Other useful ICA applications in remote sensing include feature extraction, and speckle reduction of SAR images. Chapter 20 [now Chapter 8] presents two different methods of SAR image speckle reduction using ICA, both making use of the FastICA algorithm. In two-dimensional time series modeling, Chapter 21 [now Chapter 9] makes use of a fractionally integrated autoregressive moving average (FARIMA) analysis to model the mean radial power spectral density of the sea SAR imagery. Long-range dependence models are used in addition to the fractional sea surface models for the simulation of the sea SAR image spectra at different sea states, with and without oil slicks at low computational cost.

Returning to the image classification problem, Chapter 22 [now Chapter 10] deals with the topics of pixel classification using Bayes classifier, region segmentation guided by morphology and split-and-merge algorithm, region feature extraction, and region classification.

Chapter 23 [now Chapter 11] provides a tutorial presentation of different issues of data fusion for remote sensing applications. Data fusion can improve classification and for the decision level fusion strategies, four multisensor classifiers are presented. Beyond the currently popular transform techniques, Chapter 24 [now Chapter 12] demonstrates that Hermite transform can be very useful for noise reduction and image fusion in remote sensing. The Hermite transform is an image representation model that mimics some of the important properties of human visual perception, namely local orientation analysis and the Gaussian derivative model of early vision. Chapter 25 [now Chapter 13] is another chapter that demonstrates the importance of image fusion to improving sea ice classification performance, using backpropagation trained neural network and linear discrimination analysis and texture features. Chapter 26 [now Chapter 14] is on the issue of accuracy assessment for which the Bradley–Terry model is adopted. Chapter 27 [now Chapter 15] is on land map classification using support vector machine, which has been increasingly popular as an effective classifier. The land map classification classifies the surface of the Earth into categories such as water area, forests, factories or cities. Finally, with lossless data compression in mind, Chapter 28 [now Chapter 16] focuses on information-theoretic measure of the quality of multi-band remotely sensed digital images. The procedure relies on the estimation of parameters of the noise model. Results on image sequences acquired by AVIRIS and ASTER imaging sensors offer an estimation of the information contents of each spectral band.

With rapid technological advances in both sensor and processing technologies, a book of this nature can only capture certain amount of current progress and results. However, if past experience offers any indication, the numerous mathematical techniques presented will give this volume a long lasting value.

The sister volumes of this book are the other two books edited by myself. One is *Information Processing for Remote Sensing* and the other is *Frontiers of Remote Sensing Information Processing*, both published by World Scientific in 1999 and 2003, respectively. I am grateful to all contributors of this volume for their important contribution and, in particular, to Dr. J.S. Lee, S. Serpico, L. Bruzzone and S. Omatu for chapter contributions to all three volumes. Readers are advised to go over all three volumes for a more complete information on signal and image processing for remote sensing.

C. H. Chen

Editor

Chi Hau Chen was born on December 22nd, 1937. He received his Ph.D. in electrical engineering from Purdue University in 1965, M.S.E.E. degree from the University of Tennessee, Knoxville, in 1962, and B.S.E.E. degree from the National Taiwan University in 1959.

He is currently chancellor professor of electrical and computer engineering at the University of Massachusetts, Dartmouth, where he has taught since 1968. His research areas are in statistical pattern recognition and signal/image processing with applications to remote sensing, geophysical, underwater acoustics, and nondestructive testing problems, as well as computer vision for video surveillance, time series analysis, and neural networks.

Dr. Chen has published 25 books in his area of research. He is the editor of *Digital Waveform Processing and Recognition* (CRC Press, 1982) and *Signal Processing Handbook* (Marcel Dekker, 1988). He is the chief editor of *Handbook of Pattern Recognition and Computer Vision*, volumes 1, 2, and 3 (World Scientific Publishing, 1993, 1999, and 2005, respectively). He is the editor of *Fuzzy Logic and Neural Network Handbook* (McGraw-Hill, 1966). In the area of remote sensing, he is the editor of *Information Processing for Remote Sensing* and *Frontiers of Remote Sensing Information Processing* (World Scientific Publishing, 1999 and 2003, respectively).

He served as the associate editor of the *IEEE Transactions on Acoustics Speech and Signal Processing* for 4 years, *IEEE Transactions on Geoscience and Remote Sensing* for 15 years, and since 1986 he has been the associate editor of the *International Journal of Pattern Recognition and Artificial Intelligence*.

Dr. Chen has been a fellow of the Institutue of Electrical and Electronic Engineers (IEEE) since 1988, a life fellow of the IEEE since 2003, and a fellow of the International Association of Pattern Recognition (IAPR) since 1996.

Contributors

J. van Aardt Group of Geomatics Engineering, Department of Biosystems, Katholieke Universiteit Leuven, Leuven, Belgium

Ranjan Acharyya Florida Institute of Technology, Melbourne, Florida

Nicolas Le Bihan Laboratory of Images and Signals, CNRS UMR 5083, INP of Grenoble, France

William J. Blackwell Lincoln Laboratory, Massachusetts Institute of Technology, Lexington, Massachusetts

Chi Hau Chen Department of Electrical and Computer Engineering, University of Massachusetts Dartmouth, North Dartmouth, Massachusetts

Frederick W. Chen Lincoln Laboratory, Massachusetts Institute of Technology, Lexington, Massachusetts

Leslie M. Collins Department of Electrical and Computer Engineering, Duke University, Durham, North Carolina

Fengyu Cong State Key Laboratory of Vibration, Shock & Noise, Shanghai Jiaotong University, Shanghai, China

P. Coppin Group of Geomatics Engineering, Department of Biosystems, Katholieke Universiteit Leuven, Leuven, Belgium

Fredric M. Ham Florida Institute of Technology, Melbourne, Florida

Norden E. Huang NASA Goddard Space Flight Center, Greenbelt, Maryland

Shaoling Ji State Key Laboratory of Vibration, Shock & Noise, Shanghai Jiaotong University, Shanghai, China

Peng Jia State Key Laboratory of Vibration, Shock & Noise, Shanghai Jiaotong University, Shanghai, China

I. Jonckheere Group of Geomatics Engineering, Department of Biosystems, Katholieke Universiteit Leuven, Leuven, Belgium

P. Jönsson Teachers Education, Malnö University, Sweden

S. Lhermitte Group of Geomatics Engineering, Department of Biosystems, Katholieke Universiteit Leuven, Leuven, Belgium

Steven R. Long NASA Goddard Space Flight Center, Wallops Island, Virginia

Jérôme I. Mars Laboratory of Images and Signals, CNRS UMR 5083, INP of Grenoble, France

Enders A. Robinson Columbia University, Newsburyport, Massachusetts

Xizhi Shi State Key Laboratory of Vibration, Shock & Noise, Shanghai Jiaotong University, Shanghai, China

Yingyi Tan Applied Research Associates; Inc., Raleigh, North Carolina

Stacy L. Tantum Department of Electrical and Computer Engineering, Duke University, Durham, North Carolina

J. Verbesselt Group of Geomatics Engineering, Department of Biosystems, Katholieke Universiteit Leuven, Leuven, Belgium

Valeriu Vrabie CReSTIC, University of Reims, Reims, France

Zhenhai Wang Department of Electrical and Computer Engineering, University of Massachusetts Dartmouth, North Dartmouth, Massachusetts

Sang-Ho Yun Department of Geophysics, Stanford University, Stanford, California

Howard Zebker Department of Geophysics, Stanford University, Stanford, California

Contents

1. **On the Normalized Hilbert Transform and Its Applications in Remote Sensing** 1
   Steven R. Long and Norden E. Huang

2. **Statistical Pattern Recognition and Signal Processing in Remote Sensing** 23
   Chi Hau Chen

3. **A Universal Neural Network–Based Infrasound Event Classifier** 31
   Fredric M. Ham and Ranjan Acharyya

4. **Construction of Seismic Images by Ray Tracing** 53
   Enders A. Robinson

5. **Multi-Dimensional Seismic Data Decomposition by Higher Order SVD and Unimodal ICA** 73
   Nicolas Le Bihan, Valeriu Vrabie, and Jérôme I. Mars

6. **Application of Factor Analysis in Seismic Profiling** 101
   Zhenhai Wang and Chi Hau Chen

7. **Kalman Filtering for Weak Signal Detection in Remote Sensing** 127
   Stacy L. Tantum, Yingyi Tan, and Leslie M. Collins

8. **Relating Time-Series of Meteorological and Remote Sensing Indices to Monitor Vegetation Moisture Dynamics** 151
   J. Verbesselt, P. Jönsson, S. Lhermitte, I. Jonckheere, J. van Aardt, and P. Coppin

9. **Use of a Prediction-Error Filter in Merging High- and Low-Resolution Images** 171
   Sang-Ho Yun and Howard Zebker

10. **Blind Separation of Convolutive Mixtures for Canceling Active Sonar Reverberation** 187
    Fengyu Cong, Chi Hau Chen, Shaoling Ji, Peng Jia, and Xizhi Shi

11. **Neural Network Retrievals of Atmospheric Temperature and Moisture Profiles from High-Resolution Infrared and Microwave Sounding Data** 203
    William J. Blackwell

12. **Satellite-Based Precipitation Retrieval Using Neural Networks, Principal Component Analysis, and Image Sharpening** 231
    Frederick W. Chen

Index 263

1

On the Normalized Hilbert Transform and Its Applications in Remote Sensing

Steven R. Long and Norden E. Huang

CONTENTS
1.1 Introduction 1
1.2 Review of Processing Advances 2
    1.2.1 The Normalized Empirical Mode Decomposition 2
    1.2.2 Amplitude and Frequency Representations 4
    1.2.3 Instantaneous Frequency 7
1.3 Application to Image Analysis in Remote Sensing 10
    1.3.1 The IR Digital Camera and Setup 11
    1.3.2 Experimental IR Images of Surface Processes 11
    1.3.3 Volume Computations and Isosurfaces 16
1.4 Conclusion 19
Acknowledgment 20
References 20

1.1 Introduction

The development of this new approach was motivated by the need to describe nonlinear distorted waves in detail, along with the variations of these signals that occur naturally in nonstationary processes (e.g., ocean waves). As has been often noted, natural physical processes are mostly nonlinear and nonstationary. Yet, there have historically been very few options in the available analysis methods to examine data from such nonlinear and nonstationary processes. The available methods have usually been for either linear but nonstationary, or nonlinear but stationary, and statistically deterministic processes. The need to examine data from nonlinear, nonstationary, and stochastic processes in the natural world is due to the nonlinear processes which require special treatment. The past approach of imposing a linear structure (by assumptions) on the nonlinear system is not adequate. Other than periodicity, the detailed dynamics in the processes from the data also need to be determined. This is needed because one of the typical characteristics of nonlinear processes is its intrawave frequency modulation (FM), which indicates the instantaneous frequency (IF) changes within one oscillation cycle.

In the past, when the analysis was dependent on linear Fourier analysis, there was no means of depicting the frequency changes within one wavelength (the intrawave frequency variation) except by resorting to the concept of harmonics. The term "bound harmonics" was often used in this connection. Thus, the distortions of any nonlinear waveform have often been referred to as "harmonic distortions." The concept of harmonic distortion is a mathematical artifact resulting from imposing a linear structure (through assumptions) on a nonlinear system. The harmonic distortions may thus have mathematical meaning, but there is no physical meaning associated with them, as discussed by Huang et al. [1,2]. For example, in the case of water waves, such harmonic components do not have any of the real physical characteristics of a water wave as it occurs in nature. The physically meaningful way to describe such data should be in terms of its IF, which will reveal the intrawave FMs occurring naturally.

It is reasonable to suggest that any such complicated data should consist of numerous superimposed modes. Therefore, to define only one IF value for any given time is not meaningful (see Ref. [3], for comments on the Wigner–Ville distribution). To fully consider the effects of multicomponent data, a decomposition method should be used to separate the naturally combined components completely and nearly orthogonally. In the case of nonlinear data, the orthogonality condition would need to be relaxed, as discussed by Huang et al. [1]. Initially, Huang et al. [1] proposed the empirical mode decomposition (EMD) approach to produce intrinsic mode functions (IMF), which are both monocomponent and symmetric. This was an important step toward making the application truly practical. With the EMD satisfactorily determined, an important roadblock to truly nonlinear and nonstationary analysis was finally removed. However, the difficulties resulting from the limitations stated by the Bedrosian [4] and Nuttall [5] theorems must also be addressed in connection with this approach. Both limitations have firm theoretical foundations and must be considered; IMFs satisfy only the necessary condition, but not the sufficient condition. To improve the performance of the processing as proposed by Huang et al. [1], the normalized empirical mode decomposition (NEMD) method was developed as a further improvement on the earlier processing methods.

1.2 Review of Processing Advances

1.2.1 The Normalized Empirical Mode Decomposition

The NEMD method was developed to satisfy the specific limitations set by the Bedrosian theorem while also providing a sharper measure of the local error when the quadrature differs from the Hilbert transform (HT) result.

From an example data set of a natural process, all the local maxima of the data are first determined. These local maxima are then connected with a cubic spline curve, which gives the local amplitude of the data, $A(t)$, as shown together in Figure 1.1. The envelope obtained through spline fitting is used to normalize the data by

$$y(t) = \frac{a(t)\cos\theta(t)}{A(t)} = \left(\frac{a(t)}{A(t)}\right)\cos\theta(t). \tag{1.1}$$

Here $A(t)$ represents the cubic spline fit of all the maxima from the example data, and thus $a(t)/A(t)$ should normalize $y(t)$ with all maxima then normalized to unity, as shown in Figure 1.2. As is apparent from Figure 1.2, a small number of the normalized data points can still have an amplitude in excess of unity. This is because the cubic spline is through the maxima only, so that at locations where the amplitudes are changing rapidly, the line representing the envelope spline can pass under some of the data points. These occasional

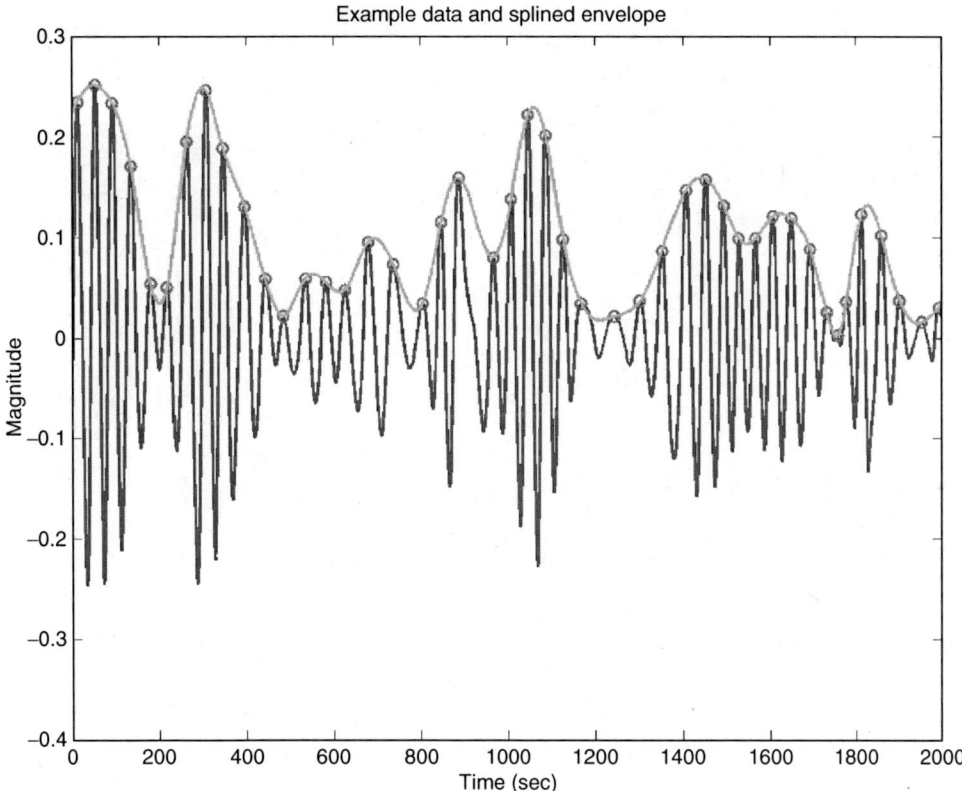

FIGURE 1.1
The best possible cubic spline fit to the local maxima of the example data. The spline fit forms an envelope as an important first step in the process. Note also how the frequency can change within a wavelength, and that the oscillations can occur in groups.

misses are unavoidable, yet the normalization scheme has effectively separated the amplitude from the carrier oscillation. The IF can then be computed from this normalized carrier function $y(t)$, just obtained. Owing to the nearly uniform amplitude, the limitations set by the Bedrosian theorem are effectively satisfied. The IF computed in this way from the normalized data from Figure 1.2 is shown in Figure 1.3, together with the original example data. With the Bedrosian theorem addressed, what of the limitations set by the Nuttall theorem?

If the HT can be considered to be the quadrature, then the absolute value of the HT performed on the perfectly normalized example data should be unity. Then any deviation from the absolute value of the HT from unity would be an indication of a difference between the quadrature and the HT results. An error index can thus be defined simply as

$$E(t) = [\text{abs}(\text{Hilbert transform } (y(t))) - 1]^2. \tag{1.2}$$

This error index would be not only an energy measure as given in the Nuttall theorem but also a function of time as shown in Figure 1.4. Therefore, it gives a local measure of the error resulting from the IF computation. This local measure of error is both logically and practically superior to the integrated error bound established by the Nuttall theorem. If the quadrature and the HT results are identical, then it follows that the error should be

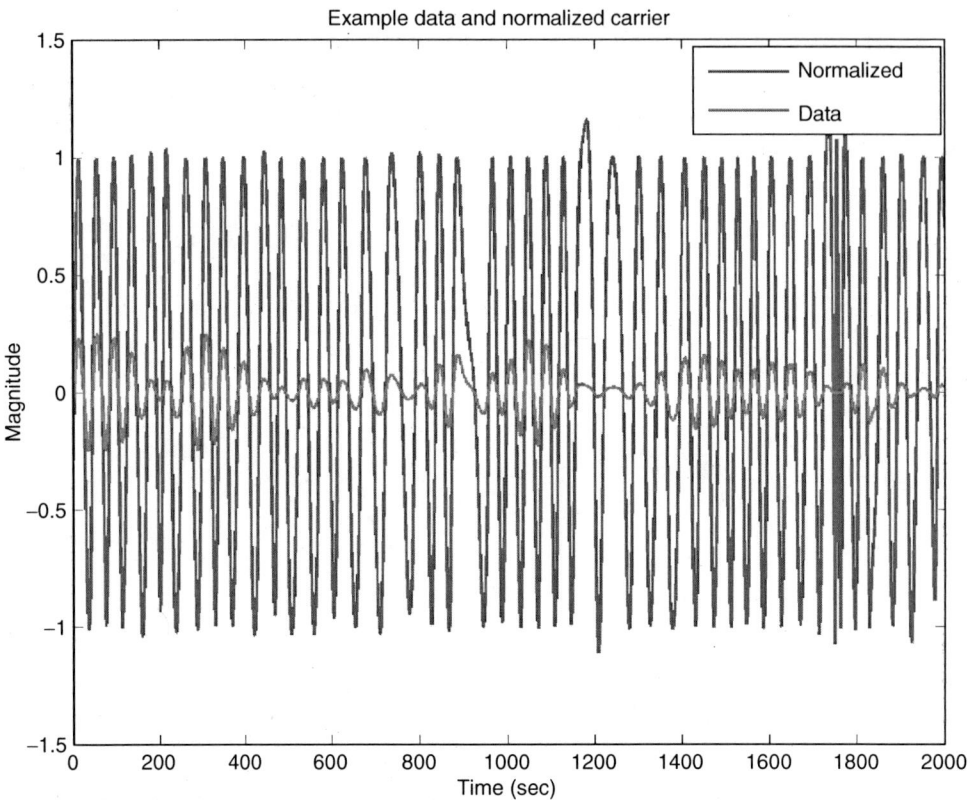

FIGURE 1.2
Normalized example data of Figure 1.1 with the cubic spline envelope. The occasional value beyond unity is due to the spline fit slightly missing the maxima at those locations.

zero. Based on experience with various natural data sets, the majority of the errors encountered here result from two sources. The first source is due to an imperfect normalization occurring at locations close to rapidly changing amplitudes, where the envelope spline-fitting is unable to turn sharply or quickly enough to cover all the data points. This type of error is even more pronounced when the amplitude is also locally small, thus amplifying any errors. The error index from this condition can be extremely large. The second source is due to nonlinear waveform distortions, which will cause corresponding variations of the phase function $\theta(t)$. As discussed by Huang et al. [1], when the phase function is not an elementary function, the differentiation of the phase determined by the HT is not identical to that determined by the quadrature. The error index from this condition is usually small (see Ref. [6]).

Overall, the NEMD method gives a more consistent, stable IF. The occasionally large error index values offer an indication where the method failed simply because the spline misses and cuts through the data momentarily. All such locations occur at the minimum amplitude with a resulting negligible energy density.

1.2.2 Amplitude and Frequency Representations

In the initial methods [1,2,6], the main result of Hilbert spectral analysis (HSA) always emphasized the FM. In the original methods, the data were first decomposed into IMFs, as

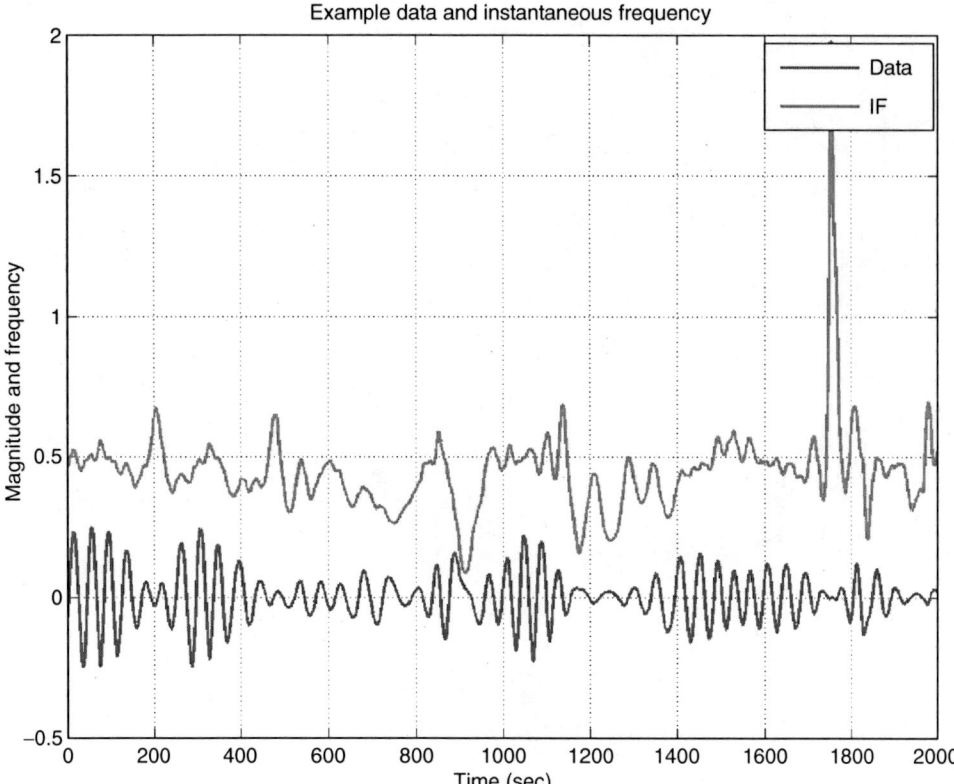

FIGURE 1.3
The instantaneous frequency determined from the normalized carrier function is shown with the example data. Data is about zero, and the instantaneous frequency varies about the horizontal 0.5 value.

defined in the initial work. Then, through the HT, the IF and the amplitude of each IMF were computed to form the Hilbert spectrum. This continues to be the method, especially when the data are normalized. The information on the amplitude or envelope variation is not examined. In the NEMD and HSA approach, it is justifiable not to pay too much attention to the amplitude variations. This is because if there is a mode mixing, the amplitude variation from such mixed mode IMFs does not reveal any true underlying physical processes. However, there are cases when the envelope variation does contain critical information. An example of this is when there is no mode mixing in any given IMF, when a beating signal representing the sum of two coexisting sinusoidal ones is encountered. In an earlier paper, Huang et al. [1] attempted to extract individual components out of the sum of two linear trigonometric functions such as

$$x(t) = \cos at + \cos bt. \tag{1.3}$$

Two seemingly separate components were recovered after over 3000 sifting steps. Yet the obtained IMFs were not purely trigonometric functions anymore, and there were obvious aliases in the resulting IMF components as well as in the residue. The approach proposed then was unnecessary and unsatisfactory. The problem, in fact, has a much simpler solution: treating the envelope as an amplitude modulation (AM), and then processing just the envelope data. The function $x(t)$, as given in Equation 1.3, can then be rewritten as

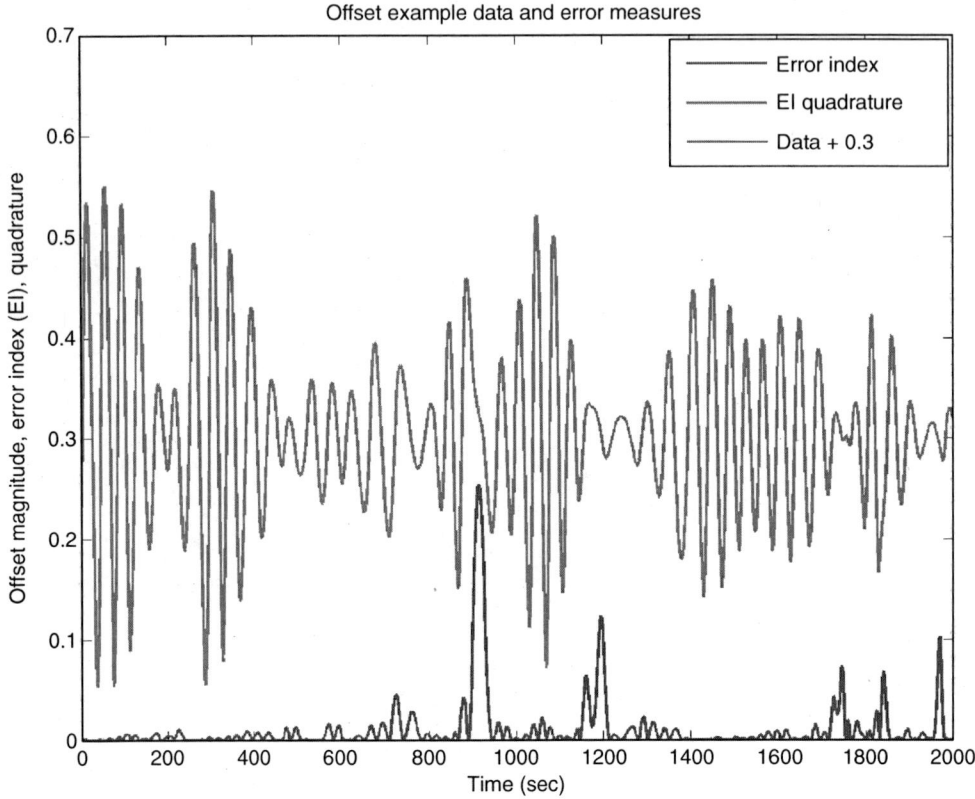

FIGURE 1.4
The error index as it changes with the data location in time. The original example data offset by 0.3 vertically for clarity is also shown. The quadrature result is not visible on this scale.

$$x(t) = \cos at + \cos bt = 2\cos\left(\frac{a+b}{2}t\right)\cos\left(\frac{a-b}{2}t\right). \quad (1.4)$$

There is no difference between the sum of the individual components and the modulating envelope form; they are trigonometric identities. If both the frequency of the carrier wave, $(a+b)/2$, and the frequency of the envelope, $(a-b)/2$, can be obtained, then all the information in the signal can be extracted. This indicates the reason to look for a new approach to extracting additional information from the envelope. In this example, however, the envelope becomes a rectified cosine wave. The frequency would be easier to determine from the simple period counting than from the Hilbert spectral result. For a more general case when the amplitudes of the two sinusoidal functions are not equal, the modulation is not simple anymore. For even more complicated cases, when there are more than two coexisting sinusoidal components with different amplitudes and frequencies, there is no general expression for the envelope and carrier. The final result could be represented as more than one frequency-modulated band in the Hilbert spectrum. It is then impossible to describe the individual components under this situation. In such cases, representing the signal as a carrier and envelope, variation should still be meaningful, for the dual representations of frequency arise from the different definitions of frequency. The Hilbert-inspired view of amplitude and FMs still renders a correct representation of the signal, but this view is very different from that of Fourier analysis. In such cases, if one is sure of the stationarity and regularity of the signal, Fourier analysis could be used, which will give more familiar results as suggested by Huang et al. [1]. The judgment for these cases is not on which one is correct, as both are correct; rather, it is on which one is more familiar and more revealing.

When more complicated data are present, such as in the case of radar returns, tsunami wave records, earthquake data, speech signals, and so on (representing a frequency "chirp"), the amplitude variation information can be found by processing the envelope and treating the data as an approximate carrier. When the envelope of frequency chirp data, such as the example given in Figure 1.5, is decomposed through the NEMD process, the IMF components are obtained as shown in Figure 1.6. Using these components (or IMFs), the Hilbert spectrum can be constructed as given in Figure 1.7, together with its FM counterpart. The physical meaning of the AM spectrum is not as clearly defined in this case. However, it serves to illustrate the AM contribution to the variability of the local frequency.

1.2.3 Instantaneous Frequency

It must be emphasized that IF is a very different concept from the frequency content of the data derived from Fourier-based methods, as discussed in great detail by Huang

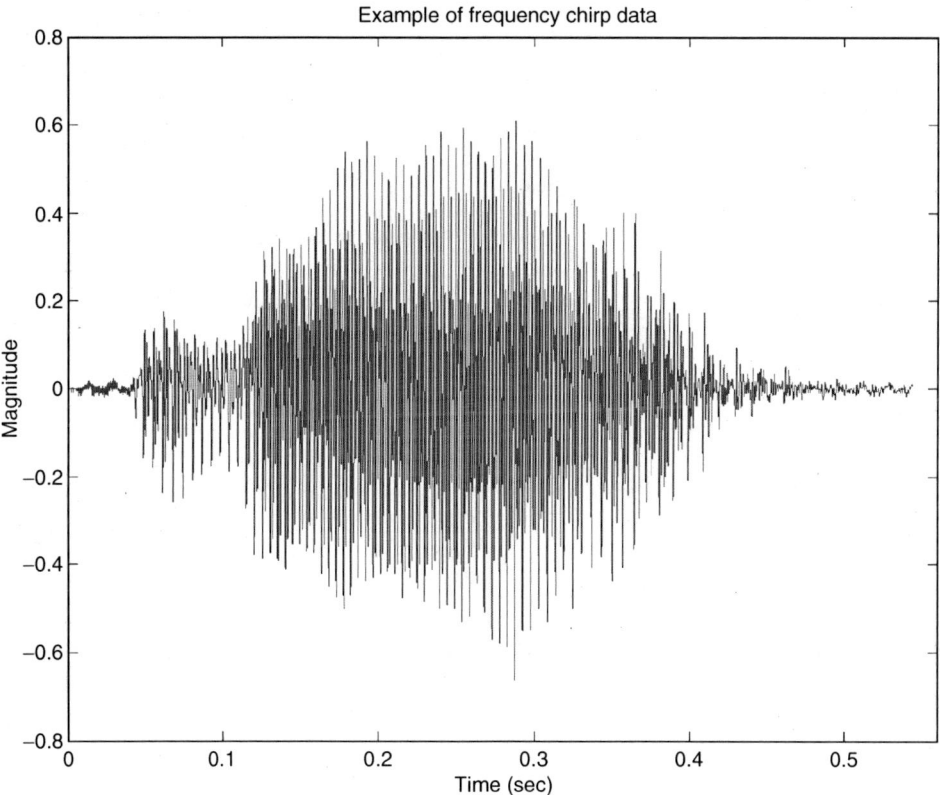

FIGURE 1.5
A typical example of complex natural data, illustrating the concept of frequency "chirps."

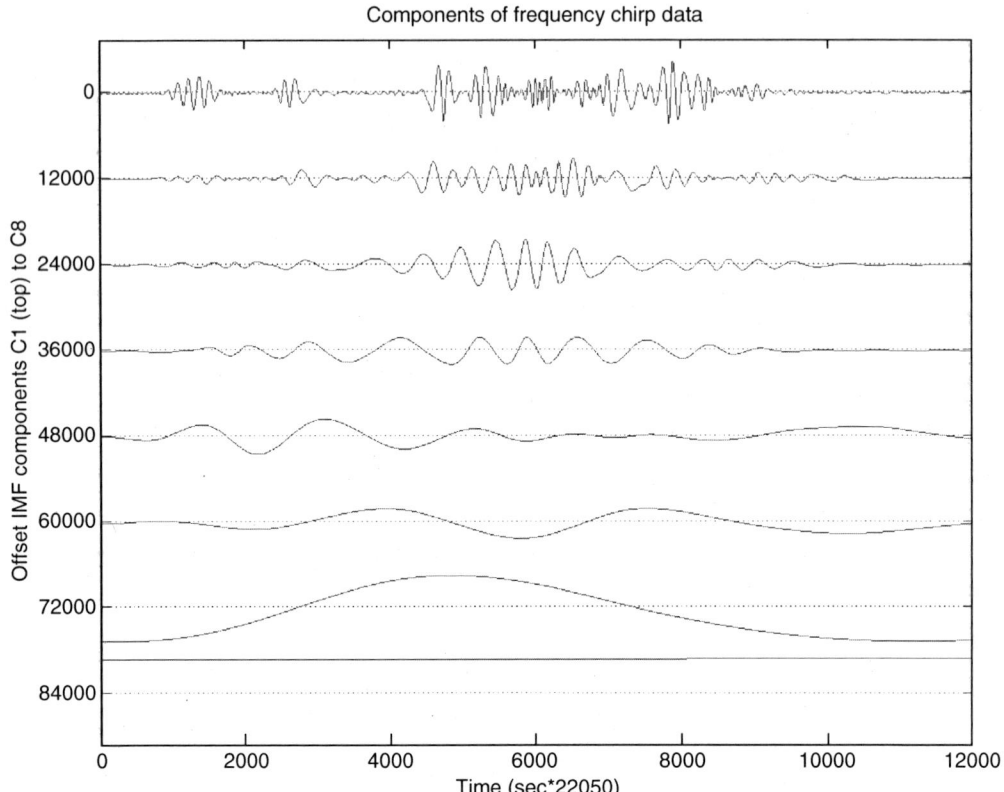

FIGURE 1.6
The eight IMF components obtained by processing the frequency chirp data of Figure 1.5, offset vertically from C1 (top) to C8 (bottom).

et al. [1]. The IF, as discussed here, is based on the instantaneous variation of the phase function from the HT of a data-adaptive decomposition, while the frequency content in the Fourier approach is an averaged frequency on the basis of a convolution of data with an *a priori* basis. Therefore, whenever the basis changes, the frequency content also changes. Similarly, when the decomposition changes, the IF also has to change. However, there are still persistent and common misconceptions on the IF computed in this manner.

One of the most prevailing misconceptions about IF is that, for any data with a discrete line spectrum, IF can be a continuous function. A variation of this misconception is that IF can give frequency values that are not one of the discrete spectral lines. This dilemma can be resolved easily. In the nonlinear cases, when the IF approach treats the harmonic distortions as continuous intrawave FMs, the Fourier-based methods treat the frequency content as discrete harmonic spectral lines. In the case of two or more beating waves, the IF approach treats the data as AM and FM modulation, while the frequency content from the Fourier method treats each constituting wave as a discrete spectral line, if the process is stationary. Although they appear perplexingly different, they represent the same data.

Another misconception is on negative IF values. According to Gabor's [7] approach, the HT is implemented through two Fourier transforms: the first transforms the data into frequency space, while the second performs an inverse Fourier transform after discarding all the negative frequency parts [3]. Therefore, according to this argument, all the negative frequency content has been discarded, which then raises the question, how can there still

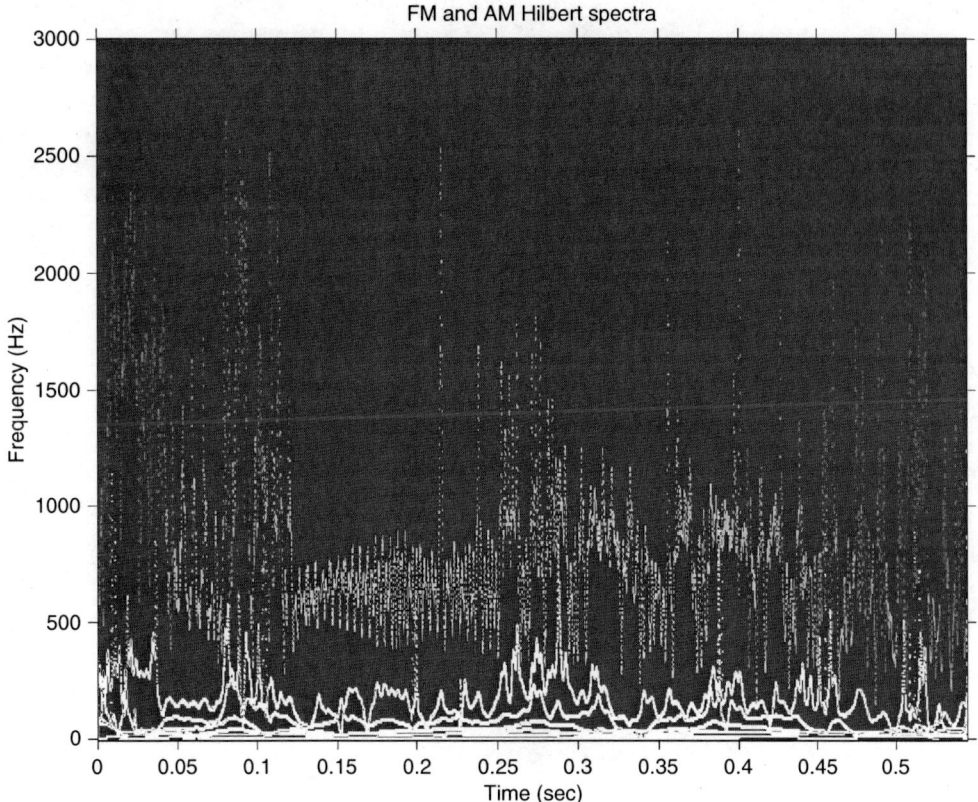

FIGURE 1.7
The AM and FM Hilbert spectral results from the frequency chirp data of Figure 1.5.

be negative frequency values? This question arises due to a misunderstanding of the nature of negative IF from the HT. The direct cause of negative frequency in the HT is the consequence of multiple extrema between two zero-crossings. Then, there are local loops not centered at the origin of the coordinate system, as discussed by Huang et al. [1]. Negative frequency can also occur even if there are no multiple extrema. For example, this would happen when there are large amplitude fluctuations, which cause the Hilbert-transformed phase loop to miss the origin. Therefore, the negative frequency does not influence the frequency content in the process of the HT through Gabor's [7] approach. Both these causes are removed by the NEMD and the normalized Hilbert transform (NHT) methods presented here.

The latest versions of these methods (NEMD/NHT) consistently give more stable IF values. They satisfy the limitation set by the Bedrosian theorem and offer a local measure of error sharper than the Nuttall theorem. Note here that in the initial spline of the amplitude done in the NEMD approach, the end effects again become important. The method used here is just to assign the end points as a maximum equal to the very last value. Other improvements using characteristic waves and linear predictions, as discussed in Ref. [1], can also be employed. There could be some improvement, but the resulting fit will be very similar.

Ever since the introduction of the EMD and HSA by Huang et al. [1,2,8], these methods have attracted increasing attention. Some investigators, however, have expressed certain reservations. For example, Olhede and Walden [9] suggested that the idea of computing

IF through the Hilbert transform is good, but that the EMD approach is not rigorous. Therefore, they have introduced the wavelet projection as the method for decomposition and adopt only the IF computation from the Hilbert transform. Flandrin et al. [10], however, suggest that the EMD is equivalent to a bank of dyadic filters, but refrain from using the HT. From the analysis presented here, it can be concluded that caution when using the HT is fully justified. The limitations imposed by Bedrosian and Nuttall certainly have solid theoretical foundations. The normalization procedure shown here will remove any reservations about further applications of the improved HT methods in data analysis. The method offers relatively little help to the approach advanced by Olhede and Walden [9] because the wavelet decomposition definitely removes the nonlinear distortions from the waveform. The consequence of this, however, is that their approach should also be limited to nonstationary, but linear, processes. It only serves the limited purpose of improving the poor frequency resolution of the continuous wavelet analysis.

As clearly shown in Equation 1.1, to give a good representation of actual wave data or other data from natural processes by means of an analytical wave profile, the analytical profile will need to have IMFs, and also obey the limitations imposed by the Bedrosian and Nuttall theorems. In the past, such a thorough examination of the data has not been done. As reported by Huang et al. [2,8], most of the actual wave data recorded are not composed of single components. Consequently, the analytical representation of a given wave profile in the form of Equation 1.1 poses a challenging problem theoretically.

1.3 Application to Image Analysis in Remote Sensing

Just as much of the data from natural phenomena are either nonlinear or nonstationary, or both, so it is also with the data that form images of natural processes. The methods of image processing are already well advanced, as can be seen in reviews such as by Castleman [11] or Russ [12]. The NEMD/NHT methods can now be added to the available tools for producing new and unique image products. Nunes et al. [13] and Linderhed [14–16], among others, have already done significant work in this new area. Because of the nonlinear and nonstationary nature of natural processes, the NEMD/NHT approach is especially well suited for image data, giving frequencies, inverse distances, or wave numbers as a function of time or distance, along with the amplitudes or energy values associated with these, as well as a sharp identification of imbedded structures. The various possibilities and products of this new analysis approach include, but are not limited to, joint and marginal distributions, which can be viewed as isosurfaces, contour plots, and surfaces that contain information on frequency, inverse wavelength, amplitude, energy and location in time, space, or both. Additionally, the concept of component images representing the intrinsic scales and structures imbedded in the data is now possible, along with a technique for obtaining frequency variations of structures within the images.

The laboratory used for producing the nonlinear waves, used as an example here, is the NASA Air–Sea Interaction Research Facility (NASIRF) located at the NASA Goddard Space Flight Center/Wallops Flight Facility, at Wallops Island, Virginia, within the Ocean Sciences Branch. The test section of the main wave tank is 18.3 m long and 0.9 m wide, filled to a depth of 0.76 m of water, leaving a height of 0.45 m over the water for airflow,

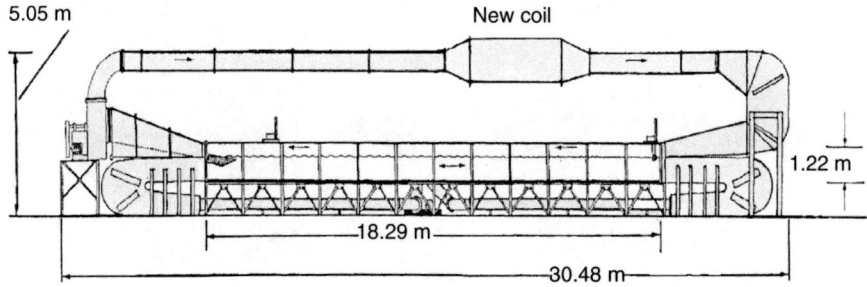

FIGURE 1.8
The NASA Air–Sea Interaction Research Facility's (NASIRF) main wave tank at Wallops Island, VA. The new coils shown were used to provide cooling and humidity control in the airflow overheated water.

if needed. The facility can produce wind and paddle-generated waves over a water current in either direction, and its capabilities, instruments and software have been described in detail by Long and colleagues [17–21]. The basic description is shown with an additional new feature indicated as *new coil* in Figure 1.8. These were recently installed to provide cold air of controlled temperature and humidity for experiments using cold air overheated water during the Flux Exchange Dynamics Study of 2004 (FEDS4) experiments, a joint experiment involving the University of Washington/Applied Physics Laboratory (UW/APL), The University of Alberta, the Lamont-Doherty Earth Observatory of Columbia University, and NASA GSFC/Wallops Flight Facility. The cold airflow overheated water optimized conditions for the collection of infrared (IR) video images.

1.3.1 The IR Digital Camera and Setup

The camera used to acquire the laboratory image presented here as an example was provided by UW/APL as part of FEDS4. The experimental setup is shown in Figure 1.9. For the example shown here, the resolution of the IR image was 640 × 512 pixels. The camera was mounted to look upwind at the water surface, so that its pixel image area covered a physical rectangle on the water surface on the order of 10 cm per side. The water within the wave tank was heated by four commercial spa heaters, while the air in the airflow was cooled and humidity controlled by NASIRF's new cooling and reheating coils. This produced a very thin layer of surface water that was cooled, so that whenever wave spilling and breaking occurred, it could be immediately seen by the IR camera.

1.3.2 Experimental IR Images of Surface Processes

With this imaging system in place, steps were taken to acquire interesting images of wave breaking and spilling due to wind and wave interactions. One such image is illustrated in Figure 1.10. To help the eyes visualize the image data, the IR camera intensity levels have been converted to a grey scale.

Using a horizontal line that slices through the central area of the image at the value of 275, Figure 1.11 illustrates the details contained in the actual array of data values obtained from the IR camera. This gives the IR camera intensity values stored in the pixels along the horizontal line. These can then be converted to actual temperatures when needed. A complex structure is evident here. Breaking wave fronts are evident in the crescent-

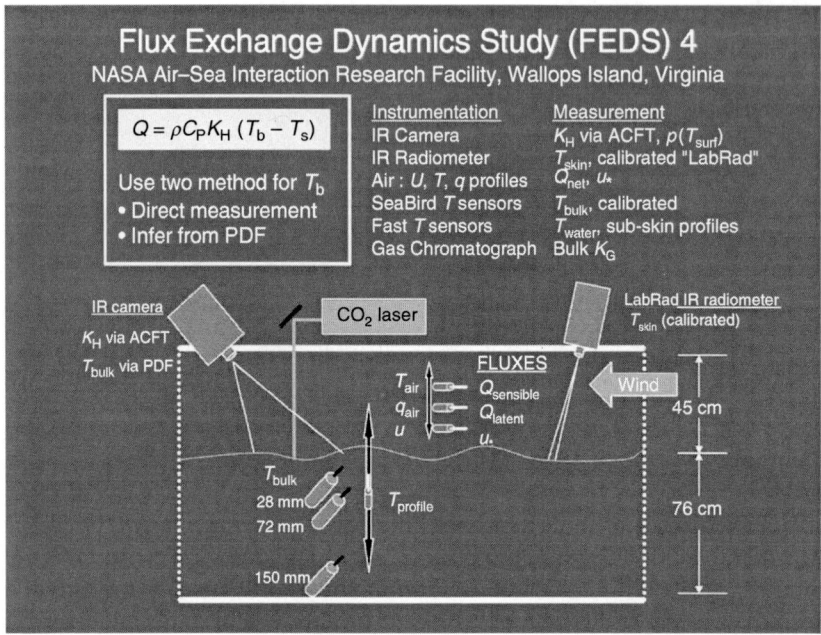

FIGURE 1.9
The experimental arrangement of FEDS4 (Flux Exchange Dynamics Study of 2004) used to capture IR images of surface wave processes. (Courtesy of A. Jessup and K. Phadnis of UW/APL.)

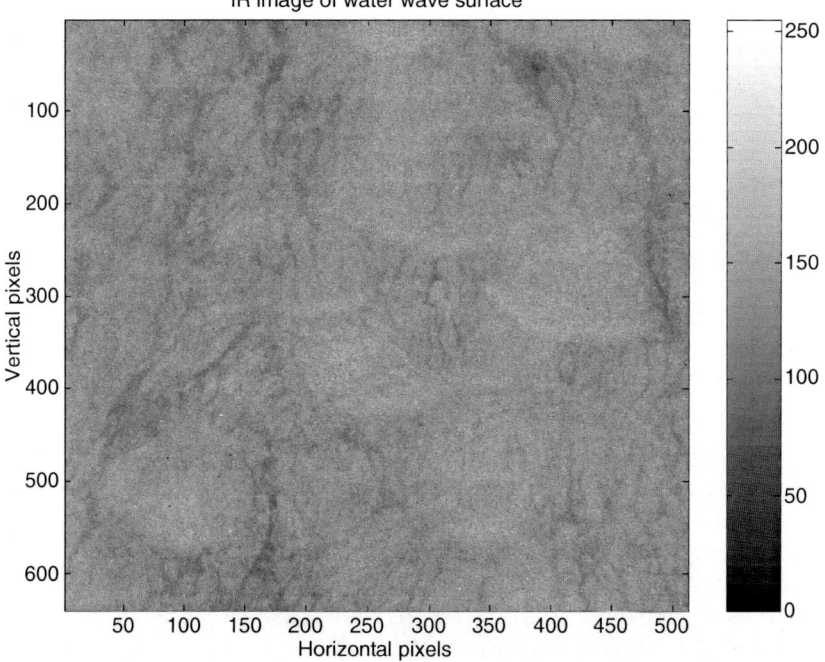

FIGURE 1.10
Surface IR image from the FEDS4 experiment. Grey bar gives the IR camera intensity levels. (Data courtesy of A. Jessup and K. Phadnis of UW/APL.)

shaped structures, where spilling and breaking brings up the underlying warmer water. After processing, the resulting components produced from the horizontal row of Figure 1.11 are shown in Figure 1.12. As can be seen, the component with the longest scale, C9, contains the bulk of the intensity values. The shorter, riding scales are fluctuations about the levels shown in component C9. The sifting was done via the extrema approach discussed in the foundation articles, and produced a total of nine components.

Using this approach, the IR image was first divided into 640 horizontal rows of 512 values each. The rows were then processed to produce the components, each of the 640 rows producing a component set similar to that shown in Figure 1.12. From these basic results, component images can be assembled. This is done by taking the first component representing the shortest scale from each of the 640 component sets. These first components are then assembled together to produce an array that is also 640 rows by 512 columns and can also be visualized as an image. This is the first component image. This production of component images is then continued in a similar fashion with the remaining components representing progressively longer scales. To visualize the shortest component scales, component images 1 through 4 were added together, as shown in Figure 1.13. Throughout the image, streaks of short wavy structures can be seen to line up in the wind direction (along the vertical axis). Even though the image is formed in the IR camera by measuring heat at many different pixel locations over a rectangular area, the surface waves have an effect that can be thus remotely sensed in the image, either as streaks of warmer water exposed by breaking or as more wavelike structures. If the longer scale components are now combined using the 5th and 6th component images, a composite image is obtained as shown in Figure 1.14. Longer scales can be seen throughout the image area where breaking and mixing occur. Other wavelike

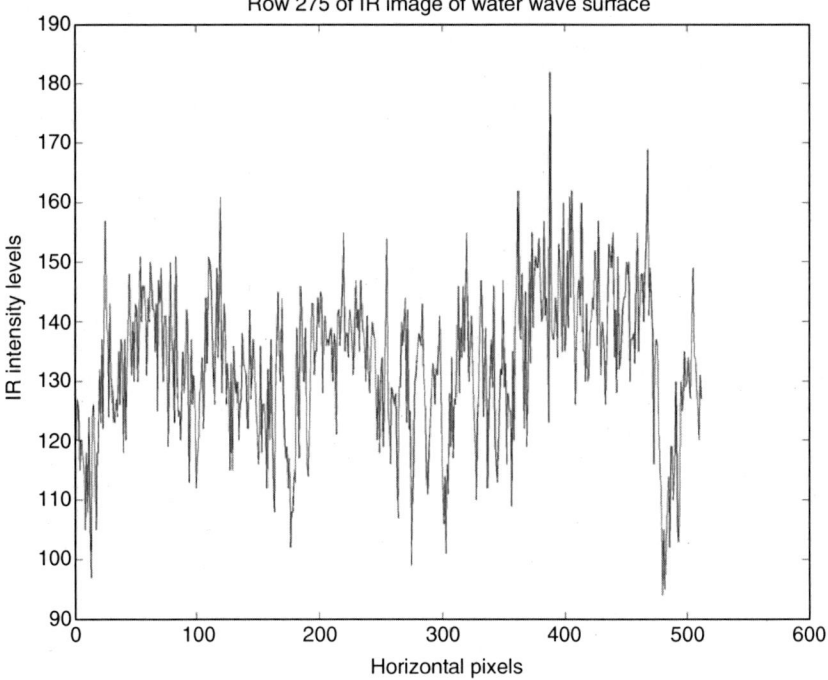

FIGURE 1.11
A horizontal slice of the raw IR image given in Figure 1.10, taken at row 275. Note the details contained in the IR image data, showing structures containing both short and longer length scales.

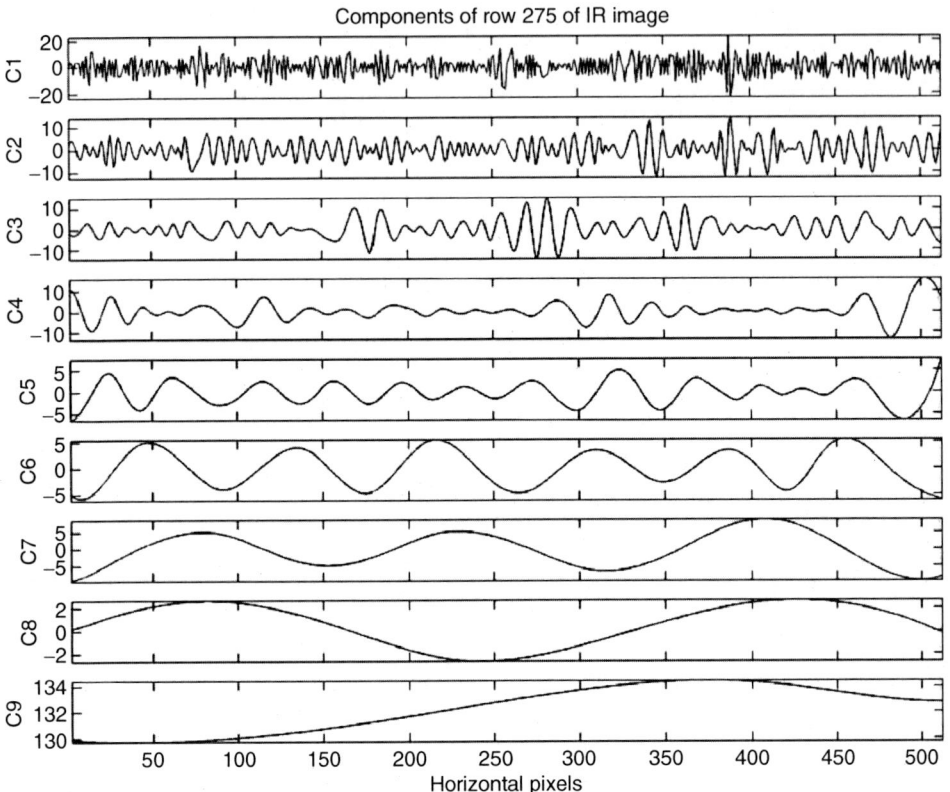

FIGURE 1.12
Components obtained by processing data from the slice shown in Figure 1.11. Note that component C9 carries the bulk of the intensity scale, while the other components with shorter scales record the fluctuations about these base levels.

structures of longer wavelengths are also visible. To produce a true wave number from images like these, one only has to convert using

$$k = 2\pi/\lambda, \quad (1.5)$$

where k is wave number (in 1/cm) and λ is wavelength (in cm). This would only require knowing the physical size of the image in centimeters or some other unit and its equivalent in pixels from the array analyzed.

Another approach to the raw image of Figure 1.10 is to separate the original image into columns instead of rows. This would make the analysis more sensitive to structures that were better aligned with that direction, and also with the direction of wind and waves. By repeating the steps leading to Figure 1.13, the shortest scale component images in component images 3 to 5 can be combined to form Figure 1.15. Component images 1 and 2 developed from the vertical column analysis were not included here, after they were found to contain results of such a short scale uniformly spread throughout the image, and without structure. Indeed, they had the appearance of uniform noise. It is apparent that more structures at these scales can be seen by analyzing along the column direction. Figure 1.16 represents the longer scale in component image 6. By the 6th component image, the lamination process starts to fail somewhat in reassembling the image from the components. Further processing is needed to better match the results at these longer scales.

On the Normalized Hilbert Transform and Its Applications in Remote Sensing 15

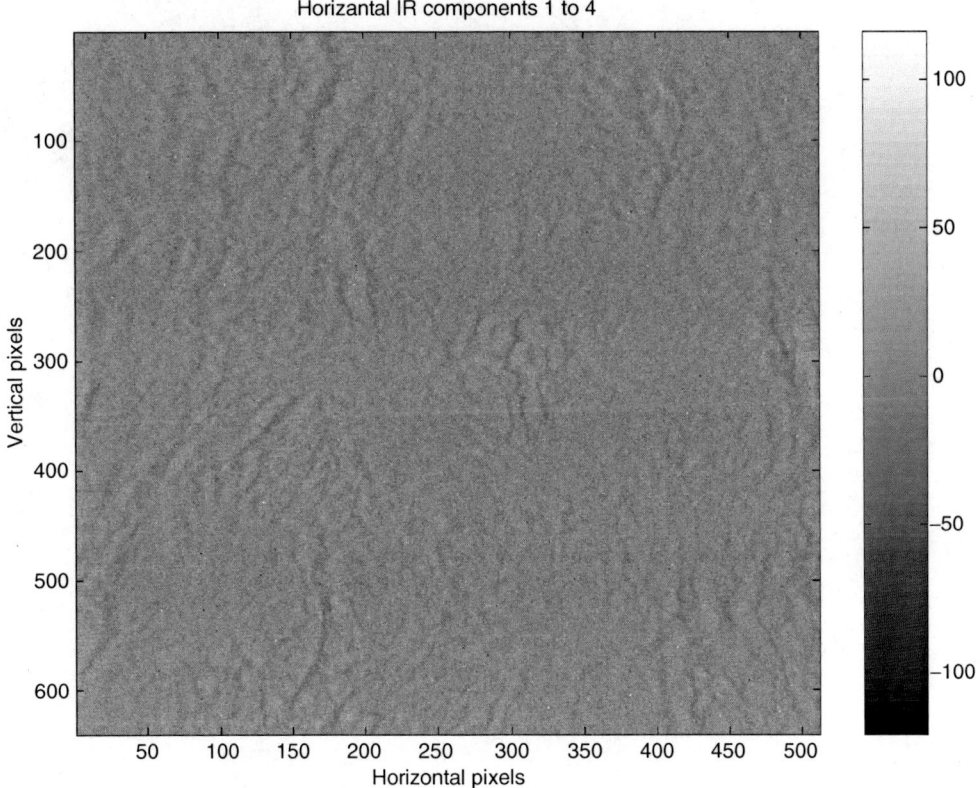

FIGURE 1.13 (See color insert following page 178.)
Component images 1 to 4 from the horizontal rows used to produce a composite image representing the shortest scales.

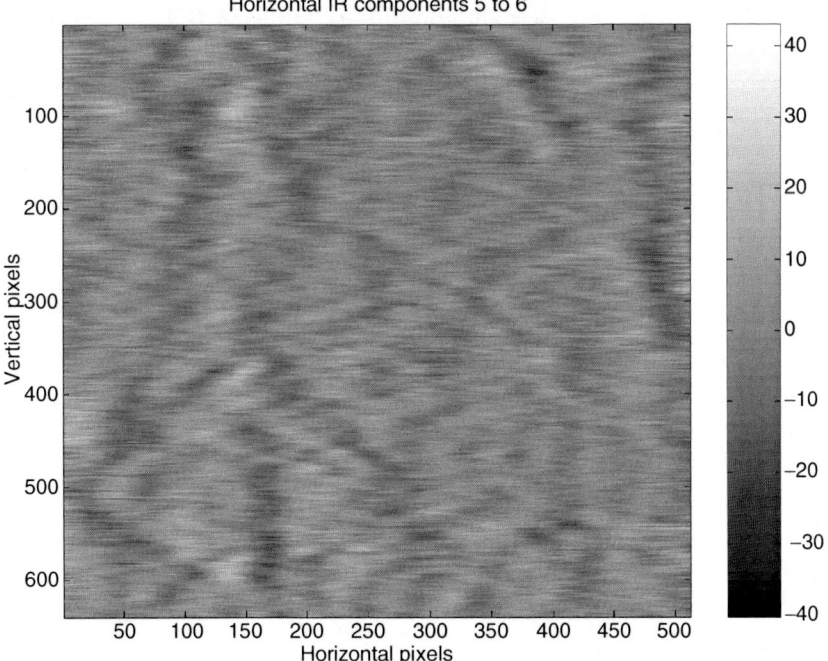

FIGURE 1.14 (See color insert following page 178.)
Component images 5 to 6 from the horizontal rows used to produce a composite image representing the longer scales.

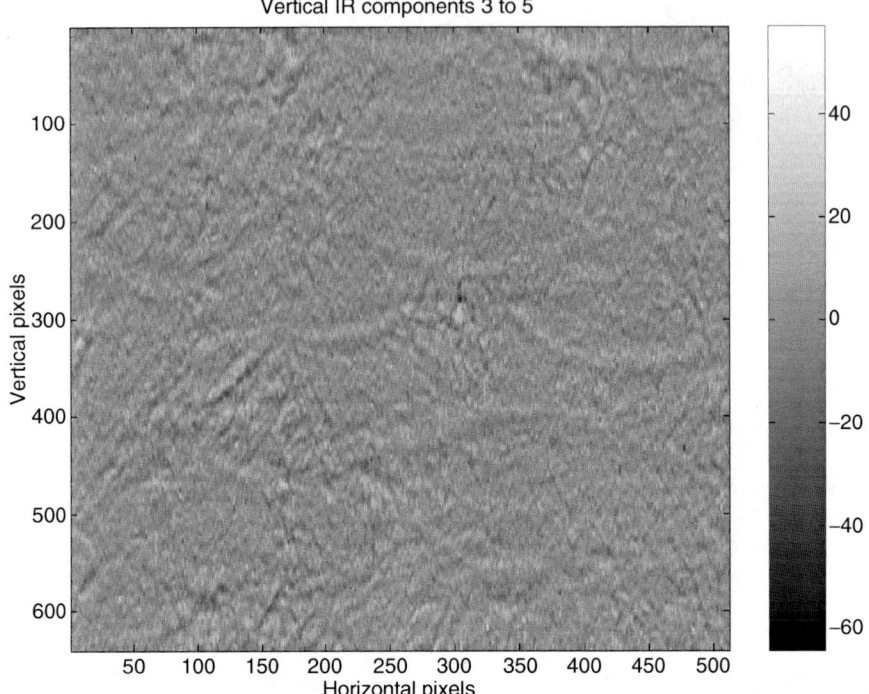

FIGURE 1.15 (See color insert following page 178.)
Component images 3 to 5 from the vertical rows here combined to produce a composite image representing the midrange scales.

When the original data are a function of time, this new approach can produce the IF and amplitude as functions of time. Here, the original data are from an IR image, so that any slice through the image (horizontal or vertical) would be a set of camera values (ultimately temperature) representing the temperature variation over a physical length. Thus, instead of producing frequency (inverse time scale), the new approach here initially produces an inverse length scale. In the case of water surface waves, this is the familiar scale of the wave number, as given in Equation 1.5. To illustrate this, consider Figure 1.17, which shows the changes of scale along the selected horizontal row 400. The largest measures of IR energy can be seen to be at the smaller inverse length scales, which imply that it came from the longer scales of components 3 and 4. Figure 1.18 repeats this for the even longer length scales in components 5 and 6.

Returning to the column-wise processing at column 250 of Figure 1.15 and Figure 1.16, further processing gives the contour plot of Figure 1.19, for components 3 through 5, and Figure 1.20, for components 4 through 6.

1.3.3 Volume Computations and Isosurfaces

Many interesting phenomena happen in the flow of time, and thus it is interesting to note how changes occur with time in the images. To include time in the analysis, a sequence of images taken at uniform time steps can be used.

By starting with a single horizontal or vertical line from the image, a contour plot can be produced, as was shown in Figure 1.7 through Figure 1.20. Using a set of sequential images covering a known time period and a pixel line of data from each (horizontal or vertical), a set of numerical arrays can be obtained from the NEMD/NHT analysis. Each

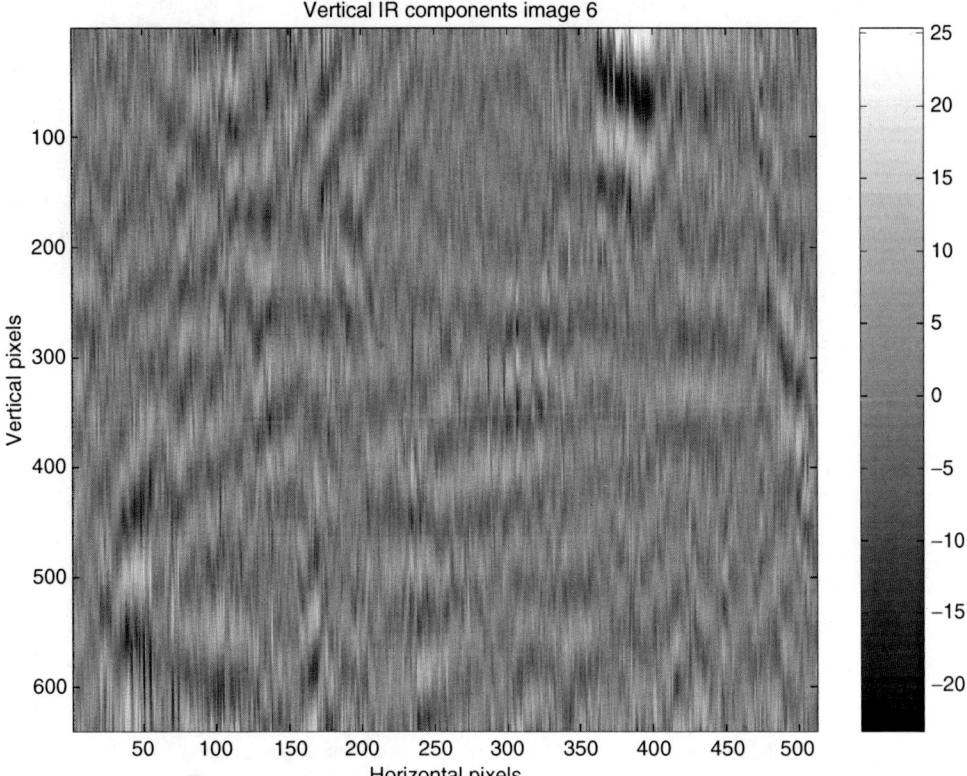

FIGURE 1.16 (See color insert following page 178.)
Component image 6 from the vertical row used to produce a composite image representing the longer scale.

array can be visualized by means of a contour plot, as already shown. The entire set of arrays can also be combined in sequence to form an array volume, or an array of dimension 3. Within the volume, each element of the array contains the amplitude or intensity of the data from the image sequence. The individual element location within the three-dimensional array specifies values associated with the stored data. One axis (call it x) of the volume can represent horizontal or vertical distance down the data line taken from the image. Another axis (call it y) can represent the resulting inverse length scale associated with the data. The additional axis (call it z) is produced by laminating the arrays together, and represents time, because each image was acquired in repetitive time steps. Thus, the position of the element in the volume gives location x along the horizontal or vertical slice, inverse length along the y-axis, and time along the z-axis.

Isosurface techniques would be needed to visualize this. This could be compared to peeling an onion, except that the different layers, or spatial contour values, are not bound in spherical shells. After a value of data intensity is specified, the isosurface visualization makes all array elements transparent outside of the level of the value chosen, while shading in the chosen value so that the elements inside that level (or behind it) cannot be seen. Some examples of this procedure can be seen in Ref. [21].

Another approach with the analysis of images is to reassemble lines from the image data using a different format. A sequence of images in units of time is needed, and using the same horizontal or vertical line from each image in the time sequence, each line can be laminated to its predecessor to build up an array that is the image length along the chosen line along one edge, and the number of images along the other axis, in units of

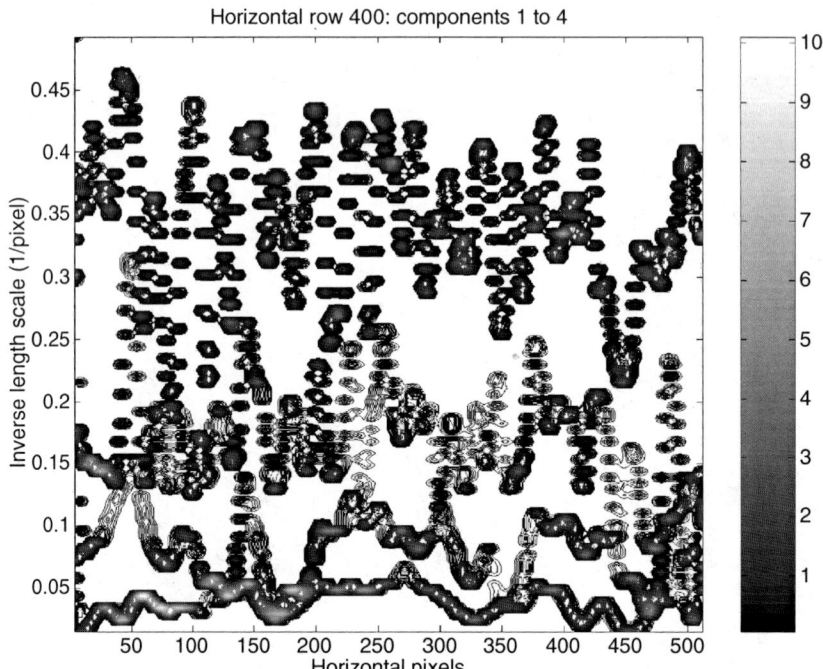

FIGURE 1.17 (See color insert following page 178.)
The results from the NEMD/NHT computation on horizontal row 400 for components 1 to 4, which resulted from Figure 1.13. Note the apparent influence of surface waves on the IR information. The most intense IR radiation can be seen at the smaller values of inverse length scale, denoting the longer scales in components 3 and 4. A wavelike influence can be seen at all scales.

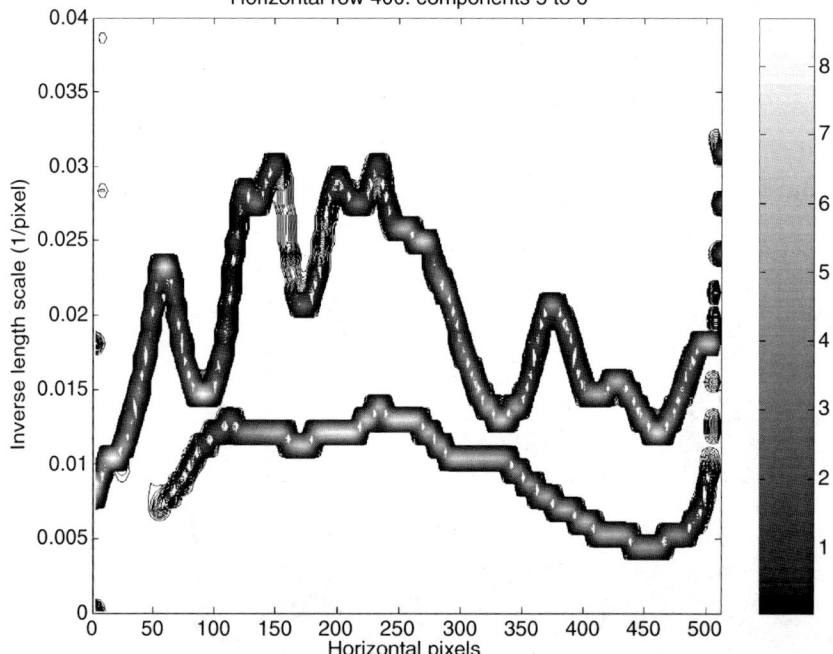

FIGURE 1.18 (See color insert following page 178.)
The results from the NEMD/NHT computation on horizontal row 400 for components 5 to 6, which resulted from Figure 1.14. Even at the longer scales, an apparent influence of surface waves on the IR information can still be seen.

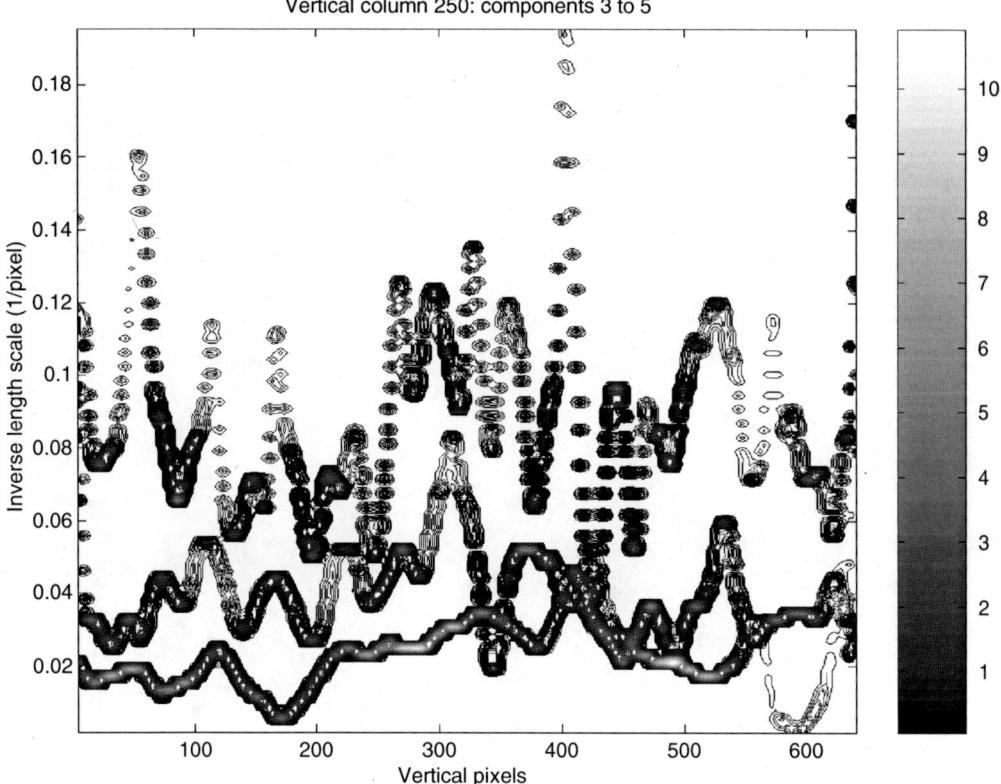

FIGURE 1.19
The contour plot developed from the vertical slice at column 250, using the components 3 to 5. The larger IR values can be seen at longer length scales.

time. Once complete, this two-dimensional array can be split into slices along the time axis. Each of these time slices, representing the variation in data values with time at a single-pixel location, can then be processed with the new NEMD/NHT technique. An example of this can also be seen in Ref. [21]. The NEMD/NHT techniques can thus reveal variations in frequency or time in the data at a specific location in the image sequence.

1.4 Conclusion

With the introduction of the normalization procedure, one of the major obstacles for NEMD/NHT analysis has been removed. Together with the establishment of the confidence limit [6] through the variation of stoppage criterion, and the statistically significant test of the information content for IMF [10,22], and the further development of the concept of IF [23], the new analysis approach has indeed approached maturity for applications empirically, if not mathematically (for a recent overview of developments, see Ref. [24]). The new NEMD/NHT methods provide the best overall approach to determine the IF for nonlinear and nonstationary data. Thus, a new tool is available to aid in further understanding and gaining deeper insight into the wealth of data now possible by remote

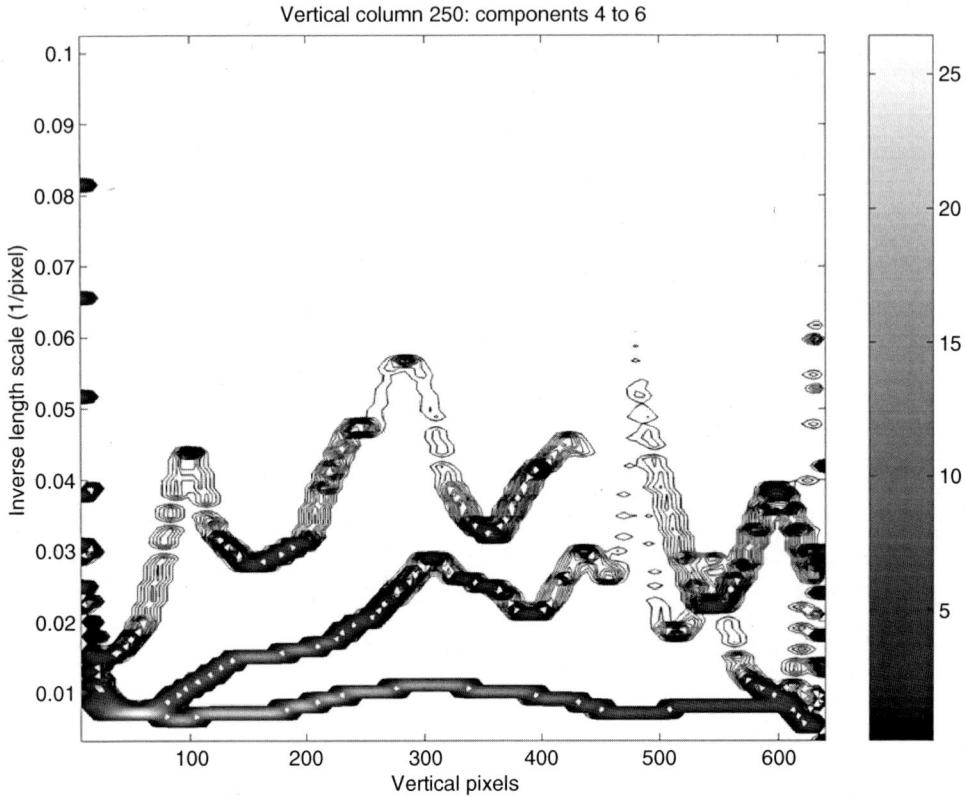

FIGURE 1.20
The contour plot developed from the vertical slice at column 250, using the components 4 through 6, as in Figure 1.19.

sensing and other means. Specifically, the application of the new method to data images was demonstrated.

This new approach is covered by several U.S. Patents held by NASA, as discussed by Huang and Long [25]. Further information on obtaining the software can be found at the NASA authorized commercial site: http://www.fuentek.com/technologies/hht.htm

Acknowledgment

The authors wish to express their continuing gratitude and thanks to Dr. Eric Lindstrom of NASA headquarters for his encouragement and support of the work.

References

1. Huang, N.E., Shen, Z., Long, S.R., Wu, M.C., Shih, S.H., Zheng, Q., Tung, C.C., and Liu, H.H., The empirical mode decomposition method and the Hilbert spectrum for non-stationary time series analysis, *Proc. Roy. Soc. London*, A454, 903–995, 1998.

2. Huang, N.E., Shen, Z., and Long, S.R., A new view of water waves—the Hilbert spectrum, *Ann. Rev. Fluid Mech.*, 31, 417–457, 1999.
3. Flandrin, P., *Time–Frequency/Time–Scale Analysis*, Academic Press, San Diego, 1999.
4. Bedrosian, E., On the quadrature approximation to the Hilbert transform of modulated signals, *Proc. IEEE*, 51, 868–869, 1963.
5. Nuttall, A.H., On the quadrature approximation to the Hilbert transform of modulated signals, *Proc. IEEE*, 54, 1458–1459, 1966.
6. Huang, N.E., Wu, M.L., Long, S.R., Shen, S.S.P., Qu, W.D., Gloersen, P., and Fan, K.L., A confidence limit for the empirical mode decomposition and the Hilbert spectral analysis, *Proc. Roy. Soc. London*, A459, 2317–2345, 2003.
7. Gabor, D., Theory of communication, *J. IEEE*, 93, 426–457, 1946.
8. Huang, N.E., Long, S.R., and Shen, Z., The mechanism for frequency downshift in nonlinear wave evolution, *Adv. Appl. Mech.*, 32, 59–111, 1996.
9. Olhede, S. and Walden, A.T., The Hilbert spectrum via wavelet projections, *Proc. Roy. Soc. London*, A460, 955–975, 2004.
10. Flandrin, P., Rilling, G., and Gonçalves, P., Empirical mode decomposition as a filterbank, *IEEE Signal Proc. Lett.*, 11(2), 112–114, 2004.
11. Castleman, K.R., *Digital Image Processing*, Prentice-Hall, Englewood Cliffs, NJ, 1996.
12. Russ, J.C., *The Image Processing Handbook*, 4th Edition, CRC Press, Boca Raton, 2002.
13. Nunes, J.C., Guyot, S., and Deléchelle, E., Texture analysis based on local analysis of the bidimensional empirical mode decomposition, *Mach. Vision Appl.*, 16(3), 177–188, 2005.
14. Linderhed, A., Compression by image empirical mode decomposition, *IEEE Int. Conf. Image Process.*, 1, 553–556, 2005.
15. Linderhed, A., Variable sampling of the empirical mode decomposition of two-dimensional signals, *Int. J. Wavelets, Multi-resolut. Inform. Process.*, 3, 2005.
16. Linderhed, A., 2D empirical mode decompositions in the spirit of image compression, *Wavelet Independ. Compon. Analy. Appl. IX, SPIE Proc.*, 4738, 1–8, 2002.
17. Long, S.R., NASA Wallops Flight Facility Air–Sea Interaction Research Facility, *NASA Reference Publication*, No. 1277, 1992, 29 pp.
18. Long, S.R., Lai, R.J., Huang, N.E., and Spedding, G.R., Blocking and trapping of waves in an inhomogeneous flow, *Dynam. Atmos. Oceans*, 20, 79–106, 1993.
19. Long, S.R., Huang, N.E., Tung, C.C., Wu, M.-L.C., Lin, R.-Q., Mollo-Christensen, E., and Yuan, Y., The Hilbert techniques: An alternate approach for non-steady time series analysis, *IEEE GRSS*, 3, 6–11, 1995.
20. Long, S.R. and Klinke, J., A closer look at short waves generated by wave interactions with adverse currents, *Gas Transfer at Water Surfaces*, Geophysical Monograph 127, American Geophysical Union, 121–128, 2002.
21. Long, S.R., Applications of HHT in image analysis, *Hilbert–Huang Transform and Its Applications*, Interdisciplinary Mathematical Sciences, 5, 289–305, World Scientific, Singapore, 2005.
22. Wu, Z. and Huang, N.E., A study of the characteristics of white noise using the empirical mode decomposition method, *Proc. Roy. Soc. London*, A460, 1597–1611, 2004.
23. Huang, N.E., Wu, Z., Long, S.R., Arnold, K.C., Blank, K., and Liu, T.W., On instantaneous frequency, *Proc. Roy. Soc. London* 2006, in press.
24. Huang, N.E., Introduction to the Hilbert–Huang transform and its related mathematical problems, *Hilbert–Huang Transform and Its Applications*, Interdisciplinary Mathematical Sciences, 5, 1–26, World Scientific, Singapore, 2005.
25. Huang, N.E. and Long, S.R., A generalized zero-crossing for local frequency determination, US Patent pending, 2003.

2

Statistical Pattern Recognition and Signal Processing in Remote Sensing

Chi Hau Chen

CONTENTS
2.1 Introduction 23
2.2 Introduction to Statistical Pattern Recognition in Remote Sensing 24
2.3 Using Self-Organizing Maps and Radial Basis Function Networks
    for Pixel Classification 26
2.4 Introduction to Statistical Signal Processing in Remote Sensing 26
2.5 Conclusions 28
References 28

2.1 Introduction

Basically, statistical pattern recognition deals with the correct classification of a pattern into one of several available pattern classes. Basic topics in statistical pattern recognition include: preprocessing, feature extraction and selection, parametric or nonparametric probability density, decision-making processes, performance evaluation, post-processing as needed, supervised and unsupervised learning, or training, and cluster analysis.

The large amount of data available makes remote-sensing data uniquely suitable for statistical pattern recognition. Signal processing is needed not only to reduce the undesired noises and interferences but also to extract desired information from the data as well as to perform the preprocessing task for pattern recognition.

Remote-sensing data considered include those from multispectral, hyperspectral, radar, optical, and infrared sensors. Statistical signal-processing methods, as used in remote sensing, include transform methods such as principal component analysis (PCA), independent component analysis (ICA), factor analysis, and the methods using high-order statistics.

This chapter is presented as a brief overview of the statistical pattern recognition and statistical signal processing in remote sensing. The views and comments presented, however, are largely those of this author. The chapter introduces the pattern recognition and signal-processing topics dealt in this book. The readers are highly recommended to refer the book by Landgrebe [1] for remote-sensing pattern classification issues and the article by Duin and Tax [2] for a survey on statistical pattern recognition.

Although there are many applications of statistical pattern recognition, its theory has been developed only during the last half century. A list of some major theoretical developments includes the following:

- Formulation of pattern recognition as a Bayes decision theory problem [3]
- Nearest neighbor decision rules (NNDRs) and density estimation [4]
- Use of Parzen density estimate in nonparametric pattern recognition [5]
- Leave-one-out method of error estimation [6]
- Use of statistical distance measures and error bounds in feature evaluation [7]
- Hidden Markov models as one way to deal with contextual information [8]
- Minimization of the perceptron criterion function [9]
- Fisher linear discriminant and multicategory generlizations [10]
- Link between backpropagation trained neural networks and the Bayes discriminant [11]
- Cover's theorem on the separability of patterns [12]
- Unsupervised learning by decomposition of mixture densities [13]
- K-mean algorithm [14]
- Self-organizing map (SOM) [15]
- Statistical learning theory and VC dimension [16,17]
- Support vector machine for pattern recognition [17]
- Combining classifiers [18]
- Nonlinear mapping [19]
- Effect of finite sample size (e.g., [13])

In the above discussion, the role of artificial neural networks on statistical classification and clustering has been taken into account. The above list is clearly not complete and is quite subjective. However, these developments clearly have a significant impact on information processing in remote sensing.

We now examine briefly the performance measures in statistical pattern recognition.

- *Error probability*. This is most popular as the Bayes decision rule is optimum for minimum error probability. It is noted that an average classification accuracy was proposed by Wilkinson [20] for remote sensing.
- *Ratio of interclass distance to within-class distance*. This is most popular for discriminant analysis that seeks to maximize such a ratio.
- *Mean square error*. This is most popular mainly in error correction learning and in neural networks.
- *ROC (receiver operating characteristics) curve*, which is a plot of the probability of correct decision versus the probability of false alarm, with other parameters given.

Other measures, like error-reject tradeoff, are often used in character recognition.

2.2 Introduction to Statistical Pattern Recognition in Remote Sensing

Feature extraction and selection is still a basic problem in statistical pattern recognition for any application. Feature measurements constructed from multiple bands of the remote-sensing data as a vector are still most commonly used in remote-sensing pattern recognition. Transform methods are useful to reduce the redundancy in vector measurements. The dimensionality reduction has been a particularly important topic in remote sensing in view of the hyperspectral image data, which normally has several hundred spectral bands.

The parametric classification rules include the Bayes or maximum likelihood decision rule and discriminant analysis. The nonparametric (or distribution free) method of classification includes NNDR and its modifications, and the Parzen density estimation. In the early 1970s, the multivariate Gaussian assumption was most popular in the multispectral data classification problem. It was demonstrated that the empirical data follows the Gaussian distribution reasonably well [21]. Even with the use of new sensors and the expanded application of remote sensing, the Gaussian assumption remains to be a good approximation. The traditional multivariate analysis still plays a useful role in remote-sensing pattern recognition [22] and, because of the importance of covariance matrix, methods to use unsupervised samples to "enhance" the data statistics have also been considered. Indeed, for good classification, data statistics must be carefully examined. An example is the synthetic aperture radar (SAR) image data. Chapter 1 of the companion volume (*Image Processing for Remote Sensing*) presents a discussion on the physical and statistical characteristics of the SAR image data.

Without making use of the Gaussian assumption, the NNDR is the most popular nonparametric classification method. It works well even with a moderate size data set and promises an error rate that is upper-bounded by twice the Bayes error rate. However, its performance is limited in remote-sensing data classification, while neural networks–based classifiers can reach the performance nearly equal to that of the Bayes classifier. Extensive study has been done in the statistical pattern recognition community to improve the performance of NNDR. We would like to mention the work of Grabowski et al. [23] here, which introduces the k-near surrounding neighbor (k-NSN) decision rule with application to remote-sensing data classification.

Some unique problem areas of statistical pattern recognition in remote sensing are the use of contextual information and the "Hughes phenomenon." The use of Markov random field model for contextual information is presented in Chapter 2 of the companion volume. While the classification performance generally improves with increases in the feature dimension, the performance reaches a peak without a proportional increase in the training sample size, beyond which the performance degrades. This is the so-called "Hughes phenomenon." Methods to reduce this phenomenon are well presented in Ref. [1].

Data fusion is important in remote sensing as different sensors, which have different strengths, are often used. The subject is treated in Chapter 11 of the companion volume. Though the approach is not limited to statistical methodology [24], the approaches in combining classifiers in statistical pattern recognition and neural networks can be quite useful in providing effective utilizations of information from different sensors or sources to achieve the best-available classification performance. Chapter 3 and Chapter 4 of the companion volume present two approaches in statistical combing of classifiers.

The recent development in support vector machine appears to present an ultimate classifier that may provide the best classification performance. Indeed, the design of the classifier is fundamental to the classification performance. There is, however, a basic question: "Is there a best classifier?" [25]. The answer is "No" as, among other reasons, it is evident that the classification process is data-dependent. Theory and practice are often not consistent in pattern recognition. The preprocessing and feature extraction and selection are important and can influence the final classification performance. There are no clear steps to be taken in preprocessing, and the optimal feature extraction and selection is still an unsolved problem. A single feature derived from the genetic algorithm may perform better than several original features. There is always a choice to be made between using a complex feature set followed by a simple classifier and a simple feature set followed by a complex classifier. Chapter 15 of the companion volume deals with the classification by support vector machine. Among many other publications on the subject, Melgani and Bruzzone [26] provide an informative comparison of the performance of several support vector machines.

2.3 Using Self-Organizing Maps and Radial Basis Function Networks for Pixel Classification

In this section, some experimental results are presented to illustrate the importance of preprocessing before classification. The data set, which is now available at the IEEE Geoscience and Remote Sensing Society database, consists of 250×350 pixel images. They were acquired by two imaging sensors installed on a Daedalus 1268 Airborne Thematic Mapper (ATM) scanner and a PLC-band, fully polarimetric NASA/JPL SAR sensor of an agricultural area near the village of Feltwell, U.K. The original SAR images include nine channels. Figure 2.1 shows the original nine channels of image data.

The radial basis function (RBF) neural network is used for classification [27]. However, preprocessing is performed by the SOM that performs preclustering. The weights of the SOM are chosen as centers for RBF neurons. RBF has five output nodes for five pattern classes on the image data considered (SAR and ATM images in an agricultural area). Weights of the "n" most-frequently-fired neurons, when each class was presented to the SOM, were separately taken as the center for the $5 \times n$ RBF neurons.

The weights between the hidden-layer neurons and the output-layer neurons were computed by a procedure for a generalized radial-basis function networks. Pixel classification using SOM alone (unsupervised) is 62.7% correct. Pixel classification using RBF alone (supervised) is 89.5% correct, at best. Pixel classification using both SOM and RBF is 95.2% correct. This result is better than the reported results on the same data set using RBF [28] at 90.5% correct or ICA-based features with nearest neighbor classification rule [29] at 86% correct.

2.4 Introduction to Statistical Signal Processing in Remote Sensing

Signal and image processing is needed in remote-sensing information processing to reduce the noise and interference with the data, to extract the desired signal and image component, or to derive useful measurements for input to the classifier. The classification problem is, in fact, very closely linked to signal and image processing [30,31].

Transform methods have been most popular in signal and image processing [32]. Though the popular wavelet transform method for remote sensing is treated elsewhere [33], we have included Chapter 1 in this volume, which presents the popular Hilbert–Huang transform; Chapter 12 of the companion volume, which deals with the use of Hermite transform in the multispectral image fusion; and Chapter 10 of this volume, Chapter 7, and Chapter 8 of the companion volume, which make use of the methods of ICA. Although there is a constant need for better sensors, the signal-processing algorithm such as the one presented in Chapter 7 of this volume demonstrates well the role of Kalman filtering in weak signal detection. Time series modeling as used in remote sensing is the subject of Chapter 8 of this volume and Chapter 9 of the companion volume. Chapter 6 of this volume makes use of the factor analysis.

Statistical Pattern Recognition and Signal Processing in Remote Sensing

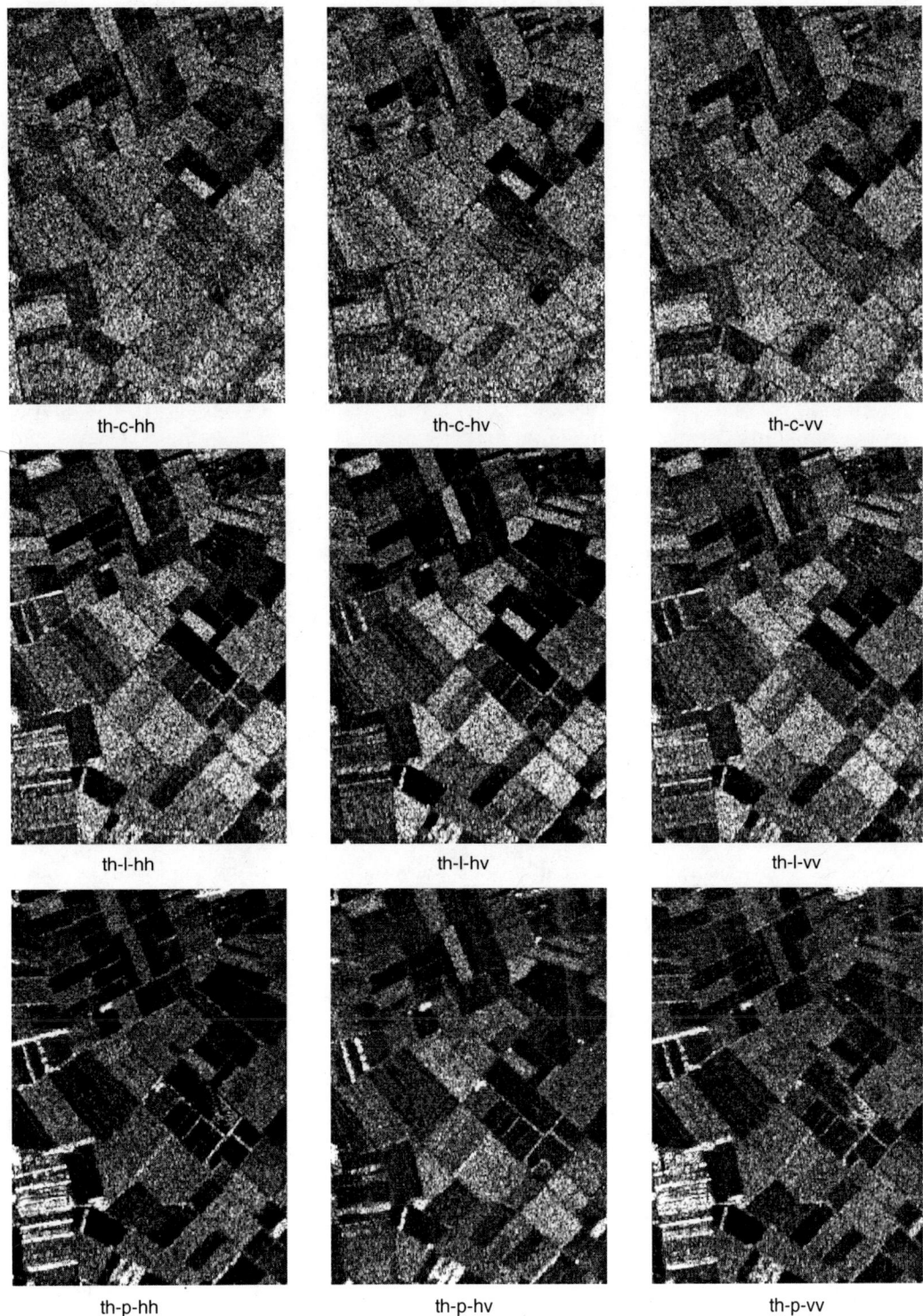

FIGURE 2.1
A nine-channel SAR image data set.

In spite of the numerous efforts with the transform methods, the basic method of PCA always has its useful role in remote sensing [34]. Signal decomposition and the use of high-order statistics can potentially offer new solutions to the remote-sensing information processing problems. Considering the nonlinear nature of the signal and image processing problems, it is necessary to point out the important roles of artificial neural networks in signal processing and classification, as presented in Chapter 3, Chapter 11, and Chapter 12 of this volume.

A lot of effort has been made in the last two decades to derive effective features in signal classification through signal processing. Such efforts include about two dozen mathematical features for use in exploration seismic pattern recognition [35], multi-dimensional attribute analysis that includes both physically and mathematically significant features or attributes for seismic interpretation [36], time domain, frequency domain, and time–frequency domain extracted features for transient signal analysis [37] and classification, and about a dozen features for active sonar classification [38]. Clearly, the feature extraction method for one type of signal cannot be transferred to other signals. To use a large number of features derived from signal processing is not desirable as there is significant information overlap among features and the resulting feature selection process can be tedious. It has not been verified that features extracted from the time–frequency representation can be more useful than the features from time–domain analysis and frequency domain alone. Ideally, a combination of a small set of physically significant and mathematically significant features should be used. Instead of looking for the optimal feature set, a small, but effective, feature set should be considered. It is doubtful that an optimal feature set for any given pattern recognition application can be developed in the near future in spite of many advances in signal and image processing.

2.5 Conclusions

Remote-sensing sensors have been able to deliver abundant information [39]. The many advances in statistical pattern recognition and signal processing can be very useful in remote-sensing information processing, either to supplement the capability of sensors or to effectively utilize the enormous amount of sensor data. The potentials and opportunities of using statistical pattern recognition and signal processing in remote sensing are thus unlimited.

References

1. Landgrebe, D.A., *Signal Theory Methods in Multispectral Remote Sensing*, John Wiley & Sons, New York, 2003.
2. Duin, R.P.W. and Tax, D.M.J., Statistical pattern recognition, in *Handbook of Pattern Recognition and Computer Vision*, 3rd ed., Chen, C.H. and Wang, P.S.P., Eds., World Scientific Publishing, Singapore, Jan. 2005, Chap. 1.
3. Chow, C.K., An optimum character recognition system using decision functions, *IEEE Trans. Electron. Comput.*, EC6, 247–254, 1957.
4. Cover, T.M. and Hart, P.E., Nearest neighbor pattern classification, *IEEE Trans. Inf. Theory*, IT-13(1), 21–27, 1967.

5. Parzen, E., On estimation of a probability density function and mode, *Annu. Math. Stat.*, 33(3), 1065–1076, 1962.
6. Lachenbruch, P.S. and Mickey, R.M., Estimation of error rates in discriminant analysis, *Technometrics*, 10, 1–11, 1968.
7. Kailath, T., The divergence and Bhattacharyya distance measures in signal selection, *IEEE Trans. Commn. Technol.*, COM-15, 52–60, 1967.
8. Duda, R.C., Hart, P.E., and Stork, D.G., *Pattern Classification*, John Wiley & Sons, New York, 2003.
9. Ho, Y.C. and Kashyap, R.L., An algorithm for linear inequalities and its applications, *IEEE Trans. Electron. Comput.*, Ece-14(5), 683–688, 1965.
10. Duda, R.C. and Hart, P.E., *Pattern Classification and Scene Analysis*, John Wiley & Sons, New York, 1972.
11. Bishop, C., *Neural Networks for Pattern Recognition*, Oxford University Press, Oxford, 1995.
12. Cover, T.M., Geometrical and statistical properties of systems of linear inequalities with applications in pattern recognition, *IEEE Trans. Electron. Comput.*, EC-14, 326–334, 1965.
13. Fukanaga, K., *Introduction to Statistical Pattern Recognition*, 2nd ed., Academic Press, New York, 1990.
14. MacQueen, J., Some methods for classification and analysis of multivariate observations, *Proc. Fifth Berkeley Symp. Probab. Stat.*, 281–297, 1997.
15. Kohonen, T., *Self-Organization and Associative Memory*, 3rd ed., Springer-Verlag, Heidelberg, 1988 (1st ed., 1980).
16. Vapnik, V.N., and Chervonenkis, A.Ya., On the uniform convergence of relative frequencies of events to their probabilities, *Theor. Probab. Appl.*, 17, 264–280, 1971.
17. Vapnik, V.N., *Statistical Learning Theory*, John Wiley & Sons, New York, 1998.
18. Kittler, J., Hatef, M., Duin, R.P.W., and Matas, J., On combining classifiers, *IEEE Trans. Pattern Anal. Mach. Intellig.*, 20(3), 226–239, 1998.
19. Sammon, J.W., Jr., A nonlinear mapping for data structure analysis, *IEEE Trans. Comput.*, C-18, 401–409, 1969.
20. Wilkinson, G., Results and implications of a study of fifteen years of satellite image classification experiments, *IEEE Trans. Geosci. Remote Sens.*, 43, 433–440, 2005.
21. Fu, K.S., Application of pattern recognition to remote sensing, in *Applications of Pattern Recognition*, Fu, K.S., Ed., CRC Press, Boca Raton, FL, 1982, Chap. 4.
22. Benediktsson, J.A., Statistical and neural network pattern recognition methods for remote sensing applications, in *Handbook of Pattern Recognition and Computer Vision*, 2nd ed., Chen, C.H. et al., Eds., World Scientific Publishing, Singapore, 1999, Chap. 3.2.
23. Graboswki, S., Jozwik, A., and Chen, C.H., Nearest neighbor decision rule for pixel classification in remote sensing, in *Frontiers of Remote Sensing Information Processing*, Chen, C.H., Ed., World Scientific Publishing, Singapore, 2003, Chap. 13.
24. Stevens, M.R. and Snorrason, M., Multisensor automatic target segmentation, in *Frontiers of Remote Sensing Information Processing*, Chen, C.H., Ed., World Scientific Publishing, Singapore, 2003, Chap. 15.
25. Richards, J., Is there a best classifier? *Proc. SPIE*, 5982, 2005.
26. Melgani, F. and Bruzzone, L., Classification of hyperspectral remote sensing images with support vector machines, *IEEE Trans. Geosci. Remote Sens.*, 42, 1778–1790, 2004.
27. Chen, C.H. and Shrestha, B., Classification of multi-sensor remote sensing images using self-organizing feature maps and radial basis function networks, *Proc. IGARSS*, Hawaii, 2000.
28. Bruzzone, L. and Prieto, D., A technique of the selection of kernel-function parameters in RBF neural networks for classification of remote sensing images, *IEEE Trans. Geosci. Remote Sens.*, 37(2), 1179–1184, 1999.
29. Zhang, X. and Chen, C.H., Independent component analysis by using joint cumulants and its application to remote sensing images, *J. VLSI Signal Process. Systems*, 37(2/3), 2004.
30. Chen, C.H., Ed., *Digital Waveform Processing and Recognition*, CRC Press, Boca Raton, FL, 1982.
31. Chen, C.H., Ed., *Signal Processing Handbook*, Marcel-Dekker, New York, 1988.
32. Chen, C.H., Transform methods in remote sensing information processing, in *Frontiers of Remote Sensing Information Processing*, Chen, C.H., Ed., World Scientific Publishing, Singapore, 2003, Chap. 2.

33. Chen, C.H., Eds., Wavelet analysis and applications, in *Frontiers of Remote Sensing Information Processing*, World Scientific Publishing, Singapore, 2003, Chaps. 7–9, pp. 139–224.
34. Lee, J.B., Woodyatt, A.S., and Berman, M., Enhancement of high spectral resolution remote sensing data by a noise-adjusted principal component transform, *IEEE Trans. Geosci. Remote Sens.*, 28, 295–304, 1990.
35. Li, Y., Bian, Z., Yan, P., and Chang, T., Pattern recognition in geophysical signal processing and interpretation, in *Handbook of Pattern Recognition and Computer Vision*, 1st ed., Chen, C.H. et al., Eds., World Scientific Publishing, Singapore, 1993, Chap. 3.2.
36. Olson, R.G., Signal transient analysis and classification techniques, in *Handbook of Pattern Recognition and Computer Vision*, 1st ed., Chen, C.H., Ed., World Scientific Publishing, Singapore, 1993, Chap. 3.3.
37. Justice, J.H., Hawkins, D.J., and Wong, G., Multidimensional attribute analysis and pattern recognition for seismic interpretation, *Pattern Recognition*, 18(6), 391–407, 1985.
38. Chen, C.H., Neural networks in active sonar classification, *Neural Networks in Ocean Environments*, IEE conference, Washington; DC; 1991.
39. Richard, J., Remote sensing sensors: capabilities and information processing requirements, in *Frontiers of Remote Sensing Information Processing*, Chen, C.H., Ed., World Scientific Publishing, Singapore, 2003, Chap. 1.

3

A Universal Neural Network–Based Infrasound Event Classifier

Fredric M. Ham and Ranjan Acharyya

CONTENTS
3.1 Overview of Infrasound and Why Classify Infrasound Events?................................ 31
3.2 Neural Networks for Infrasound Classification 32
3.3 Details of the Approach.. 33
    3.3.1 Infrasound Data Collected for Training and Testing 34
    3.3.2 Radial Basis Function Neural Networks 34
3.4 Data Preprocessing 38
    3.4.1 Noise Filtering.. 38
    3.4.2 Feature Extraction Process 38
    3.4.3 Useful Definitions.. 42
    3.4.4 Selection Process for the Optimal Number of Feature Vector Components... 44
    3.4.5 Optimal Output Threshold Values and 3-D ROC Curves 44
3.5 Simulation Results 47
3.6 Conclusions.. 51
Acknowledgments 51
References 51

3.1 Overview of Infrasound and Why Classify Infrasound Events?

Infrasound is a longitudinal pressure wave [1–4]. The characteristics of these waves are similar to audible acoustic waves but the frequency range is far below what the human ear can detect. The typical frequency range is from 0.01 to 10 Hz (Figure 3.1). Nature is an incredible creator of infrasonic signals that can emanate from sources such as volcano eruptions, earthquakes, severe weather, tsunamis, meteors (bolides), gravity waves, microbaroms (infrasound radiated from ocean waves), surf, mountain ranges (mountain associated waves), avalanches, and auroral waves to name a few. Infrasound can also result from man-made events such as mining blasts, the space shuttle, high-speed aircraft, artillery fire, rockets, vehicles, and nuclear events. Because of relatively low atmospheric absorption at low frequencies, infrasound waves can travel long distances in the Earth's atmosphere and can be detected with sensitive ground-based sensors.

    An integral part of the comprehensive nuclear test ban treaty (CTBT) international monitoring system (IMS) is an infrasound network system [3]. The goal is to have 60

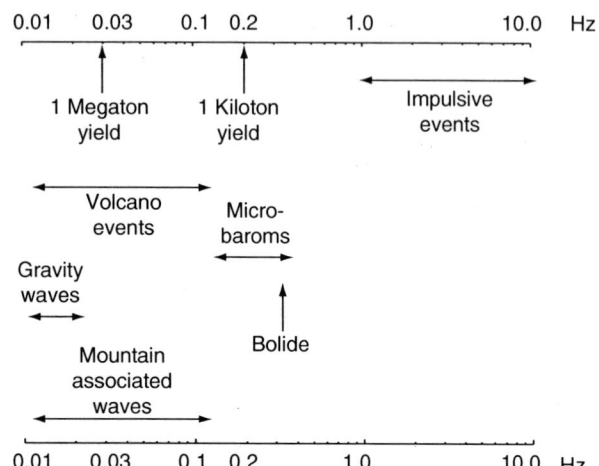

FIGURE 3.1 Infrasound spectrum.

infrasound arrays operational worldwide over the next several years. The main objective of the infrasound monitoring system is the detection and verification, localization, and classification of nuclear explosions as well as other infrasonic signals-of-interest (SOI). Detection refers to the problem of detecting an SOI in the presence of all other unwanted sources and noises. Localization deals with finding the origin of a source, and classification deals with the discrimination of different infrasound events of interest. This chapter concentrates on the classification part only.

3.2 Neural Networks for Infrasound Classification

Humans excel at the task of classifying patterns. We all perform this task on a daily basis. Do we wear the checkered or the striped shirt today? For example, we will probably select from a *group* of checkered shirts versus a *group* of striped shirts. The grouping process is carried out (probably at a near subconscious level) by our ability to discriminate among all shirts in our closet and we group the striped ones in the *striped class* and the checkered ones in the *checkered class* (that is, without physically moving them around in the closet, only in our minds). However, if the closet is dimly lit, this creates a potential problem and diminishes our ability to make the right selection (that is, we are working in a "noisy" environment). In the case of using an artificial neural network for classification of patterns (or various "events") the same problem exists with noise. Noise is everywhere.

In general, a common problem associated with event classification (or detection and localization for that matter) is environmental noise. In the infrasound problem, many times the distance between the source and the sensors is relatively large (as opposed to region infrasonic phenomena). Increases in the distance between sources and sensors heighten the environmental dependence of the signals. For example, the signal of an infrasonic event that takes place near an ocean may have significantly different characteristics as compared to the same event that occurs in a desert. A major contributor of noise for the signal near an ocean is microbaroms. As mentioned above, microbaroms are generated in the air from large ocean waves. One important characteristic of neural networks is their noise rejection capability [5]. This, and several other attributes, makes them highly desirable to use as classifiers.

3.3 Details of the Approach

Our approach of classifying infrasound events is based on a parallel bank neural network structure [6–10]. The basic architecture is shown in Figure 3.2. There are several reasons for using such an architecture; however, one very important advantage of dedicating one module to perform the classification of one event class is that the architecture is fault tolerant (i.e., if one module fails, the rest of the individual classifiers will continue to function). However, the overall performance of the classifier is enhanced when the parallel bank neural network classifier (PBNNC) architecture is used. Individual banks (or modules) within the classifier architecture are radial basis function neural networks (RBF NNs) [5]. Also, each classifier has its own dedicated preprocessor. Customized feature vectors are computed optimally for each classifier and are based on cepstral coefficients and a subset of their associated derivatives (differences) [11]. This will be explained in detail later. The different neural modules are trained to classify one and only one class; however, for the requisite module responsible for one of the classes, it is also trained not to recognize all other classes (negative reinforcement). During the training process, the output is set to a "1" for a correct class and a "0" for all the other signals associated with all the other classes. When the training process is complete the final output thresholds will be set to an optimal value based on a three-dimensional receiver operating characteristic (3-D ROC) curve for each one of the neural modules (see Figure 3.2).

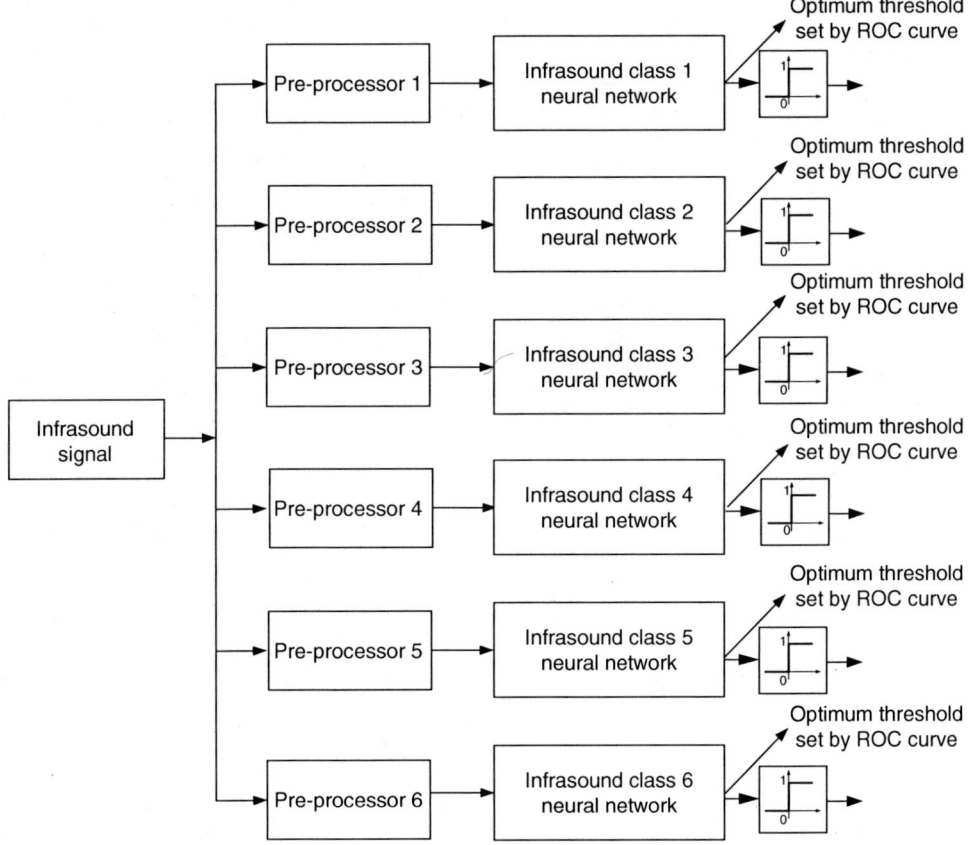

FIGURE 3.2
Basic parallel bank neural network classifier (PBNNC) architecture.

3.3.1 Infrasound Data Collected for Training and Testing

The data used for training and testing the individual networks are obtained from multiple infrasound arrays located in different geographical regions with different geometries. The six infrasound classes used in this study are shown in Table 3.1, and the various array geometries are shown in Figure 3.3(a) through Figure 3.3(e) [12,13]. Table 3.2 shows the various classes, along with the array numbers where the data were collected, and the associated sampling frequencies.

3.3.2 Radial Basis Function Neural Networks

As previously mentioned, each of the neural network modules in Figure 3.2 is an RBF NN. A brief overview of RBF NNs will be given here. This is not meant to be an exhaustive discourse on the subject, but only an introduction to the subject. More details can be found in Refs. [5,14].

Earlier work on the RBF NN was carried out for handling multivariate interpolation problems [15,16]. However, more recently they have been used for probability density estimation [17–19] and approximations of smooth multivariate functions [20]. In principle, the RBF NN makes adjustments of its weights so that the error between the actual and the desired responses is minimized relative to an optimization criterion through a defined learning algorithm [5]. Once trained, the network performs the interpolation in the output vector space, thus the generalization property.

Radial basis functions are one type of positive-definite kernels that are extensively used for multivariate interpolation and approximation. Radial basis functions can be used for problems of any dimension, and the smoothness of the interpolants can be achieved to any desirable extent. Moreover, the structures of the interpolants are very simple. However, there are several challenges that go along with the aforementioned attributes of RBF NNs. For example, many times an ill-conditioned linear system must be solved, and the complexity of both time and space increases with the number of interpolation points. But these types of problems can be overcome.

The interpolation problem may be formulated as follows. Assume M distinct data points $X = \{x_1, \ldots, x_M\}$. Also assume the data set is bounded in a region Ω (for a specific class). Each observed data point $x \in \Re^u$ (u corresponds to the dimension of the input space) may correspond to some function of x. Mathematically, the interpolation problem may be stated as follows. Given a set of M points, i.e., $\{x_i \in \Re^u | i = 1, 2, \ldots, M\}$ and a corresponding set of M real numbers $\{d_i \in \Re | i = 1, 2, \ldots, M\}$ (desired outputs or the targets), find a function $F: \Re^M \to \Re$ that satisfies the interpolation condition

$$F(x_i) = d_i, \quad i = 1, 2, \ldots, M \tag{3.1}$$

TABLE 3.1

Infrasound Classes Used for Training and Testing

Class Number	Event	No. SOI ($n = 574$)	No. SOI Used for Training ($n = 351$)	No. SOI Used for Testing ($n = 223$)
1	Vehicle	8	4	4
2	Artillery fire (ARTY)	264	132	132
3	Jet	12	8	4
4	Missile	24	16	8
5	Rocket	70	45	25
6	Shuttle	196	146	50

A Universal Neural Network–Based Infrasound Event Classifier

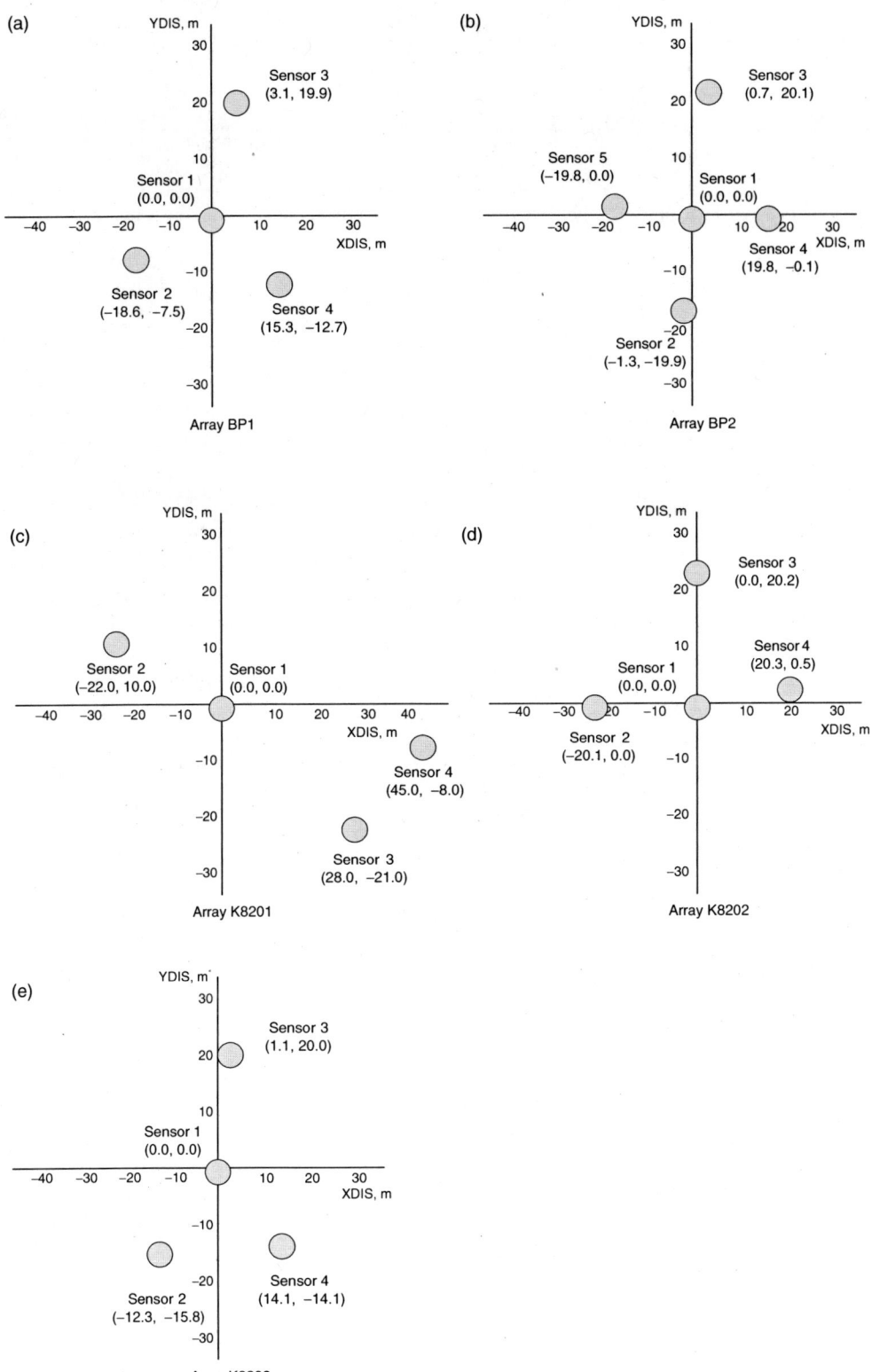

FIGURE 3.3
Five different array geometries.

TABLE 3.2

Array Numbers Associated with the Event Classes and the Sampling Frequencies Used to Collect the Data

Class Number	Event	Array	Sampling Frequency, Hz
1	Vehicle	K8201	100
2	Artillery fire (ARTY) (K8201: Sites 1 and 2)	K8201; K8203	(K8201: Sites 1 and 100; Sites 2 and 50); 50
3	Jet	K8201	50
4	Missile	K8201; K8203	50; 50
5	Rocket	BP1; BP2	100; 100
6	Shuttle	BP2; BP103[a]	100; 50

[a] Array geometry not available.

Thus, all the points must pass through the interpolating surface. A radial basis function may be a special interpolating function of the form

$$F(x) = \sum_{i=1}^{M} w_i \phi_i(\|x - x_i\|_2) \quad (3.2)$$

where $\phi(\bullet)$ is known as the radial basis function and $\|\bullet\|_2$ denotes the Euclidean norm. In general, the data points x_i are the centers of the radial basis functions and are frequently written as c_i.

One of the problems encountered when attempting to fit a function to data points is over-fitting of the data, that is, the value of M is too large. However, generally speaking, this is less a problem the RBF NN that it is with, for example, a multi-layer perceptron trained by backpropagation [5]. The RBF NN is attempting to construct the hyperspace for a particular problem when given a limited number of data points.

Let us take another point of view concerning how an RBF NN performs its construction of a hypersurface. Regularization theory [5,14] is applied to the construction of the hypersurface. A geometrical explanation follows.

Consider a set of input data obtained from several events from a single class. The input data may be from temporal signals or defined features obtained from these signals using an appropriate transformation. The input data would be transformed by a nonlinear function in the hidden layer of the RBF NN. Each event would then correspond to a point in the feature space. Figure 3.4 depicts a two-dimensional (2-D) feature set, that is, the dimension of the output of the hidden layer in the RBF NN is two. In Figure 3.4, "(a)", "(b)", and "(c)" correspond to three separate events. The purpose here is to construct a surface (shown by the dotted line in Figure 3.4) such that the dotted region encompasses events of the same class. If the RBF network is to classify four different classes, there must be four different regions (four dotted contours), one for each class. Ideally, each of these regions should be separate with no overlap. However, because there is always a limited amount of observed data, perfect reconstruction of the hyperspace is not possible and it is inevitable that overlap will occur.

To overcome this problem it is necessary to incorporate global information from Ω (i.e., the class space) in approximating the unknown hyperspace. One choice is to introduce a smoothness constraint on the targets. Mathematical details will not be given here, but for an in-depth development see Refs. [5,14].

Let us now turn our attention to the actual RBF NN architecture and how the network is trained. In its basic form, the RBF NN has three layers: an input layer, one hidden

A Universal Neural Network–Based Infrasound Event Classifier

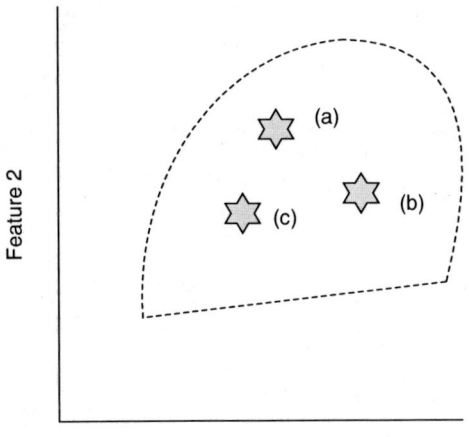

FIGURE 3.4
Example of a two-dimensional feature set.

layer, and one output layer. Referring to Figure 3.5, the source nodes (or the input components) make up the input layer. The hidden layer performs a nonlinear transformation (i.e., the radial basis functions residing in the hidden layer perform this transformation) of the input to the network and is generally of a higher dimension than the input. This nonlinear transformation of the input in the hidden layer may be viewed as a basis for the construction of the input in the transformed space. Thus, the term radial basis function.

In Figure 3.5, the output of the RBF NN (i.e., at the output layer) is calculated according to

$$y_i = f_i(x) = \sum_{k=1}^{N} w_{ik}\phi_k(x,c_k) = \sum_{k=1}^{N} w_{ik}\phi_k(\|x - c_k\|_2), \quad i = 1, 2, \ldots, m \text{ (no. outputs)} \quad (3.3)$$

where $x \in \Re^{u \times 1}$ is the input vector, $\phi_k(\bullet)$ is a (RBF) function that maps $\Re^+$ (set of all positive real numbers) to $\Re$ (field of real numbers), $\|\bullet\|_2$ denotes the Euclidean norm, w_{ik} are the weights in the output layer, N is the number of neurons in the hidden layer, and $c_k \in \Re^{u \times 1}$ are the RBF centers that are selected based on the input vector space. The Euclidean distance between the center of each neuron in the hidden layer and the input to the network is computed. The output of the neuron in a hidden layer is a nonlinear

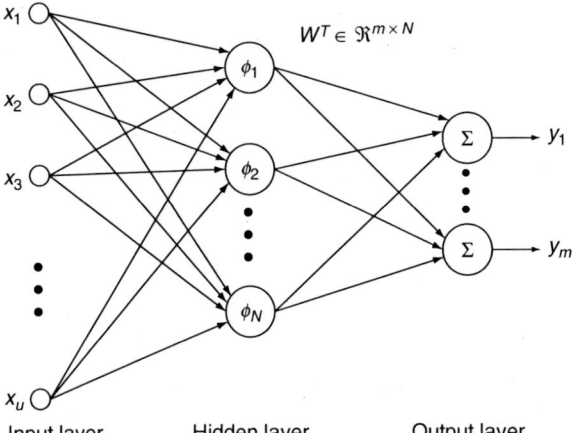

FIGURE 3.5
RBF NN architecture.

function of this distance, and the output of the network is computed as a weighted sum of the hidden layer outputs.

The functional form of the radial basis function, $\phi_k(\bullet)$, can be any of the following:

- Linear function: $\phi(x) = x$
- Cubic approximation: $\phi(x) = x^3$
- Thin-plate-spline function: $\phi(x) = x^2 \ln(x)$
- Gaussian function: $\phi(x) = \exp(-x^2/\sigma^2)$
- Multi-quadratic function: $\phi(x) = \sqrt{x^2 + \sigma^2}$
- Inverse multi-quadratic function: $\phi(x) = 1/(\sqrt{x^2 + \sigma^2})$

The parameter σ controls the "width" of the RBF and is commonly referred to as the spread parameter. In many practical applications the Gaussian RBF is used. The centers, c_k, of the Gaussian functions are points used to perform a sampling of the input vector space. In general, the centers form a subset of the input data.

3.4 Data Preprocessing

3.4.1 Noise Filtering

Microbaroms, as previously defined, are a persistently present source of noise that resides in most collected infrasound signals [21–23]. Microbaroms are a class of infrasonic signals characterized by narrow-band, nearly sinusoidal waveforms, with a period between 6 and 8 sec. These signals can be generated by marine storms through a non-linear interaction of surface waves [24]. The frequency content of the microbaroms often coincides with that of small-yield nuclear explosions. This could be bothersome in many applications; however, simple band-pass filtering can alleviate the problem in many cases. Therefore, a band-pass filter with a pass band between 1 and 49 Hz (for signals sampled at 100 Hz) is used here to eliminate the effects of the microbaroms. Figure 3.6 shows how band-pass filtering can be used to eliminate the microbaroms problem.

3.4.2 Feature Extraction Process

Depicted in each of the six graphs in Figure 3.7 is a collection of eight signals from each class, that is, $y_{ij}(t)$ for $i = 1, 2, \ldots, 6$ (classes) and $j = 1, 2, \ldots, 8$ (number of signals) (see Table 3.1 for total number of signals in each class). A feature extraction process is desired that will capture the salient features of the signals in each class and at the same time be invariant relative to the array geometry, the geographical location of the array, the sampling frequency, and the length of the time window. The overall performance of the classifier is contingent on the data that is used to train the neural network in each of the six modules shown in Figure 3.2. Moreover, the neural network's ability to distinguish between the various events (presented the neural networks as feature vectors) is the distinctiveness of the features between the classes. However, within each class it is desirable to have the feature vectors as similar to each other as possible.

There are two major questions to be answered: (1) What will cause the signals in one class to have markedly different characteristics? (2) What can be done to minimize these

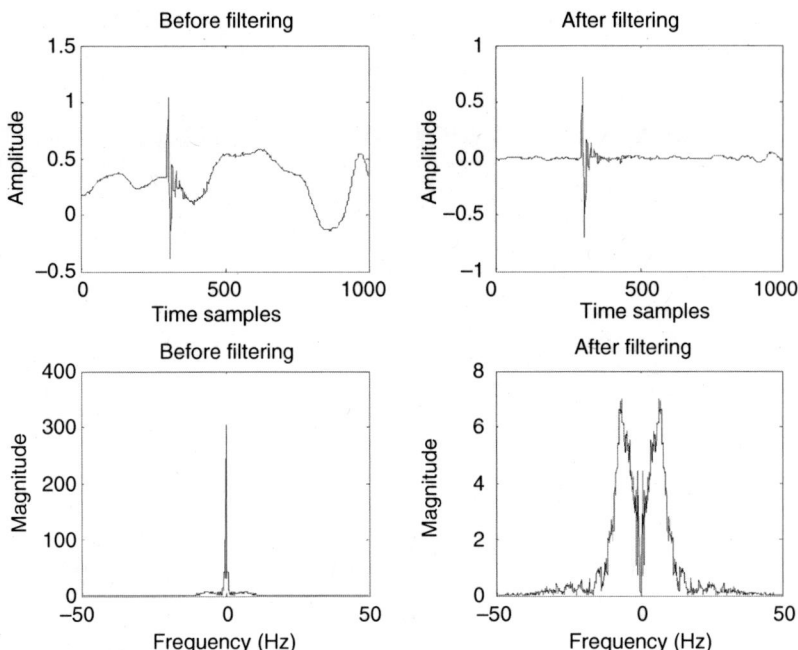

FIGURE 3.6
Results of band-pass filtering to eliminate the effects of microbaroms (an artillery signal).

differences and achieve uniformity within a class and distinctively different feature vector characteristics between classes?

The answer to the first question is quite simple—noise. This can be noise associated with the sensors, the data acquisition equipment, or other unwanted signals that are not of interest. The answer to the second question is also quite simple (once you know the answer)—using a feature extraction process based on computed cepstral coefficients and a subset of their associated derivatives (differences) [10,11,25–28].

As mentioned in Section 3.3, each classifier has its own dedicated preprocessor (see Figure 3.2). Customized feature vectors are computed optimally for each classifier (or neural module) and are based on the aforementioned cepstral coefficients and a subset of their associated derivatives (or differences). The preprocessing procedure is as follows.

Each time-domain signal is first normalized and then its mean value is computed and removed. Next, the power spectral density (PSD) is calculated for each signal, which is a mixture of the desired component and possibly other unwanted signals and noise. Therefore, when the PSDs are computed for a set of signals in a defined class there will be spectral components associated with noise and other unwanted signals that need to be suppressed. This can be systematically accomplished by first computing the average PSD (i.e., PSD_{avg}) over the suite of PSDs for a particular class. The spectral components are defined as μ_i for $i = 1, 2, \ldots$ for PSD_{avg}. The maximum spectral component, μ_{max}, of PSD_{avg} is then determined. This is considered the dominant spectral component within a particular class and its value is used to suppress selected components in the resident PSDs for any particular class according to the following:

$$\text{if } \mu_i > \varepsilon_1 \mu_{max} \text{ (typically } \varepsilon_1 = 0.001\text{)}$$
$$\text{then } \mu_i \leftarrow \mu_i$$
$$\text{else } \varepsilon_2 \leftarrow \mu_i \text{ (typically } \varepsilon_2 = 0.00001\text{)}$$

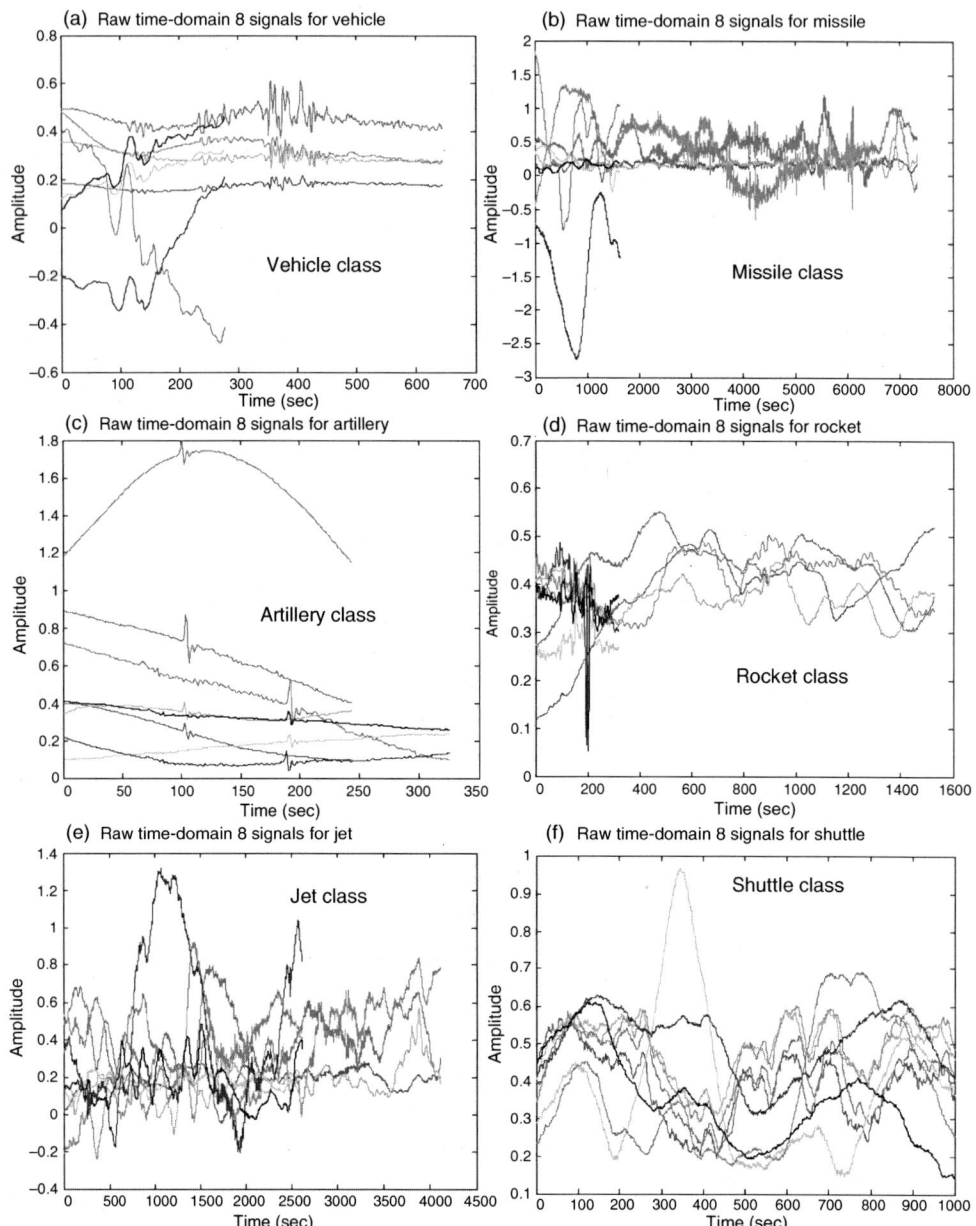

FIGURE 3.7 (See color insert following page 178.)
Infrasound signals for six classes.

To some extent, this will minimize the effects of any unwanted components that may reside in the signals and at the same time minimize the effects of noise. However, another step can be taken to further minimize the effects of any unwanted signals and noise that may reside in the data. This is based on a minimum variance criterion applied to the spectral components of the PSDs in a particular class after the previously described step is completed. The second step is carried out by taking the first 90% of the spectral components that are rank-ordered according to the smallest variance. The rest of the components in the power spectral densities within a particular class are set to a small value, that is, ε_3 (typically 0.00001). Therefore, the number of spectral components greater than ε_3 will dictate the number of components in the cepstral domain (i.e., the number of cepstral coefficients and associated differences). Depending on the class, the number of coefficients and differences will vary. For example, in the simulations that were run, the largest number of components was 2401 (artillery class) and the smallest number was 543 (vehicle class). Next, the mel-frequency scaling step is carried out with defined values for α and β [10], then the inverse discrete cosine transform is taken and the derivatives (differences) are computed.

From this set of computed cepstral coefficients and differences, it is desired to select those components that will constitute a feature vector that is consistent within a particular class. That is, there is minimal variation among similar components across the suite of feature vectors. So the approach taken here is to think in terms of minimum variance of these similar components within the feature set.

Recall, the time-domain infrasound signals are assumed to be band-pass filtered to remove any effects of microbaroms as described previously. For each discrete-time infrasound signal, $y(k)$, where k is the discrete time index (an integer), the specific preprocessing steps are (dropping the time dependence k):

(1) Normalize (i.e., divide each sample in the signal $y(k)$ by the absolute value of the maximum amplitude, $|y_{max}|$, and also divide by the square root of the computed variance of the signal, σ_y^2, and then remove the mean:

$$y \leftarrow y/\{|y_{max}|, \sigma_y\} \tag{3.4}$$

$$y \leftarrow y - \text{mean}(y) \tag{3.5}$$

(2) Compute the PSD, $S_{yy}(k_\omega)$, of the signal y:

$$S_{yy}(k_\omega) = \sum_{\tau=0}^{\infty} R_{yy}(\tau) e^{-jk_\omega \tau} \tag{3.6}$$

where $R_{yy}(\tau)$ is the autocorrelation of the infrasound signal y.

(3) Find the average of the entire set of PSDs in the class, i.e., PSD_{avg}
(4) Retain only those spectral components whose contributions will maximize the overall performance of the global classifier:

if $\mu_i > \varepsilon_1 \mu_{max}$ (typically $\varepsilon_1 = 0.001$)
then $\mu_i \leftarrow \mu_i$
else $\varepsilon_2 \leftarrow \mu_i$ (typically $\varepsilon_2 = 0.00001$)

(5) Compute variances of the components selected in Step (4). Then take the first 90% of the spectral components that are rank-ordered according to the smallest variance. Set the remaining components to a small value, i.e., ε_3 (typically 0.00001).

(6) Apply mel-frequency scaling to $S_{yy}(k_\omega)$:

$$S_{mel}(k_\omega) = \alpha \log_e [\beta S_{yy}(k_\omega)] \tag{3.7}$$

where $\alpha = 11.25$, $\beta = 0.03$.

(7) Take the inverse discrete cosine transform:

$$x_{mel}(n) = \frac{1}{n} \sum_{k_\omega=0}^{N-1} S_m(k_\omega) \cos(2\pi k_\omega n/N) \quad \text{for } n = 0, 1, 2, \ldots, N-1 \quad (3.8)$$

(8) Take the consecutive differences of the sequence $x_{mel}(n)$ to obtain $x'_{mel}(n)$.

(9) Concatenate the sequence of differences, $x'_{mel}(n)$, with the cepstral coefficient sequence, $x_{mel}(n)$, to form the augmented sequence:

$$x^a_{mel} = [x'_{mel}(i)|x_{mel}(j)] \quad (3.9)$$

where i and j are determined experimentally. As mentioned previously, $i = 400$ and $j = 600$.

(10) Take the absolute value of the elements in the sequence x^a_{mel} yielding:

$$x^a_{mel,abs} = |x^a_{mel}| \quad (3.10)$$

(11) Take the $\log_e$ of $x^a_{mel,abs}$ from the previous step to give:

$$x^a_{mel,abs,log} = \log_e [x^a_{mel,abs}] \quad (3.11)$$

Applying this 11-step feature extraction process to the infrasound signals in the six different classes results in the feature vectors shown in Figure 3.8. The length of each feature vector is 34. This will be explained in the next section. If these sets of feature vectors are compared to their time-domain signal counterparts (see Figure 3.7), it is obvious that the feature extraction process produces feature vectors that are much more consistent than the time-domain signals. Moreover, comparing the feature vectors between classes reveals that the different sets of feature vectors are markedly distinct. This should result in improved classification performance.

3.4.3 Useful Definitions

Before we go on, let us define some useful quantities that apply to the assessment of performance for classifiers. The confusion matrix [29] for a two-class classifier is shown in Table 3.3.

In Table 3.3 we have the following:

p: number of correct predictions that an occurrence is positive
q: number of incorrect predictions that an occurrence is positive
r: number of incorrect of predictions that an occurrence is negative
s: number of correct predictions that an occurrence is negative

With this, the correct classification rate (CCR) is defined as

$$\begin{aligned} CCR &= \frac{\text{No. correct predictions} - \text{No.} \times \text{classifications}}{\text{No. predictions}} \\ &= \frac{p + s - \text{No. multiple classifications}}{p + q + r + s} \end{aligned} \quad (3.12)$$

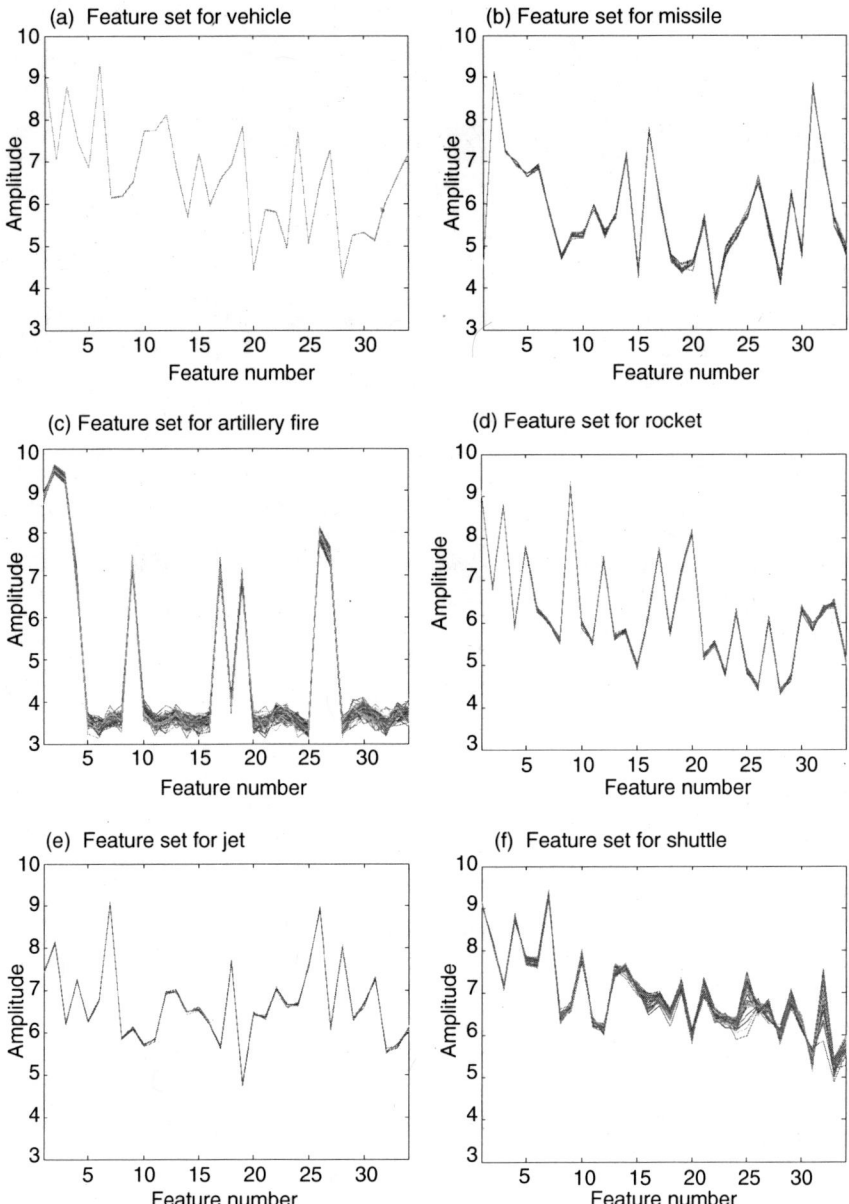

FIGURE 3.8 (See color insert following page 178.)
Infrasound signals for six class different classes.

Multiple classifications refer to more than one of the neural modules showing a "positive" at the output of the RBF NN indicating that the input to the global classifier belongs to more than one class (whether this is true or not). So there could be double, triple, quadruple, etc., classifications for one event.

The accuracy (ACC) is given by

$$\text{ACC} = \frac{\text{No. correct predictions}}{\text{No predictions}} = \frac{p+s}{p+q+r+s} \qquad (3.13)$$

TABLE 3.3

Confusion Matrix for a Two-Class Classifier

		Predicted Value	
Actual Value		Positive	Negative
	Positive	p	q
	Negative	r	s

As seen from Equation 3.12 and Equation 3.13, if multiple classifications occur, the CCR is a more conservative performance measure than the ACC. However, if no multiple classifications occur, the CCR = ACC.

The true positive (TP) rate is the proportion of positive cases that are correctly identified. This is computed using

$$\text{TP} = \frac{p}{p+q} \tag{3.14}$$

The false positive (FP) rate is the proportion of negative cases that are incorrectly classified as positive occurrences. This is computed using

$$\text{FP} = \frac{r}{r+s} \tag{3.15}$$

3.4.4 Selection Process for the Optimal Number of Feature Vector Components

From the set of computed cepstral coefficients and differences generated using the feature extraction process given above, an optimal subset of these is desired that will constitute the feature vectors used to train and test the PBNNC shown in Figure 3.2. The optimal subset (i.e., the optimal feature vector length) is determined by taking a minimum variance approach. Specifically, a 3-D graph is generated that plots the performance; that is, CCR versus the RBF NN spread parameter and the feature vector number (see Figure 3.9). From this graph, mean values and variances are computed across the range of spread parameters for each of the defined number of components in the feature vector. The selection criterion is defined as simultaneously maximizing the mean and at the same time minimizing the variance. Maximization of the mean ensures maximum performance; that is, maximizing the CCR and at the same time minimizing the variance to minimize variation in the feature set within each of the classes. The output threshold at each of the neural modules (i.e., the output of the single output neuron of each RBF NN) is set optimally according to a 3-D ROC curve. This will be explained next.

Figure 3.10 shows the two plots used to determine the maximum mean and the minimum variance. The table insert between the two graphs shows that even though the mean value for 40 elements in the feature vector is (slightly) larger than that for 34 elements, the variance for 40 is nearly three times that for 34 elements. Therefore, a length of 34 elements for the feature vectors is the best choice.

3.4.5 Optimal Output Threshold Values and 3-D ROC Curves

At the output of the RBF NN for each of the six neural modules, there is a single output neuron with hard-limiting binary values used during the training process (see Figure 3.2). After training, to determine whether a particular SOI belongs to one of the six classes, the

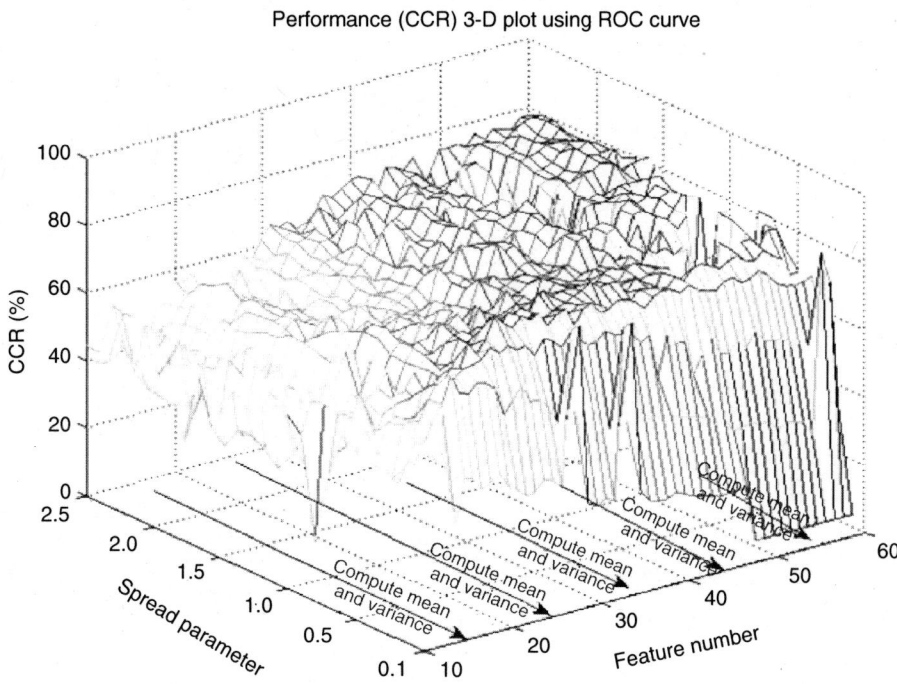

FIGURE 3.9 (See color insert following page 178.)
Performance plot used to determine the optimal number components in the feature vector. Ill conditioning occurs for the feature number less than 10, and for the feature number greater than 60, the CCR dramatically declines.

threshold value of the output neurons is optimally set according to an ROC curve [30–32] for that individual neural module (i.e., one particular class). Before an explanation of the 3-D ROC curve is given, let us first review 2-D ROC curves and see how they are used to optimally set threshold values.

An ROC curve is a plot of the TP rate versus the FP rate, or the sensitivity versus (1 − specificity); a sample ROC curve is shown in Figure 3.11. The optimal threshold value corresponds to a point nearest the ideal point (0, 1) on the graph. The point (0, 1) is considered ideal because in this case there would be no false positives and only true positives. However, because of noise and other undesirable effects in the data, the point closest to the (0, 1) point (i.e., the minimum Euclidean distance) is the best that we can do. This will then dictate the optimal threshold value to be used at the output of the RBF NN.

Since there are six classifiers, that is, six neural modules in the global classifier, six ROC curves must be generated. However, using 2-D ROC curves to set the thresholds at the outputs of the six RBF NN classifiers will not result in optimal thresholds. This is because *misclassifications* are not taken into account when setting the threshold for a particular neural module that is responsible for classifying a particular set of infrasound signals. Recall that one neural module is associated with one infrasonic class, and each neural module acts as its own classifier. Therefore, it is necessary to account for the misclassifications that can occur and this can be accomplished by adding a third dimension to the ROC curve. When the misclassifications are taken into account the (0, 1, 0) point now becomes the optimal point, and the smallest Euclidean distance to this point is directly related to

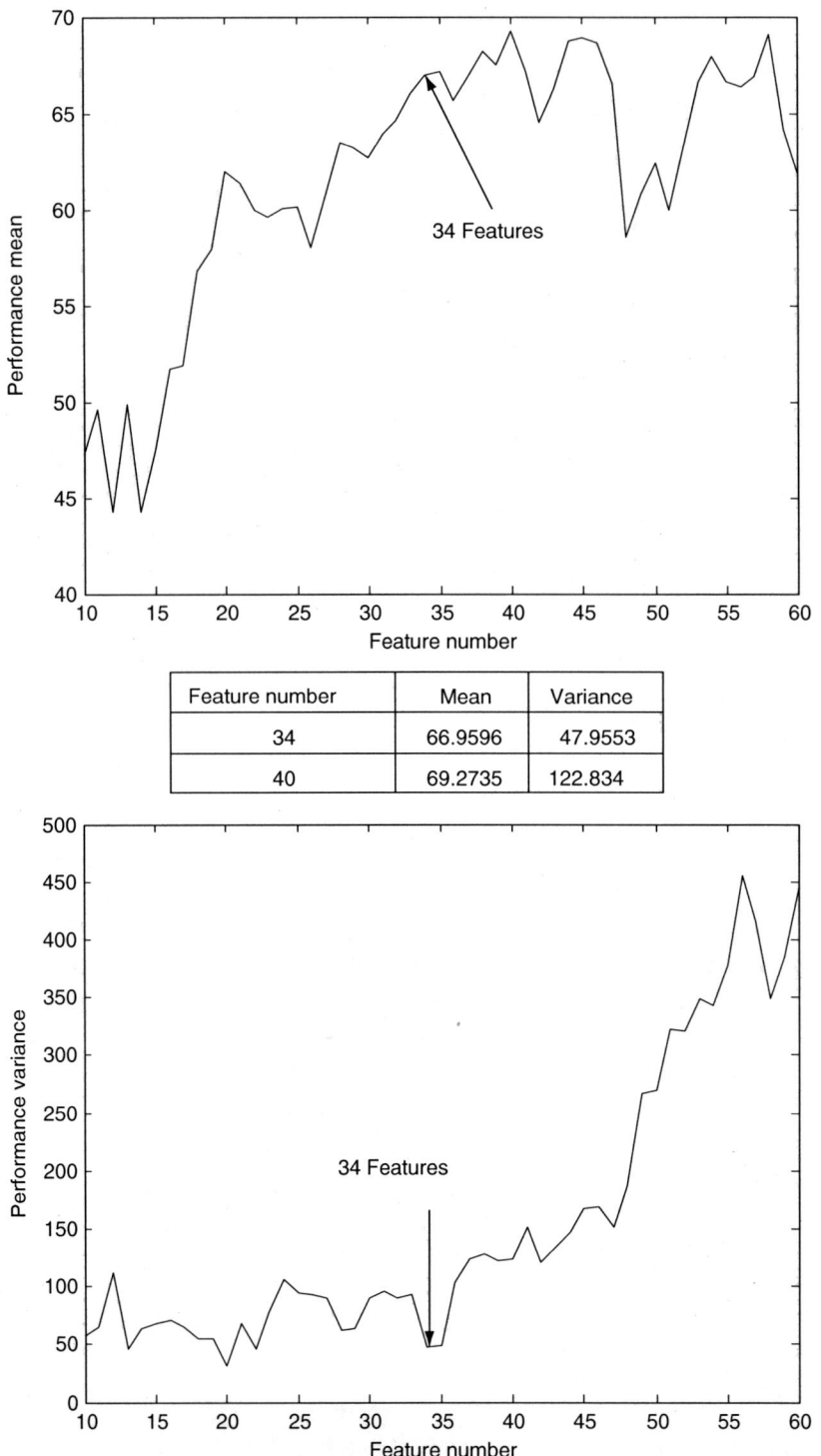

FIGURE 3.10
Performance means and performance variances versus feature number used to determine the optimal length of the feature vector.

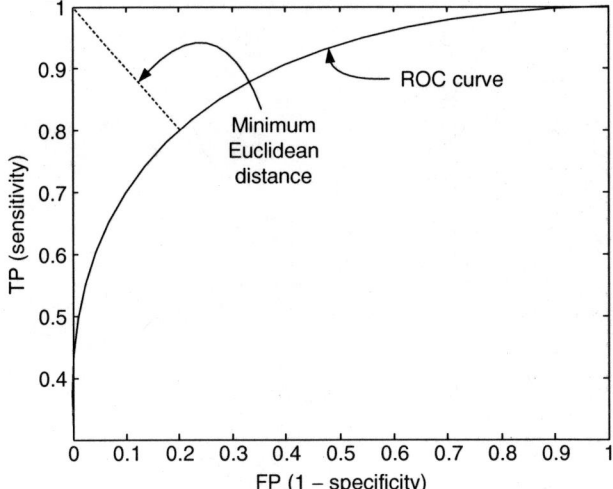

FIGURE 3.11
Example ROC curve.

the optimal threshold value for each neural module. Figure 3.12 shows the six 3-D ROC curves associated with the classifiers.

3.5 Simulation Results

The four basic parameters that are to be optimized in the process of training the neural network classifier (i.e., the bank of six RBF NNs) are the RBF NN spread parameters, the output thresholds of each neural module, the combination of 34 components in the feature vectors for each class (note again in Figure 3.2, each neural module has its own custom preprocessor) and of course the resulting weights of each RBF NN. The MATLAB neural networks toolbox was used to design the six RBF NNs [33].

Table 3.1 shows the specific classes and the associated number of signals used to train and test the RBF NNs. Of the 574 infrasound signals, 351 were used for training the remaining 223 were used for testing. The criterion used to divide the data between the training and testing sets was to maintain independence. Hence, the four array signals from any one event are always kept together, either in the training set or the test set.

After the optimal number components for each feature vector was determined, i.e., 34 elements, and the optimal combination of the 34 components for each preprocessor, the optimal RBF spread parameters are determined along with the optimal threshold value (the six graphs in Figure 3.12 were used for this purpose). For both the RBF spread parameters and the output thresholds, the selection criterion is based on maximizing the CCR of the local network and the overall (global) classifier CCR.

The RBF spread parameter and the output threshold for each neural module was determined one by one by fixing the spread parameter, i.e., σ, for all other neural modules to 0.3, and holding the threshold value at 0.5. Once the first neural module's spread

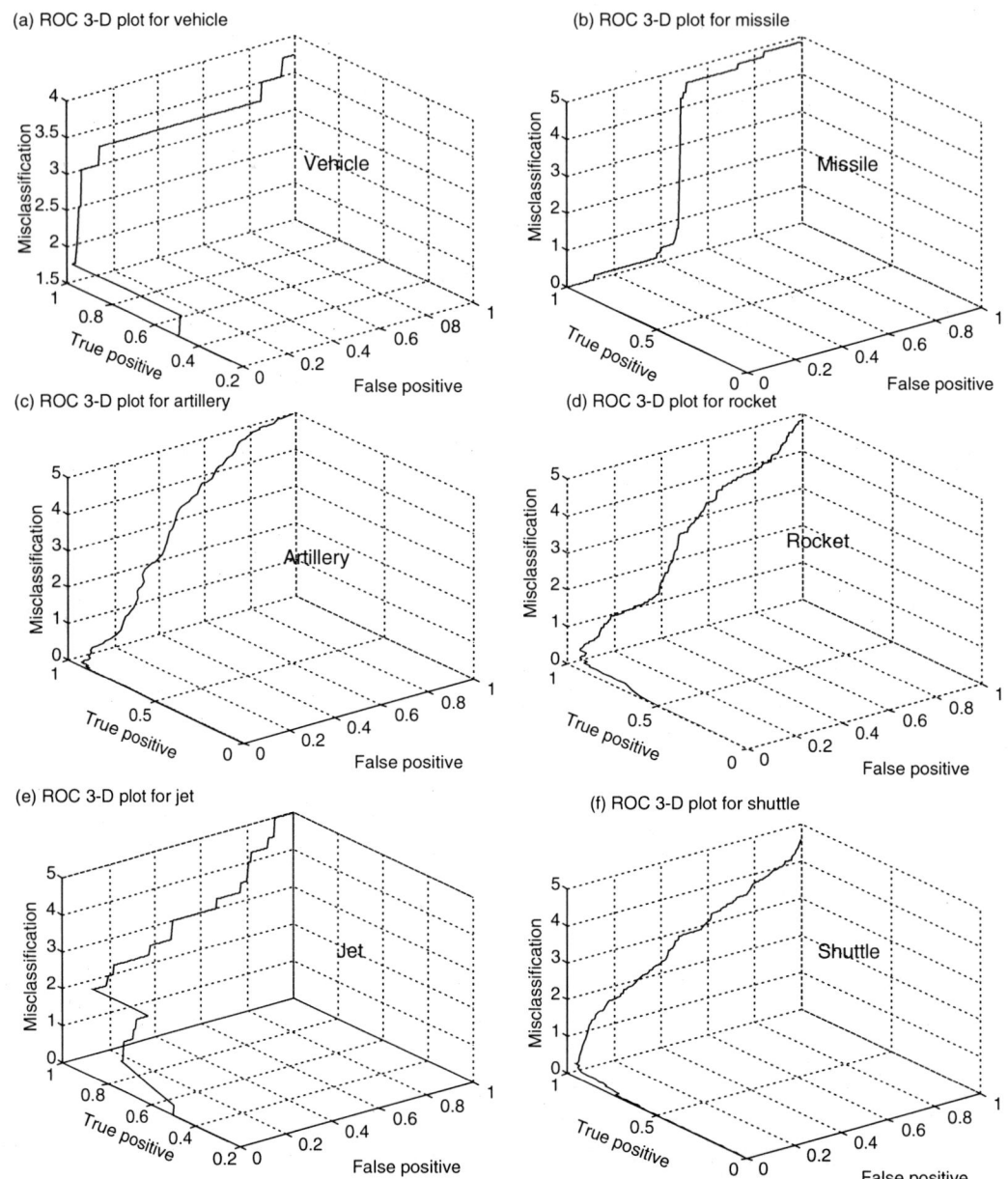

FIGURE 3.12
3-D ROC curves for the six classes.

parameter and threshold is determined, then the spread parameter and output threshold of the second neural module is computed while holding all other neural modules' (except the first one) spread parameters and output thresholds fixed at 0.3 and 0.5, respectively.

Table 3.4 gives the final values of the spread parameter and the output threshold for the global classifier. Figure 3.13 shows the classifier architecture with the final values indicated for the RBF NN spread parameters and the output thresholds.

Table 3.5 shows the confusion matrix for the six-classifier. Concentrating on the 6×6 portion of the matrix for each of the defined classes, the diagonal elements correspond to

A Universal Neural Network–Based Infrasound Event Classifier

TABLE 3.4

Spread Parameter and Threshold of Six-Class Classifier

	Spread Parameter	Threshold Value	True Positive	False Positive
Vehicle	0.2	0.3144	0.5	0
Artillery	2.2	0.6770	0.9621	0.0330
Jet	0.3	0.6921	0.5	0
Missile	1.8	0.9221	1	0
Rocket	0.2	0.4446	0.9600	0.0202
Shuttle	0.3	0.6170	0.9600	0.0289

the correct predictions. The trace of this 6×6 matrix divided by the total number of signals tested (i.e., 223) gives the accuracy of the global classifier. The formula for the accuracy is given in Equation 3.13, and here ACC = 94.6%. The off-diagonal elements indicate the misclassifications that occurred and those in parentheses indicate double classifications (i.e., the actual class was identified correctly, but there was another one of the output thresholds for another class that was exceeded). The off-diagonal element that is in square

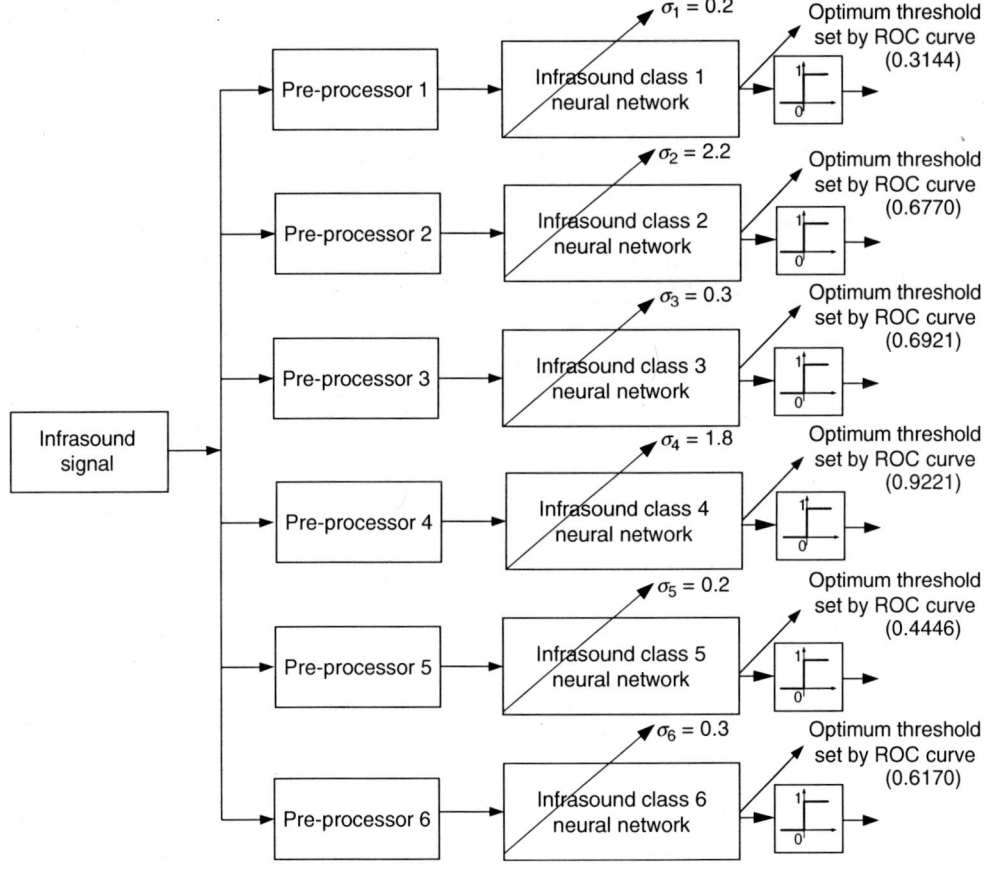

FIGURE 3.13

Parallel bank neural network classifier architecture with the final values for the RBF spread parameters and the output threshold values.

TABLE 3.5

Confusion Matrix for the Six-Class Classifier

		Predicted Value							
		Vehicle	Artillery	Jet	Missile	Rocket	Shuttle	Unclassified	Total (223)
	Vehicle	2	(1)	0	0	0	0	2	4
	Artillery	0	127	0	0	0	0	5	132
Actual	Jet	0	0	2	0	0	0	2	4
Value	Missile	0	0	0	8	0	0	0	8
	Rocket	0	0	0	0	24	(5)	1	25
	Shuttle	0	(1)[1]	0	0	1(3)	48	1	50

brackets is a double misclassification, that is, this event is misclassified along with another misclassified event (this is a shuttle event that is misclassified as both a "rocket" event as well as an "artillery" event).

Table 3.6 shows the final global classifier results giving both the CCR (see Equation 3.12) and the ACC (see Equation 3.13). Simulations were also run using "bi-polar" outputs instead of binary outputs. For the case of bi-polar outputs, the output is bound between −1 and +1 instead of 0 and 1. As can be seen from the table, the binary case yielded the best results. Finally, Table 3.7 shows the results for the case where the threshold levels on the outputs of the individual RBF NNs are ignored and only the output with this largest value is taken as the "winner," that is, "winner-takes-all"; this is considered to be the class that the input SOI belongs to. It should be noted that even though the CCR shows a higher level of performance for the winner-takes-all approach, this is probably not a viable method for classification. The reason being that if there were truly multiple events occurring simultaneously, they would never be indicated as such using this approach.

TABLE 3.6

Six-Class Classification Result Using a Threshold Value for Each Network

Performance Type	Binary Outputs (%)	Bi-Polar Outputs (%)
CCR	90.1	88.38
ACC	94.6	92.8

TABLE 3.7

Six-Class Classification Result Using "Winner-Takes-All"

Performance Type	Binary Method (%)	Bi-Polar Method (%)
CCR	93.7	92.4
ACC	93.7	92.4

3.6 Conclusions

Radial basis function neural networks were used to classify six different infrasound events. The classifier was built with a parallel structure of neural modules that individually are responsible for classifying one and only one infrasound event, referred to as PBNNC architecture. The overall accuracy of the classifier was found to be greater than 90%, using the CCR performance criterion. A feature extraction technique was employed that had a major impact toward increasing the classification performance over most other methods that have been tried in the past. Receiver operating characteristic curves were also employed to optimally set the output thresholds of the individual neural modules in the PBNNC architecture. This also contributed to increased performance of the global classifier. And finally, by optimizing the individual spread parameters of the RBF NN, the overall classifier performance was increased.

Acknowledgments

The authors would like to thank Dr. Steve Tenney and Dr. Duong Tran-Luu from the Army Research Laboratory for their support of this work. The authors also thank Dr. Kamel Rekab, University of Missouri–Kansas City, and Mr. Young-Chan Lee, Florida Institute of Technology, for their comments and insight.

References

1. Pierce, A.D., *Acoustics: An Introduction to Its Physical Principles and Applications*, Acoustical Society of America Publications, Sewickley, PA, 1989.
2. Valentina, V.N., *Microseismic and Infrasound Waves*, Research Reports in Physics, Springer-Verlag, New York, 1992.
3. National Research Council, *Comprehensive Nuclear Test Ban Treaty Monitoring*, National Academy Press, Washington, DC, 1997.
4. Bedard, A.J. and Georges, T.M., Atmospheric infrasound, *Physics Today*, 53(3), 32–37, 2000.
5. Ham, F.M. and Kostanic, I., *Principles of Neurocomputing for Science and Engineering*, McGraw-Hill, New York, 2001.
6. Torres, H.M. and Rufiner, H.L., Automatic speaker identification by means of mel cepstrum, wavelets and wavelet packets, In *Proc. 22nd Annu. EMBS Int. Conf.*, Chicago, IL, July 23–28, 2000, pp. 978–981.
7. Foo, S.W. and Lim, E.G., Speaker recognition using adaptively boosted classifier, In *Proc. The IEEE Region 10th Int. Conf. Electr. Electron. Technol. (TENCON 2001)*, Singapore, Aug. 19–22, 2001, pp. 442–446.
8. Moonasar, V. and Venayagamoorthy, G., A committee of neural networks for automatic speaker recognition (ASR) systems, *IJCNN*, Vol. 4, Washington, DC, 2001, pp. 2936–2940.
9. Inal, M. and Fatihoglu, Y.S., Self-organizing map and associative memory model hybrid classifier for speaker recognition, 6th Seminar on Neural Network Applications in Electrical Engineering, NEUREL-2002, Belgrade, Yugoslavia, Sept. 26–28, 2002, pp. 71–74.
10. Ham, F.M., Rekab, K., Park, S., Acharyya, R., and Lee, Y.-C., Classification of infrasound events using radial basis function neural networks, Special Session: Applications of learning and data-driven methods to earth sciences and climate modeling, *Proc. Int. Joint Conf. Neural Networks*, Montréal, Québec, Canada, July 31–August 4, 2005, pp. 2649–2654.

11. Mammone, R.J., Zhang, X., and Ramachandran, R.P., Robust speaker recognition: a feature-based approach, *IEEE Signal Process. Mag.*, 13(5), 58–71, 1996.
12. Noble, J., Tenney, S.M., Whitaker, R.W., and ReVelle, D.O., Event detection from small aperture arrays, U.S. Army Research Laboratory and Los Alamos National Laboratory Report, September 2002.
13. Tenney, S.M., Noble, J., Whitaker, R.W., and Sandoval, T., Infrasonic SCUD-B launch signatures, U.S. Army Research Laboratory and Los Alamos National Laboratory Report, October 2003.
14. Haykin, S., *Neural Networks: A Comprehensive Foundation*, 2nd ed., Prentice-Hall, Upper Saddle River, NJ, 1999.
15. Powell, M.J.D., Radial basis functions for multivariable interpolation: a review, *IMA conference on Algorithms for the Approximation of Function and Data*, RMCS, Shrivenham, U.K., 1985, pp. 143–167.
16. Powell, M.J.D., Radial basis functions for multivariable interpolation: a review, *Algorithms for the Approximation of Function and Data*, Mason, J.C. and Cox, M.G., Eds., Clarendon Press, Oxford, U.K., 1987.
17. Parzen, E., On estimation of a probability density function and mode, *Ann. Math. Stat.*, 33, 1065–1076, 1962.
18. Duda, R.O., and Hart, P.E., *Pattern Classification and Scene Analysis*, John Wiley & Sons, New York, 1973.
19. Specht, D.F., Probabilistic neural networks, *Neural Networks*, 3(1), 109–118, 1990.
20. Poggio, T. and Girosi, F., *A Theory of Networks for Approximation and Learning*, A.I. Memo 1140, MIT Press, Cambridge, MA, 1989.
21. Wilson, C.R. and Forbes, R.B., Infrasonic waves from Alaskan volcanic eruptions, *J. Geophys. Res.*, 74, 4511–4522, 1969.
22. Wilson, C.R., Olson, J.V., and Richards, R., Library of typical infrasonic signals, Report prepared for *ENSCO* (subcontract no. 269343-2360.009), Vols. 1–4, 1996.
23. Bedard, A.J., Infrasound originating near mountain regions in Colorado, *J. Appl. Meteorol.*, 17, 1014, 1978.
24. Olson, J.V. and Szuberla, C.A.L., Distribution of wave packet sizes in microbarom wave trains observed in Alaska, *J. Acoust. Soc. Am.*, 117(3), 1032–1037, 2005.
25. Ham, F.M., Leeney, T.A., Canady, H.M., and Wheeler, J.C., Discrimination of volcano activity using infrasonic data and a backpropagation neural network, In *Proc. SPIE Conf. Appl. Sci. Computat. Intell. II*, Priddy, K.L., Keller, P.E., Fogel, D.B., and Bezdek, J.C., Eds., Orlando, FL, 1999, Vol. 3722, pp. 344–356.
26. Ham, F.M., Leeney, T.A., Canady, H.M., and Wheeler, J.C., An infrasonic event neural network classifier, In *Proc. 1999 Int. Joint Conf. Neural Networks*, Session 10.7, Paper No. 77, Washington, DC, July 10–16, 1999.
27. Ham, F.M., Neural network classification of infrasound events using multiple array data, *International Infrasound Workshop 2001*, Kailua-Kona, Hawaii, 2001.
28. Ham, F.M. and Park, S., A robust neural network classifier for infrasound events using multiple array data, In *Proc. 2002 World Congress on Computational Intelligence*—International Joint Conference on Neural Networks, Honolulu, Hawaii, May 12–17, 2002, pp. 2615–2619.
29. Kohavi, R. and Provost, F., Glossary of terms, *Mach. Learn.*, 30(2/3), 271–274, 1998.
30. McDonough, R.N. and Whalen, A.D., *Detection of Signals in Noise*, 2nd ed., Academic Press, San Diego, CA, 1995.
31. Smith, S.W., *The Scientist and Engineer's Guide to Digital Signal Processing*, California Technical Publishing, San Diego, CA, 1997.
32. Hanley, J.A. and McNeil, B.J., The meaning and use of the area under a receiver operating characteristic (ROC) curve, *Radiology*, 143(1), 29–36, 1982.
33. Demuth, H. and Beale, M., *Neural Network Toolbox for Use with MATLAB*, The MathWorks, Inc., Natick, MA, 1998.

4

Construction of Seismic Images by Ray Tracing

Enders A. Robinson

CONTENTS

4.1 Introduction 53
4.2 Acquisition and Interpretation 54
4.3 Digital Seismic Processing 56
4.4 Imaging by Seismic Processing 58
4.5 Iterative Improvement 60
4.6 Migration in the Case of Constant Velocity 61
4.7 Implementation of Migration 62
4.8 Seismic Rays 64
4.9 The Ray Equations 68
4.10 Numerical Ray Tracing 69
4.11 Conclusions 71
References 72

4.1 Introduction

Reflection seismology is a method of remote imaging used in the exploration of petroleum. The seismic reflection method was developed in the 1920s. Initially, the source was a dynamite explosion set off in a shallow hole drilled into the ground, and the receiver was a geophone planted on the ground. In difficult areas, a single source would refer to an array of dynamite charges around a central point, called the *source point*, and a receiver would refer to an array of geophones around a central point, called the *receiver point*. The received waves were recorded on a photographic paper on a drum. The developed paper was the seismic record or seismogram. Each receiver accounted for a single wiggly line on the record, which is called a *seismic trace* or simply a *trace*. In other words, a seismic trace is a signal (or time series) received at a specific receiver location from a specific source location. The recordings were taken for a time span starting at the time of the shot (called time zero) until about three or four seconds after the shot. In the early days, a seismic crew would record about 10 or 20 seismic records per day, with a dozen or two traces on each record. Figure 4.1 shows a seismic record with wiggly lines as traces. Seismic crew number 23 of the Atlantic Refining Company shot the record on October 9, 1952. As written on the record, the traces were shot with a source that was laid out as a 36-hole circular array. The first circle in the array had a diameter of 130 feet with 6 holes, each hole loaded with 10 lbs of dynamite. The second circle in the array had a diameter of 215 feet with 11 holes (which should have been 12 holes, but one hole was missing), each

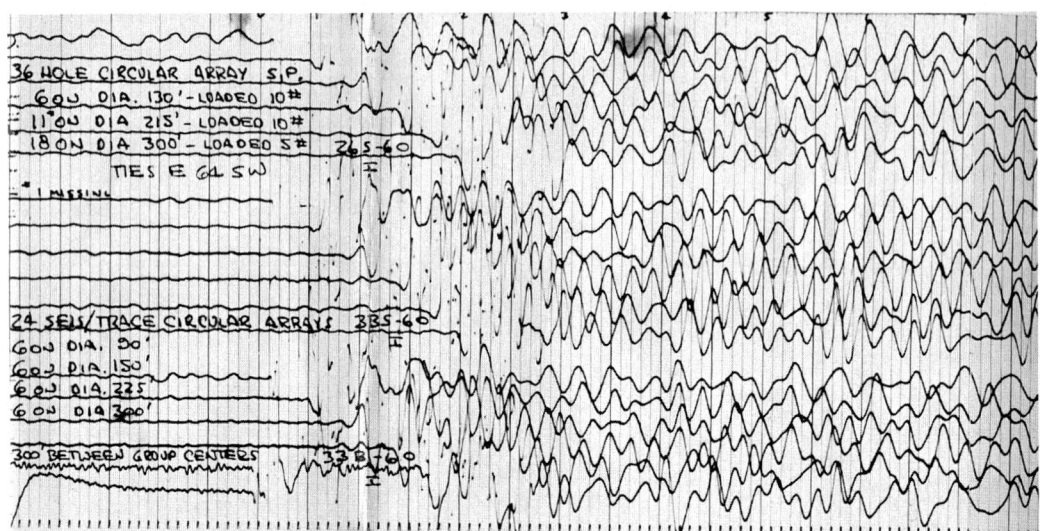

FIGURE 4.1
Seismic record taken in 1952.

hole loaded with 10 lbs of dynamite. The third circle in the array had a diameter of 300 feet with 18 holes, each hole loaded with 5 lbs of dynamite. Each charge was at a depth of 20 feet. The total charge was thus 260 lbs of dynamite, which is a large amount of dynamite for a single seismic record. The receiver for each trace was made up of a group of 24 geophones (also called seismometers) in a circular array with 6 geophones on each of 4 circles of diameters 50 feet, 150 feet, 225 feet, and 300 feet, respectively. There was a 300-feet gap between group centers (i.e., receiver points). This record is called an *NR seismogram*. The purpose of the elaborate source and receiver arrays was an effort to bring out visible reflections on the record. The effort was fruitless. Regions where geophysicists can see no reflections on the raw records are termed as no record or no reflection (NR) areas. The region in which this record was taken, as the great majority of possible oil-bearing regions in the world, was an NR area. In such areas, the seismic method (before digital processing) failed, and hence wildcat wells had to be drilled based on surface geology and a lot of guess work. There was a very low rate of discovery in the NR regions. Because of the tremendous cost of drilling to great depths, there was little chance that any oil would ever be discovered. The outlook for oil was bleak in the 1950s.

4.2 Acquisition and Interpretation

From the years of its inception up to about 1965, the seismic method involved two steps, namely *acquisition* and *interpretation*. Acquisition refers to the generation and recording of seismic data. Sources and receivers are laid out on the surface of the Earth. The objective is to probe the unknown structure below the surface. The sources are made up of vibrators (called vibroseis), dynamite shots, or air guns. The receivers are geophones on land and hydrophones at sea. The sources are activated one at a time, not all together. Suppose a single source is activated, the resulting seismic waves travel from the source into the Earth. The waves pass down through sedimentary rock strata, from which the waves are reflected upward. A reflector is an interface between layers of contrasting physical properties. A reflector might represent a change in lithology, a fault, or an unconformity. The reflected energy returns to the surface, where it is recorded. For each source activated, there are many receivers surrounding the source point. Each recorded signal, called a *seismic trace*, is associated with a particular source point and a particular receiver point. The traces, as recorded, are referred to as the raw traces. A raw trace contains all the received events. These events are produced by the subsurface structure of the Earth. The events due to primary reflections are wanted; all the other events are unwanted.

Interpretation was the next step after acquisition. Each seismic record was examined through the eye and the primary reflections that could be seen were marked by a pencil. A primary reflection is an event that represents a passage from the source to the depth point, and then a passage directly back to the receiver (Figure 4.2). At a reflection, the traces become coherent; that is, they come into phase with each other. In other words, at a reflection, the crests and troughs on adjacent traces appear to fit into one another. The arrival time of a reflection indicates the depth of the reflecting horizon below the surface, while the time differential (the so-called step-out time) in the arrivals of a given peak or trough at successive receiver positions provides information on the dip of the reflecting horizon. In favorable areas, it is possible to follow the same reflection over a distance much greater than that covered by the receiver spread for a single record. In such cases, the records are placed side-by-side. The reflection from the last trace of one record correlates with the first trace of the next record. Such a correlation can be continued on successive records as long as the reflection persists. In areas of rapid structural change, the ensemble of raw traces is unable to show the true geometry of subsurface structures. In some cases, it is possible to identify an isolated structure such as a fault or a syncline on the basis of its characteristic reflection pattern. In NR regions, the raw record section does not give a usable image of the subsurface at all.

Seismic wave propagation in three dimensions is a complicated process. The rock layers absorb, reflect, refract, or scatter the waves. Inside the different layers, the waves propagate at different velocities. The waves are reflected and refracted at the interfaces between the layers. Only elementary geometries can be treated exactly in three dimensions. If the reflecting interfaces are horizontal (or nearly so), the waves going straight down will be reflected nearly straight up. Thus, the wave motion is essentially vertical. If the time axes on the records are placed in the vertical position, time appears in the same direction as the raypaths. By using the correct wave velocity, the time axis can be converted into the depth axis. The result is that the primary reflections show the locations

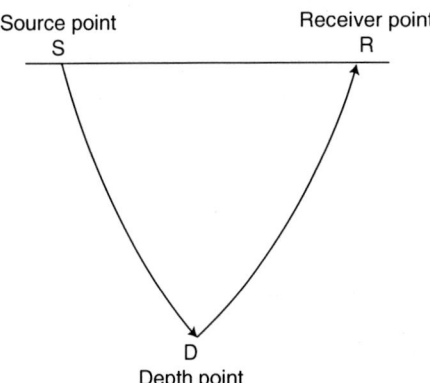

FIGURE 4.2
Primary reflection.

of the reflecting interfaces. Thus, in areas that have nearly level reflecting horizons, the primary reflections, as recorded, essentially show the correct depth positions of the subsurface interfaces. However, in areas that have a more complicated subsurface structure, the primary reflections as recorded in time do not occur at the correct depth positions in space. As a result, the primary reflections have to be moved (or migrated) to their proper spatial positions.

In areas of complex geology, it is necessary to move (or migrate) the energy of each primary reflection to the spatial position of the reflecting point. The method is similar to that used in *Huygens's construction*. Huygens articulated the principle that every point of a wavefront can be regarded as the origin of a secondary spherical wave, and the envelope of all these secondary waves constitutes the propagated wavefront. In the predigital days, migration was carried out by straightedge and compass or by a special-purpose hand-manipulated drawing machine on a large sheet of a graph paper. The arrival times of the observed reflections were marked on the seismic records. These times are the two-way traveltimes from the source point to the receiver point. If the source and the receiver were at the same point (i.e., coincident), then the raypath down would be the same as the raypath up. In such a case, the one-way time is one half of the two-way time. From the two-way traveltime data, such a one-way time was estimated for each source point. This one-way time was multiplied by an estimated seismic velocity. The travel distance to the interface was thus obtained. A circle was drawn with the surface point as center and the travel distance as radius. This process was repeated for the other source points. In Huygens's construction, the envelope of the spherical secondary waves gives the new wavefront. In a similar manner, in the seismic case, the envelope of the circles gives the reflecting interface. This method of migration was done in 1921 in the first reflection seismic survey ever taken.

4.3 Digital Seismic Processing

Historically, most seismic work fell under the category of two-dimensional (2D) imaging. In such cases, the source positions and the receiver positions are placed on a horizontal surface line called the x-axis. The time axis is a vertical line called the t-axis. Each source would produce many traces—one trace for each receiver position on the x-axis. The waves that make up each trace take a great variety of paths, each requiring a different time to travel from the source to receiver. Some waves are refracted and others scattered. Some waves travel along the surface of the Earth, and others are reflected upward from various interfaces. A primary reflection is an event that represents a passage from the source to the depth point, and then a passage directly back to the receiver. A multiple reflection is an event that has undergone three, five, or some other odd number of reflections in its travel path. In other words, a multiple reflection takes a zig-zag course with the same number of down legs as up legs. Depending on their time delay from the primary events with which they are associated, multiple reflections are characterized as short-path, implying that they interfere with the primary reflection, or as long-path, where they appear as separate events. Usually, primary reflections are simply called *reflections* or *primaries*, whereas multiple reflections are simply called *multiples*. A water-layer reverberation is a type of multiple reflection due to the multiple bounces of seismic energy back and forth between the water surface and the water bottom. Such reverberations are common in marine seismic data. A reverberation re-echoes (i.e., bounces back and forth) in the water layer for a prolonged period of time. Because of its resonant nature, a reverberation is a troublesome type of multiple. Reverberations conceal the primary reflections. The primary reflections (i.e., events that have undergone only one reflection) are needed for image formation. To make use of the primary reflected signals on the record, it is necessary to distinguish them from the other type of signals on the record. Random noise, such as wind noise, is usually minor and, in such cases, can be neglected. All the seismic signals, except primary reflections, are unwanted. These unwanted signals are due to the seismic energy introduced by the seismic source signal; hence, they are called *signal-generated noise*. Thus, we are faced with the problem of (primary reflected) signal enhancement and (signal-generated) noise suppression. In the analog days (approximately up to about 1965), the separation of signal and noise was done through the eye.

In the 1950s, a good part of the Earth's sedimentary basins, including essentially all water-covered regions, were classified as NR areas. Unfortunately, in such areas, the signal-generated noise overwhelms the primary reflections. As a result, the primary reflections cannot be picked up visually. For example, water-layer reverberations as a rule completely overwhelm the primaries in the water-covered regions such as the Gulf of Mexico, the North Sea, and the Persian Gulf. The NR areas of the world could be explored for oil in a direct way by drilling, but not by the remote detection method of reflection seismology. The decades of the 1940s and 1950s were replete with inventions, not the least of which was the modern high-speed electronic stored-program digital computer. In the years from 1952 to 1954, almost every major oil company joined the MIT Geophysical Analysis Group to use the digital computer to process NR seismograms (Robinson, 2005). Historically, the additive model (trace = $s+n$) was used. In this model, the desired primary reflections were the signal s and everything else was the noise n. Electric filers and other analog methods were used, but they failed to give the desired primary reflections. The breakthrough was the recognition that the convolutional model (trace = $s*n$) is the correct model for a seismic trace. Note that the asterisk denotes convolution. In this model, the signal-generated noise is the signal s and the unpredictable primary reflections are the noise n. Deconvolution removes the signal-generated noise (such as instrument responses, ground roll, diffractions, ghosts, reverberations, and other types of multiple reflections) so as to yield the underlying primary reflections. The MIT Geophysical Analysis Group demonstrated the success of deconvolution on many NR seismograms, including the record shown in Figure 4.1. However, the oil companies were not ready to undertake digital seismic processing at that time. They were discouraged because an excursion into digital seismic processing would require new effort that would be expensive, and still the effort might fail because of the unreliability of the existing computers. It is true that in 1954 the available digital computers were far from suitable for geophysical processing. However, each year from 1946 onward, there was a constant stream of improvements in computers, and this development was accelerating every year. With patience and time, the oil and geophysical companies would convert to digital processing. It would happen when the need for hard-to-find oil was great enough to justify the investment necessary to turn NR seismograms into meaningful data. Digital signal processing was a new idea to the seismic exploration industry, and the industry shied away from converting to digital methods until the 1960s. The conversion of the seismic exploration industry to digital was in full force by about 1965, at which time transistorized computers were generally available at a reasonable price. Of course, a reasonable price for one computer then would be in the range from hundreds of thousands of dollars to millions of dollars. With the digital computer, a whole new step in seismic exploration was added, namely digital processing. However, once the conversion to digital was undertaken in the years around 1965, it was done quickly and effectively. Reflection seismology now involves three steps, namely *acquisition, processing,* and *interpretation*. A comprehensive presentation of seismic data processing is given by Yilmaz (1987).

4.4 Imaging by Seismic Processing

The term *imaging* refers to the formation of a computer image. The purpose of seismic processing is to convert the raw seismic data into a useful image of the subsurface structure of the Earth. From about 1965 onward, most of the new oil fields discovered were the result of the digital processing of the seismic data. Digital signal processing deconvolves the data and then superimposes (migrates) the results. As a result, seismic processing is divided into two main divisions: the deconvolution phase, which produces primaries-only traces (as well as possible), and the migration phase, which moves the primaries to their true depth positions (as well as possible). The result is the desired image. The first phase of imaging (i.e., deconvolution) is carried out on the traces, either individually by means of single-channel processing or in groups by means of multi-channel processing. Ancillary signal-enhancement methods typically include such things as the analyses of velocities and frequencies, static and dynamic corrections, and alternative types of deconvolution. Deconvolution is performed on one or a few traces at a time; hence, the small capacity of the computers of the 1960s was not a severely limiting factor.

The second phase of imaging (i.e., migration) is the movement of the amplitudes of the primary-reflection events to their proper spatial locations (the depth points). Migration can be implemented by a Huygens-like superposition of the deconvolved traces. In a mechanical medium, such as the Earth, forces between the small rock particles transmit the disturbance. The disturbance at some regions of rock acts locally on nearby regions. Huygens imagined that the disturbance on a given wavefront is made up of many separate disturbances, each of which acts like a point source that radiates a spherically symmetric secondary wave, or wavelet. The superposition of these secondary waves gives the wavefront at a later time. The idea that the new wavefront is obtained by superposition is the crowning achievement of Huygens. See Figure 4.3. In a similar way, seismic migration uses superposition to find the subsurface reflecting interfaces.

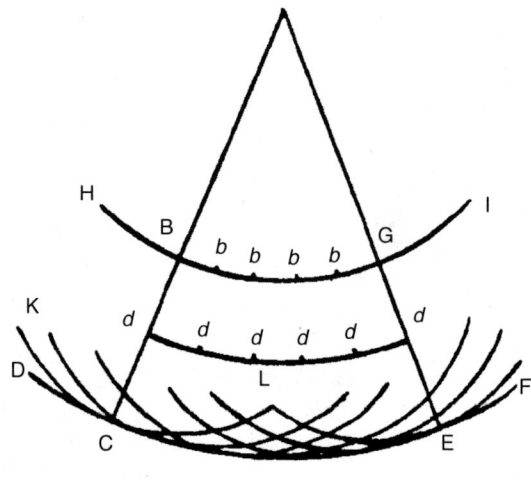

FIGURE 4.3
Huygens's construction.

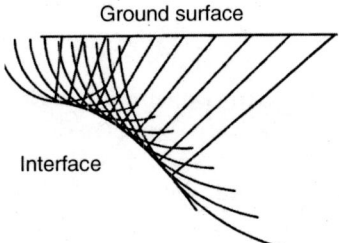

FIGURE 4.4
Construction of a reflecting interface.

See Figure 4.4. In the case of the migration, there is an added benefit of superposition, namely, superposition is one of the most effective ways to accentuate signals and suppress noise. The superposition used in migration is designed to return the primary reflections to their proper spatial locations. The remnants of the signal-generated noise on the deconvolved traces are out of step with the primary events. As a result, the remaining signal-generated noise tends to be destroyed by the superposition. The superposition used in migration provides the desired image of the underground structure of the Earth.

Historically, migration (i.e., the movement of reflected events to their proper locations in space) was carried out manually, sometimes making use of elaborate drawing instruments. The transfer of the manual processes to a digital computer involved the manipulation of a great number of traces at once. The resulting digital migration schemes all relied heavily on the superposition of the traces. This tremendous amount of data handling had a tendency to overload the limited capacities of the computers made in the 1960s and 1970s. As a result, it was necessary to simplify the migration problem and to break it down into smaller parts. Thus, migration was done by a sequence of approximate operations, such as stacking, followed by normal moveout, dip moveout, and migration after stack. The process known as *time migration* was often used, which improved the records in time, but stopped short of placing the events in their proper spatial positions. All kinds of modifications and adjustments were made to these piecemeal operations, and seismic migration in the 1970s and 1980s was a complicated discipline—an art as much as a science. The use of this art required much skill. Meanwhile, great advances in technology were taking place. In the 1990s, everything seemed to come together. Major improvements in instrumentation and computers resulted in light compact geophysical field equipment and affordable computers with high speed and massive storage. Instead of the modest number of sources and receivers used in 2D seismic processing, the tremendous number required for three-dimensional (3D) processing started to be used on a regular basis in field operations. Finally, the computers were large enough to handle the data for 3D imaging. Event movements (or migration) in three dimensions can now be carried out economically and efficiently by time-honored superposition methods such as those used in the Huygens's construction. These migration methods are generally named as *prestack migration*. This name is a relic, which implies that stacking and all the other piecemeal operations are no longer used in the migration scheme. Until the 1990s, 3D seismic imaging was rarely used because of the prohibitive costs involved. Today, 3D methods are commonly used, and the resulting subsurface images are of extraordinary quality. Three-dimensional prestack migration significantly improves seismic interpretation because the locations of geological structures, especially faults, are given much more accurately. In addition, 3D migration collapses diffractions from secondary sources such as reflector terminations against faults and corrects bow ties to form synclines. Three-dimensional seismic work gives beautiful images of the underground structure of the Earth.

4.5 Iterative Improvement

Let (x, y) represent the surface coordinates and z represent the depth coordinate. Migration takes the deconvolved records and moves (i.e., migrates) the reflected events to depth points in the 3D volume (x, y, z). In this way, seismic migration produces an image of the geologic structure $g(x, y, z)$ from the deconvolved seismic data. In other words, migration is a process in which primary reflections are moved to their correct locations in space. Thus, for migration, we need the primary reflected events. What else is required? The answer is the complete velocity function $v(x, y, z)$, which gives the wave velocity at each point in the given volume of the Earth under exploration.

At this point, let us note that the word *velocity* is used in two different ways. One way is to use the word velocity for the scalar that gives the rate of change of position in relation to time. When velocity is a scalar, the terms *speed* or *swiftness* are often used instead. The other (and more correct) way is to use the word velocity for the vector quantity that specifies both the speed of a body and its direction of motion. In geophysics, the word velocity is used for the (scalar) speed or swiftness of a seismic wave. The reciprocal of velocity v is *slowness*, $n = 1/v$. Wave velocity can vary vertically and laterally in isotropic media. In anisotropic media, it can also vary azimuthally. However, we consider only isotropic media; therefore, at a given point, the wave velocity is the same in all directions. Wave velocity tends to increase with depth in the Earth because deep layers suffer more compaction from the weight of the layers above. Wave velocity can be determined from laboratory measurements, acoustic logs, and vertical seismic profiles or from velocity analysis of seismic data. Often, we say velocity when we mean wave velocity. Over the years, various methods have been devised to obtain a sampling of the velocity distribution within the Earth. The velocity functions so determined vary from method to method. For example, the velocity measured vertically from a check-shot or vertical seismic profile (VSP) differs from the stacking velocity derived from normal moveout measurements of ommon depth point gathers. Ideally, we would want to know the velocity at each and every point in the volume of Earth of interest. In many areas, there are significant and rapid lateral or vertical changes in the velocity that distort the time image. Migration requires an accurate knowledge of vertical and horizontal seismic velocity variations. Because the velocity depends on the types of rocks, a complete knowledge of the velocity is essentially equivalent to a complete description of the geologic structure $g(x, y, z)$. However, as we have stated above, the velocity function is required to get the geologic structure. In other words, to get the answer (the geologic structure) $g(x, y, z)$, we must know the answer (the velocity function) $v(x, y, z)$.

Seismic interpretation takes the images generated as representatives of the physical Earth. In an iterative improvement scheme, any observable discrepancies in the image are used as forcing functions to correct the velocity function. Sometimes, simple adjustments can be made, and, at other times, the whole imaging process has to be redone one or more times before a satisfactory solution can be obtained. In other words, there is interplay between seismic processing and seismic interpretation, which is a manifestation of the well-accepted exchange between the disciplines of geophysics and geology. *Iterative improvement* is a well-known method commonly used in those cases where you must know the answer to find the answer. By the use of iterative improvement, the seismic inverse problem is solved. In other words, the imaging of seismic data requires a model of seismic velocity. Initially, a model of smoothly varying velocity is used. If the results are not satisfactory, the velocity model is adjusted and a new image is formed. This process is repeated until a satisfactory image is obtained. To get the image, we must know the velocity; the method of iterative improvement deals with this problem.

4.6 Migration in the Case of Constant Velocity

Consider a primary reflection. Its two-way traveltime is the time it takes for the seismic energy to travel down from the source $S = (x_S, y_S)$ to depth point $D = (x_D, y_D, z_D)$ and then back up to the receiver $R = (x_R, y_R)$. The deconvolved trace $f(S, R, t)$ gives the amplitude of the reflected signal as a function of two-way traveltime t, which is given in milliseconds from the time that the source is activated. We know S, R, t, and $f(S, R, t)$. The problem is to find D, which is the depth point at refection.

An isotropic medium is a medium whose properties at a point are the same in all directions. In particular, the wave velocity at a point is the same in all directions. Fresnel's principle of least time requires that in an isotropic medium the rays are orthogonal trajectories of the wavefronts. In other words, the rays are normal to the wavefronts. However, in an anisotropic medium, the rays need not be orthogonal trajectories of the wavefronts. A homogeneous medium is a medium whose physical properties are the same throughout. For ease of exposition, let us first consider the case of a homogenous isotropic medium. Within a homogeneous isotropic material, the velocity v has the same value at all points and in all directions. The rays are straight lines since by symmetry they cannot bend in any preferred direction, as there are none. The two-way traveltime t is the elapsed time for a seismic wave to travel from its source to a given depth point and return to a receiver at the surface of the Earth. The two-way traveltime t is thus equal to the one-way traveltime t_1 from the source point S to the depth point D plus the one-way traveltime t_2 from the depth point D to the receiver point R. Note that the traveltime from D to R is the same as the traveltime from R to D. We may write $t = t_1 + t_2$, which in terms of distance is

$$vt = \sqrt{(x_D - x_S)^2 + (y_D - y_S)^2 + (z_D - z_S)^2} + \sqrt{(x_D - x_R)^2 + (y_D - y_R)^2 + (z_D - z_R)^2}$$

We recall that an ellipse can be drawn with two pins, a loop of string, and a pencil. The pins are placed at the foci and the ends of the string are attached to the pins. The pencil is placed on the paper inside the string, so the string is taut. The string forms a triangle. If the pencil is moved around so that the string stays taut, the sum of the distances from the pencil to the pins remains constant, and the curve traced out by the pencil is an ellipse. Thus, if vt is the length of the string, then any point on the ellipse could be the depth point D that produces the reflection for that source S, receiver R, and traveltime t. We therefore take that event and move it out to each point on the ellipse.

Suppose we have two traces with only one event on each trace. Suppose both events come from the same reflecting surface. In Figure 4.5, we show the two ellipses. In the spirit of Huygens's construction, the reflector must be the common tangent to the ellipses. This example shows how migration works. In practice, we would take

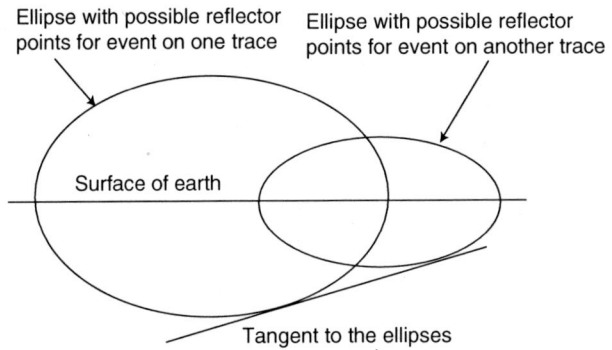

FIGURE 4.5
Reflecting interface as a tangent.

the amplitude at each digital time instant t on the trace, and scatter the amplitude on the constructed ellipsoid. In this way, the trace is spread out into a 3D volume. Then we repeat this operation for each trace, until every trace is spread out into a 3D volume. The next step is superposition. All of these volumes are added together. (In practice, each trace would be spread out and cumulatively added into one given volume.) Interference tends to destroy the noise, and we are left with the desired 3D image of the Earth.

4.7 Implementation of Migration

A raypath is a course along which wave energy propagates through the Earth. In isotropic media, the raypath is perpendicular to the local wavefront. The raypath can be calculated using ray tracing. Let the point $P = (x_P, y_P)$ be either a source location or a receiver location. The subsurface volume is represented by a 3D grid (x, y, z) of depth points D. To minimize the amount of ray tracing, we first compute a traveltime table for each and every surface location, whether the location be a source point or a receiver point. In other words, for each surface location P, we compute the one-way traveltime from P to each depth point D in the 3D grid. We put these one-way traveltimes into a 3D table that is labeled by the surface location P.

The traveltime for a primary reflection is the total two-way (i.e., down and up) time for a path originating at the source point S, reflected at the depth point D, and received at the receiver point R. Two identification numbers are associated with each trace: one for the source S and the other for the receiver R. We pull out the respective tables for these two identification numbers. We add the two tables together, element by element. The result is a 3D table for the two-way traveltimes for that seismic trace.

Let us give a 2D example. We assume that the medium has a constant velocity, which we take to be 1. Let the subsurface grid for depth points D be given by (z, x), where depth is given by $z = 1, 2, \ldots, 10$ and horizontal distance is given by $x = 1, 2, \ldots, 15$. Let the surface locations P be $(z = 1, x)$, where $x = 1, 2, \ldots, 15$. Suppose the source is $S = (1, 3)$. We want to construct a table of one-way traveltimes, where depth z denotes the row and horizontal distance x denotes the column. The one-way traveltime from the source to the depth point (z, x) is $t(z, x) = \sqrt{(z-1)^2 + (x-3)^2}$. For example, the traveltime from source to depth point $(z, x) = (4, 6)$ is

$$t(4, 6) = \sqrt{(4-1)^2 + (6-3)^2} = \sqrt{18} = 4.24$$

This number (rounded) appears in the fourth row, sixth column of the table below. The computed table (rounded) for the source is

2.0	1.0	0.0	1.0	2.0	3.0	4.0	5.0	6.0	7.0	8.0	9.0	10.0	11.0	12.0
2.2	1.4	1.0	1.4	2.2	3.2	4.1	5.1	6.1	7.1	8.1	9.1	10.0	11.0	12.0
2.8	2.2	2.0	2.2	2.8	3.6	4.5	5.4	6.3	7.3	8.2	9.2	10.2	11.2	12.2
3.6	3.2	3.0	3.2	3.6	4.2	5.0	5.8	6.7	7.6	8.5	9.5	10.4	11.4	12.4
4.5	4.1	4.0	4.1	4.5	5.0	5.7	6.4	7.2	8.1	8.9	9.8	10.8	11.7	12.6
5.4	5.1	5.0	5.1	5.4	5.8	6.4	7.1	7.8	8.6	9.4	10.3	11.2	12.1	13.0
6.3	6.1	6.0	6.1	6.3	6.7	7.2	7.8	8.5	9.2	10.0	10.8	11.7	12.5	13.4
7.3	7.1	7.0	7.1	7.3	7.6	8.1	8.6	9.2	9.9	10.6	11.4	12.2	13.0	13.9
8.2	8.1	8.0	8.1	8.2	8.5	8.9	9.4	10.0	10.6	11.3	12.0	12.8	13.6	14.4
9.2	9.1	9.0	9.1	9.2	9.5	9.8	10.3	10.8	11.4	12.0	12.7	13.5	14.2	15.0

Next, let us compute the traveltime for the receiver point. Note that the traveltime from the depth point to the receiver is the same as the traveltime from the receiver to the depth point. Suppose the receiver is $R = (11, 1)$. Then the one-way traveltime from the receiver to the depth point (z, x) is $t(z, x) = \sqrt{(z-1)^2 + (x-11)^2}$. For example, the traveltime from source to depth point $(z, x) = (4, 6)$ is $t(4, 6) = \sqrt{(4-1)^2 + (6-11)^2} = 5.85$. This number (rounded) appears in the fourth row, sixth column of the table below. The computed table (rounded) for the receiver is

10.0	9.0	8.0	7.0	6.0	5.0	4.0	3.0	2.0	1.0	0.0	1.0	2.0	3.0	4.0
10.0	9.1	8.1	7.1	6.1	5.1	4.1	3.2	2.2	1.4	1.0	1.4	2.2	3.2	4.1
10.2	9.2	8.2	7.3	6.3	5.4	4.5	3.6	2.8	2.2	2.0	2.2	2.8	3.6	4.5
10.4	9.5	8.5	7.6	6.7	5.8	5.0	4.2	3.6	3.2	3.0	3.2	3.6	4.2	5.0
10.8	9.8	8.9	8.1	7.2	6.4	5.7	5.0	4.5	4.1	4.0	4.1	4.5	5.0	5.7
11.2	10.3	9.4	8.6	7.8	7.1	6.4	5.8	5.4	5.1	5.0	5.1	5.4	5.8	6.4
11.7	10.8	10.0	9.2	8.5	7.8	7.2	6.7	6.3	6.1	6.0	6.1	6.3	6.7	7.2
12.2	11.4	10.6	9.9	9.2	8.6	8.1	7.6	7.3	7.1	7.0	7.1	7.3	7.6	8.1
12.8	12.0	11.3	10.6	10.0	9.4	8.9	8.5	8.2	8.1	8.0	8.1	8.2	8.5	8.9
13.5	12.7	12.0	11.4	10.8	10.3	9.8	9.5	9.2	9.1	9.0	9.1	9.2	9.5	9.8

The addition of the above two tables gives the two-way traveltimes. For example, the traveltime from source to depth point $(z, x) = (4, 6)$ and back to receiver is $t(4, 6) = 4.24 + 5.85 = 10.09$. This number (rounded) appears in the fourth row, sixth column of the two-way table below. The two-way table (rounded) for source and receiver is

12	10	8	8	8	8	8	8	8	8	8	10	12	14	16
12	10	9	8	8	8	8	8	8	8	9	10	12	14	16
13	11	10	10	9	9	9	9	9	10	10	11	13	15	17
14	13	12	11	10	10	10	10	10	11	12	13	14	16	17
15	14	13	12	12	11	11	11	12	12	13	14	15	17	18
17	15	14	14	13	13	13	13	13	14	14	15	17	18	19
18	17	16	15	15	15	14	15	15	15	16	17	18	19	21
19	18	18	17	16	16	16	16	16	17	18	18	19	21	22
21	20	19	19	18	18	18	18	18	19	19	20	21	22	23
23	22	21	20	20	20	20	20	20	21	22	23	24	25	

Figure 4.6 shows a contour map of the above table. The contour lines are elliptic curves of constant two-way traveltime for the given source and receiver pair.
Let the deconvolved trace (i.e., the trace with primary reflections only) be

Time	0	1	2	3	4	5	6	7	8	9	10	11	12	13	14	15	16	17	18	19
Trace Sample	0	0	0	0	0	0	0	0	0	0	-1	0	0	2	0	0	0	0	3	0

The next step is to place the amplitude of each trace at depth locations, where the traveltime of the trace sample equals the traveltime as given in the above two-way table. The trace is zero for all times except 10, 13, and 18. We now spread the trace

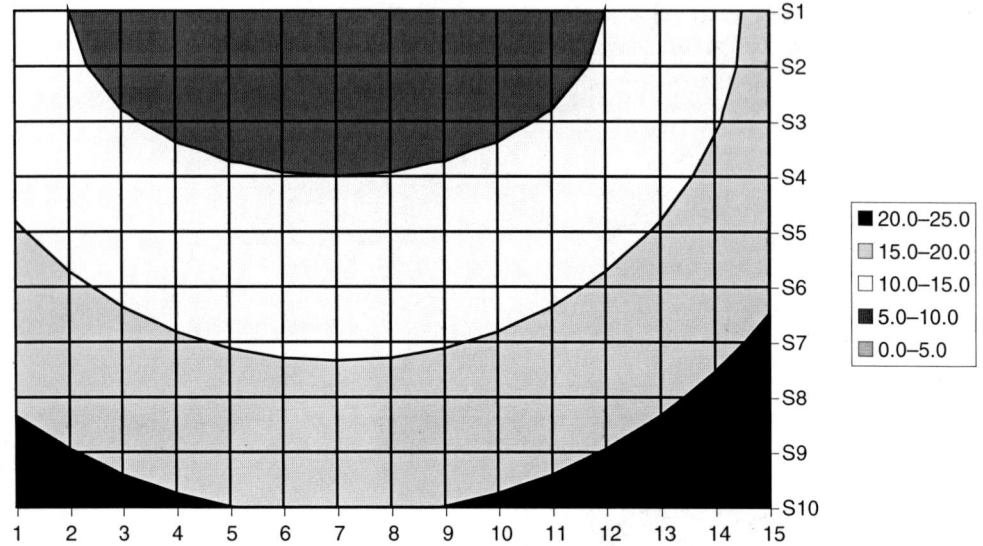

FIGURE 4.6
Elliptic contour lines.

out as follows. In the above two-way table, the entries with 10, 13, and 18 are replaced by the trace values −1, 2, 3, respectively. All other entries are replaced by zero. The result is

0	−1	0	0	0	0	0	0	0	0	0	0	0	0	0
0	−1	0	0	0	0	0	0	0	0	0	−1	0	0	0
2	0	−1	−1	0	0	0	0	0	−1	−1	0	2	0	0
0	2	0	0	−1	−1	−1	−1	−1	0	0	2	0	0	0
0	0	2	0	0	0	0	0	0	0	2	0	0	0	3
0	0	0	0	2	2	2	2	2	0	0	0	0	3	0
3	0	0	0	0	0	0	0	0	0	0	0	0	0	0
0	3	3	0	0	0	0	0	0	0	3	3	0	0	0
0	0	0	0	3	3	3	3	3	0	0	0	0	0	0
0	0	0	0	0	0	0	0	0	0	0	0	0	0	0

This operation is repeated for all traces in the survey, and the resulting tables are added together. The final image appears by the constructive and destructive interference among the individual trace contributions. The above procedure for a constant velocity is the same as with a variable velocity, except now the traveltimes are computed according to the velocity function $v(x, y, z)$. The eikonal equation can provide the means; hence, for the rest of this paper we will develop the properties of this basic equation.

4.8 Seismic Rays

To move the received reflected events back into the Earth and place their energy at the point of reflection, it is necessary to have a good understanding of ray theory. We assume the medium is isotropic. Rays are directed curves that are always perpendicular to the wavefront at any given time. The rays point along the direction of the motion of the wavefront. As time progresses, the disturbance propagates, and we obtain a family of wavefronts. We will now describe the behavior of the rays and wavefronts in media with a continuously varying velocity.

In the treatment of light as wave motion, there is a region of approximation in which the wavelength is small in comparison with the dimensions of the components of the optical system involved. This region of approximation is treated by the methods of geometrical optics. When the wave character of the light cannot be ignored, then the methods of physical optics apply. Since the wavelength of light is very small compared to ordinary objects, geometrical optics can describe the behavior of a light beam satisfactorily in many situations. Within the approximation represented by geometrical optics, light travels along lines called *rays*. The ray is essentially the path along which most of the energy is transmitted from one point to another. The ray is a mathematical device rather than a physical entity. In practice, one can produce very narrow beams of light (e.g., a laser beam), which may be considered as physical manifestations of rays. When we turn to a seismic wave, the wavelength is not particularly small in comparison with the dimensions of geologic layers within the Earth. However, the concept of a seismic ray fulfills an important need. Geometric seismics is not nearly as accurate as geometric optics, but still ray theory is used to solve many important practical problems. In particular, the most popular form of prestack migration is based on tracing the raypaths of the primary reflections.

In ancient times, Archimedes defined the straight line as the shortest path between two points. Heron explained the paths of reflected rays of light based on a principle of least distance. In the 17th century, Fermat proposed the principle of least time, which let him account for refraction as well as reflection. The Mississippi River has created most of Louisiana with sand and silt. The river could not have deposited these sediments by remaining in one channel. If it had remained in one channel, southern Louisiana would be a long narrow peninsula reaching into the Gulf of Mexico. Southern Louisiana exists in its present form because the Mississippi River has flowed here and there within an arc of about two hundred miles wide, frequently and radically changing course, surging over the left or the right bank to flow in a new direction. It is always the river's purpose to get to the Gulf in the least time. This means that its path must follow the steepest way down. The gradient is the vector that points in the direction of the steepest ascent. Thus, the river's path must follow the direction of the negatives gradient, which is the path of steepest descent. As the mouth advances southward and the river lengthens, the steepness of the path declines, the current slows, and sediment builds up the bed. Eventually, the bed builds up so much that the river spills to one side to follow what has become the steepest way down. Major shifts of that nature have occurred about once in a millennium. The Mississippi's main channel of three thousand years ago is now Bayou Teche. A few hundred years later, the channel shifted abruptly to the east. About two thousand years ago, the channel shifted to the south. About one thousand years ago, the channel shifted to the river's present course. Today, the Mississippi River has advanced past New Orleans and out into the Gulf that the channel is about to shift again to the Atchafalaya. By the route of the Atchafalaya, the distance across the delta plain is 145 miles, which is about half the length of the route of the present channel. The Mississippi River intends changing its course to this shorter and steeper route.

The concept of potential was first developed to deal with problems of gravitational attraction. In fact, a simple gravitational analogy is helpful in explaining potential. We do work in carrying an object up a hill. This work is stored as potential energy, and it can be recovered by descending in any way we choose. A topographic map can be used to visualize the terrain. Topographic maps provide information about the elevation of the surface above sea level. The elevation is represented on a topographic map by contour lines. Each point on a contour line has the same elevation. In other words, a contour line represents a horizontal slice through the surface of the land. A set of contour lines tells you the shape of the land. For example, hills are represented by concentric loops, whereas stream valleys are represented by V-shapes. The contour interval is the elevation difference between adjacent contour lines. Steep slopes have closely spaced contour lines, while gentle slopes have very widely spaced contour lines.

In seismic theory, the counterpart of gravitational potential is the wavefront $t(x, y, z) = $ constant. A vector field is a rule that assigns a vector, in our case the gradient

$$\nabla t(x, y, z) = \left(\frac{\partial t}{\partial x}, \frac{\partial t}{\partial y}, \frac{\partial t}{\partial z}\right)$$

to each point (x, y, z). In visualizing a vector field, we imagine there is a vector extending from each point. Thus, the vector field associates a direction to each point. If a hypothetical particle moves in such a manner that its direction at any point coincides with the direction of the vector field at that point, then the curve traced out is called a *flow line*. In the seismic case, the wavefront corresponds to the equipotential surface and the seismic ray corresponds to the flow line.

In 3D space, let $\mathbf{r}$ be the vector from the origin $(0, 0, 0)$ to an arbitrary point (x, y, z). A vector is specified by its components. A compact notation is to write $\mathbf{r}$ as $\mathbf{r} = (x, y, z)$. We call $\mathbf{r}$ the radius vector. A more explicit way is to write $\mathbf{r}$ as $\mathbf{r} = x\hat{\mathbf{x}} + y\hat{\mathbf{y}} + z\hat{\mathbf{z}}$, where $\hat{\mathbf{x}}, \hat{\mathbf{y}}$, and $\hat{\mathbf{z}}$ are the three orthogonal unit vectors. These unit vectors are defined as the vectors that have magnitude equal to one and have directions lying along the x, y, z axes, respectively. They are referred to as "x-hat" and so on.

Now, let the vector $\mathbf{r} = (x, y, z)$ represent a point on a given ray (Figure 4.7). Let s denote arc length along the ray. Let $\mathbf{r} + d\mathbf{r} = (x + dx, y + dy, z + dz)$ give an adjacent point on the same ray. The vector $d\mathbf{r} = (dx, dy, dz)$ is (approximately) the tangent vector to the ray. The length of this vector is $\sqrt{(dx^2 + dy^2 + dz^2)}$ which is approximately equal to the increment ds of the arc length on the ray. As a result the unit vector tangent to the ray is

$$\mathbf{u} = \frac{d\mathbf{r}}{ds} = \frac{dx}{ds}\hat{\mathbf{x}} + \frac{dy}{ds}\hat{\mathbf{y}} + \frac{dz}{ds}\hat{\mathbf{z}}$$

The unit tangent vector can also be written as

$$\mathbf{u} = \left(\frac{dx}{ds}, \frac{dy}{ds}, \frac{dz}{ds}\right)$$

The velocity along the ray is $v = ds/dt$ so $dt = ds/v = n\,ds$, where $n(\mathbf{r}) = 1/v(\mathbf{r})$ is the slowness and ds is an increment of path length along the given ray. Thus, the seismic traveltime field is

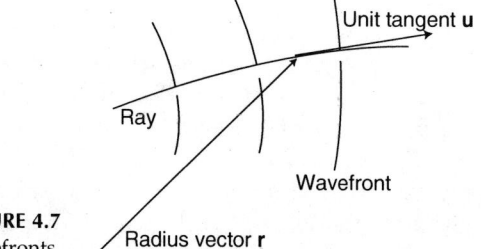

FIGURE 4.7
Raypath and wavefronts.

$$t(\mathbf{r}) = \int_{\mathbf{r}_0}^{\mathbf{r}} n(\mathbf{r}) \, ds$$

It is understood, of course, that the path of integration is along the given ray. The above equation holds for any ray. A wavefront is a surface of constant traveltime. The time difference between two wavefronts is a constant independent of the ray used to calculate the time difference.

A wave as it travels must follow the path of least time. The wavefronts are like contour lines on a hill. The height of the hill is measured in time. Take a point on a contour line. In what direction will the ray point? Suppose the ray points along the contour line (that is, along the wavefront). As the wave travels a certain distance along this hypothetical ray, it takes time. But, all time is the same along the wavefront. Thus, a wave cannot travel along a wavefront. It follows that a ray must point away from a wavefront. Suppose a ray points away from the wavefront. The wave wants to take the least time to travel to the new wavefront. By isotropy, the wave velocity is the same in all directions. Since the traveltime is velocity multiplied by distance, the wave wants to take the raypath that goes the shortest distance. The shortest distance is along the path that has no component along the wavefront; that is, the shortest distance is along the normal to the wavefront. In other words, the ray's unit tangent vector $\mathbf{u}$ must be orthogonal to the wavefront. By definition, the gradient is a vector that points in the direction orthogonal to the wavefront. Thus, the ray's unit tangent vector $\mathbf{u}$ and the gradient ∇t of the wavefront must point in the same direction.

If the given wavefront is at time t and the new wavefront is at time $t + dt$, then the traveltime along the ray is dt. If s measures the path length along the given ray, then the travel distance in time dt is ds. Along the raypath, the increments dt and ds are related by the slowness, that is, $dt = n \, ds$. Thus, the slowness is equal to the directional derivative in the direction of the raypath, that is, $n = dt/ds$. In other words, the swiftness along the raypath direction is $v = ds/dt$, and the slowness along the raypath direction is $n = dt/ds$. If we write the directional derivative in terms of its components, this equation becomes

$$n = \frac{dt}{ds} = \frac{\partial t}{\partial x}\frac{dx}{ds} + \frac{\partial t}{\partial y}\frac{dy}{ds} + \frac{\partial t}{\partial z}\frac{dz}{ds} = \nabla t \cdot \frac{d\mathbf{r}}{ds}$$

Because $d\mathbf{r}/ds = \mathbf{u}$, it follows that the above equation is $n = \nabla t \cdot \mathbf{u}$. Since $\mathbf{u}$ is a unit vector in the same direction of the gradient, it follows that

$$n = |\nabla t| \, |\mathbf{u}| \cos 0 = |\nabla t|$$

In other words, the slowness is equal to the magnitude of the gradient. Since gradient ∇t and the raypath each have the same direction $\mathbf{u}$, and the gradient has magnitude n, and $\mathbf{u}$ has magnitude unity, it follows that

$$\nabla t = n\mathbf{u}$$

This equation is the vector eikonal equation. The vector eikonal equation written in terms of its components is

$$\left(\frac{\partial t}{\partial x}, \frac{\partial t}{\partial y}, \frac{\partial t}{\partial z}\right) = n\left(\frac{dx}{ds}, \frac{dy}{ds}, \frac{dz}{ds}\right)$$

If we take the squared magnitude of each side, we obtain the eikonal equation

$$\left(\frac{\partial t}{\partial x}\right)^2 + \left(\frac{\partial t}{\partial y}\right)^2 + \left(\frac{\partial t}{\partial z}\right)^2 = n^2$$

The left-hand side involves the wavefront and the right-hand side involves the ray. The connecting link is the slowness. In the eikonal equation, the function $t(x, y, z)$ is the traveltime (also called the eikonal) from the source to the point with the coordinates (x, y, z), and $n(x, y, z) = 1/v(x, y, z)$ is the slowness (or reciprocal velocity) at that point. The eikonal equation describes the traveltime propagation as an isotropic medium. To obtain a well-posed initial value problem, it is necessary to know the velocity function $v(x, y, z)$ at all points in space. Moreover, as an initial condition, the source or some particular wavefront must be specified. Furthermore, one must choose one of the two branches of the solutions (namely, either the wave going from the source or else the wave going to the source). The eikonal equation then yields the traveltime field $t(x, y, z)$ in the heterogeneous medium, as required for migration.

What does the eikonal equation $\nabla t = n\mathbf{u}$ say? It says that, because of Fermat's principle of least time, the raypath direction must be orthogonal to the wavefront. The eikonal equation is the fundamental equation that connects the ray (which corresponds to the fuselage of the airplane) to the wavefront (which corresponds to the wings of the airplane). The wings let the fuselage feel the effects of points removed from the path of the fuselage. The eikonal equation makes a traveling wave (as envisaged by Huygens) fundamentally different from a traveling particle (as envisaged by Newton). Hamilton perceived that there is a wave–particle duality, which provides the mathematical foundation of quantum mechanics.

4.9 The Ray Equations

In this section, the position vector $\mathbf{r}$ always represents a point on a specific raypath, and not any arbitrary point in space. As time increases, $\mathbf{r}$ traces out the particular raypath in question. The seismic ray at any given point follows the direction of the gradient of the traveltime field $t(\mathbf{r})$. As before, let $\mathbf{u}$ be the unit vector along the ray. The ray, in general, follows a curved path, and $n\mathbf{u}$ is the tangent to this curved raypath. We now want to derive an equation that will tell us how $n\mathbf{u}$ changes along the curved raypath. The vector eikonal equation is written as

$$n\mathbf{u} = \nabla t$$

We now take the derivative of the vector eikonal equation with respect to distance s along the raypath. We obtain the ray equation

$$\frac{d}{ds}(n\mathbf{u}) = \frac{d}{ds}(\nabla t)$$

Interchange ∇ and (d/ds) and use $dt/ds = n$. Thus, the right-hand side becomes

$$\frac{d\nabla t}{ds} = \nabla\left(\frac{dt}{ds}\right) = \nabla n$$

Thus, the ray equation becomes

Construction of Seismic Images by Ray Tracing

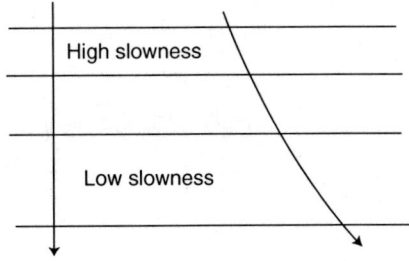

FIGURE 4.8
Straight and curved rays.

$$\frac{d}{ds}(n\mathbf{u}) = \nabla n$$

This equation, together with the equation for the unit tangent vector

$$\frac{d\mathbf{r}}{ds} = \mathbf{u}$$

are called the *ray equations*.

We need to understand how a single ray, say a seismic ray, moving along a particular path can know what is an extremal path in the variational sense. To illustrate the problem, consider a medium whose slowness n decreases with vertical depth, but is constant laterally. Thus, the gradient of the slowness at any location points straight up (Figure 4.8). The vertical line on the left depicts a raypath parallel to the gradient of the slowness. This ray undergoes no refraction. However, the path of the ray on the right intersects the contour lines of slowness at an angle. The right-hand ray is refracted and follows a curved path, even though the ray strikes the same horizontal contour lines of slowness as did the left-hand ray, where there was no refraction. This shows that the path of a ray cannot be explained solely in terms of the values of the slowness on the path. We must also consider the transverse values of the slowness along neighboring paths, that is, along paths not taken by that particular ray.

The classical wave explanation, proposed by Huygens, resolves this problem by saying that light does not propagate in the form of a single ray. According to the wave interpretation, light propagates as a wavefront possessing transverse width. Think of an airplane traveling along the raypath. The fuselage of the airplane points in the direction of the raypath. The wings of the aircraft are along the wavefront. Clearly, the wavefront propagates more rapidly on the side where the slowness is low (i.e., where the velocity is high) than on the side where the slowness is high. As a result, the wavefront naturally turns in the direction of the gradient of slowness.

4.10 Numerical Ray Tracing

Computer technology and seismic instrumentation have experienced great advances in the past few years. As a result, exploration geophysics is in a state of transition from computer-limited 2D processing to computer-intensive 3D processing. In the past, most seismic surveys were along surface lines, which yield 2D subsurface images. The wave equation acts nicely in one dimension, and in three dimensions, but not in two dimensions. In one dimension, waves (as on a string) propagate without distortion. In three dimensions, waves (as in the Earth) propagate in an undistorted way except for a spherical correction factor. However, in two dimensions, wave propagation is complicated and distorted. By its very nature, 2D processing can never account for events originating outside of the plane. As a result, 2D processing is broken up into a large number of approximate partial steps in a sequence of operations. These steps are ingenious, but they can never give a true image. However, 3D processing accounts for all of the events. It is now cost effective to lay out seismic surveys over a surface area and to do 3D processing. The third dimension is no longer missing, and, consequently, the need for a large number of piecemeal 2D approximations is gone. Prestack depth migration is a 3D imaging process that is computationally extensive, but mathematically simple. The resulting 3D images of the interior of the Earth surpass all expectations in utility and beauty.

Let us now consider the general case in which we have a spatially varying velocity function $v(x, y, z) = v(\mathbf{r})$. This velocity function represents a velocity field. For a fixed constant v_0, the equation $v(\mathbf{r}) = v_0$ specifies those positions, $\mathbf{r}$, where velocity has this fixed value. The locus of such positions makes up an *isovelocity surface*. The gradient

$$\nabla v(\mathbf{r}) = \left(\frac{\partial v}{\partial x}, \frac{\partial v}{\partial y}, \frac{\partial v}{\partial z}\right)$$

is normal to the isovelocity surface and points in the direction of the greatest increase in velocity. Similarly, the equation $n(\mathbf{r}) = n_0$ for a fixed value of slowness n_0 specifies an *isoslowness surface*. The gradient

$$\nabla n(\mathbf{r}) = \left(\frac{\partial n}{\partial x}, \frac{\partial n}{\partial y}, \frac{\partial n}{\partial z}\right)$$

is normal to the isoslowness surface and points in the direction of greatest increase in slowness. The isovelocity and isoslowness surfaces coincide, and

$$\nabla v = -n^{-2}\nabla n$$

so the respective gradients point in the opposite direction, as we would expect.

A seismic ray makes its way through the slowness field. As the wavefront progresses in time, the raypath is bent according to the slowness field. For example, suppose we have a stratified Earth in which the slowness decreases with depth, a vertical raypath does not bend, as it is pulled equally in all lateral directions. However, a nonvertical ray drags on its slow side, therefore it curves away from the vertical and bends toward the horizontal. This is the case of a diving wave, whose raypath eventually curves enough to reach the Earth's surface again. Certainly, the slowness field, together with the initial direction of the ray, determines the entire raypath. Except in special cases, however, we must determine such raypaths numerically.

Assume that we know the slowness function $n(\mathbf{r})$ and that we know the ray direction $\mathbf{u}_1$ at point $\mathbf{r}_1$. We now want to derive an algorithm for finding the ray direction $\mathbf{u}_2$ at point $\mathbf{r}_2$. We choose a small, but finite, change in path length Δs. Then we use the first ray equation, which we recall is

$$\frac{d\mathbf{r}}{ds} = \mathbf{u}$$

to compute the change $\Delta \mathbf{r} = \mathbf{r}_2 - \mathbf{r}_1$. The required approximation is

$$\Delta \mathbf{r} = \mathbf{u}_1 \Delta s$$

Construction of Seismic Images by Ray Tracing

or

$$r_2 = r_1 + u_1 \Delta s$$

We have thus found the first desired quantity r_2. Next, we use the second ray equation, which we recall is

$$\frac{d(n\mathbf{u})}{ds} = \nabla n$$

in the form

$$d(n\mathbf{u}) = \nabla n \, ds$$

The required approximation is

$$\Delta(n\mathbf{u}) = (\nabla n) \, \Delta s$$

or

$$n(r_2)\mathbf{u}_2 - n(r_1)\mathbf{u}_1 = \nabla n \, \Delta s$$

For accuracy, ∇n may be evaluated by differentiating the known function $n(\mathbf{r})$ midway between r_1 and r_2. Thus, the desired $\mathbf{u}_2$ is given as

$$\mathbf{u}_2 = \frac{n(r_1)}{n(r_2)} \mathbf{u}_1 + \frac{\Delta s}{n(r_2)} \nabla n$$

Note that the vector $\mathbf{u}_1$ is pulled in the direction of ∇n in forming $\mathbf{u}_2$, that is, the ray drags on the slow side, and so is bent in the direction of increasing slowness. The special case of no bending occurs when $\mathbf{u}_1$ and ∇n are parallel. As we have seen, a vertical wave in a horizontally stratified medium is an example of such a special case. We have thus found how to advance the wave along the ray by an incremental raypath distance. We can repeat the algorithm to advance the wave by any desired distance.

4.11 Conclusions

The acquisition of seismic data in many promising areas yields raw traces that cannot be interpreted. The reason is that signal-generated noise conceals the desired primary reflections. The solution to this problem was found in the 1950s with the introduction of the first commercially available digital computers and the signal-processing method of deconvolution. Digital signal-enhancement methods, and, in particular, the various methods of deconvolution, are able to suppress signal-generated noise on seismic records and bring out the desired primary-reflected energy. Next, the energy of the primary reflections must be moved to the spatial positions of the subsurface reflectors. This process, called *migration*, involves the superposition of all the deconvolved traces according to a scheme similar to Huygens's construction. The result gives the reflecting horizons and other features that make up the desired image. Thus, digital processing, as it is currently done, is roughly divided into two main parts, namely *signal enhancement* and *event migration*. The day is not far off, provided that research actively continues in the future as it has in the past, when the processing scheme will not be divided into two parts, but will be united as a whole. The signal-generated noise consists of physical signals that future processing should not destroy, but utilize. When all the seismic information is used in an integrated way, then the images produced will be even more excellent.

References

Robinson, E.A., The MIT Geophysical Analysis Group from inception to 1954, *Geophysics*, 70, 7JA–30JA, 2005.

Yilmaz, O., *Seismic Data Processing*, Society of Exploration Geophysicists, Tulsa, 1987.

5

Multi-Dimensional Seismic Data Decomposition by Higher Order SVD and Unimodal ICA

Nicolas Le Bihan, Valeriu Vrabie, and Jérôme I. Mars

CONTENTS

5.1 Introduction 74
5.2 Matrix Data Sets 74
    5.2.1 Acquisition 75
    5.2.2 Matrix Model 75
5.3 Matrix Processing 76
    5.3.1 SVD 76
        5.3.1.1 Definition 76
        5.3.1.2 Subspace Method 76
    5.3.2 SVD and ICA 77
        5.3.2.1 Motivation 77
        5.3.2.2 Independent Component Analysis 77
        5.3.2.3 Subspace Method Using SVD–ICA 79
    5.3.3 Application 80
5.4 Multi-Way Array Data Sets 83
    5.4.1 Multi-Way Acquisition 84
    5.4.2 Multi-Way Model 84
5.5 Multi-Way Array Processing 85
    5.5.1 HOSVD 85
        5.5.1.1 HOSVD Definition 85
        5.5.1.2 Computation of the HOSVD 86
        5.5.1.3 The (r_c, r_x, r_t)-rank 87
        5.5.1.4 Three-Mode Subspace Method 88
    5.5.2 HOSVD and Unimodal ICA 88
        5.5.2.1 HOSVD and ICA 89
        5.5.2.2 Subspace Method Using HOSVD–Unimodal ICA 89
    5.5.3 Application to Simulated Data 90
    5.5.4 Application to Real Data 95
5.6 Conclusions 98
References 98

5.1 Introduction

This chapter describes multi-dimensional seismic data processing using the higher order singular value decomposition (HOSVD) and partial (unimodal) independent component analysis (ICA). These techniques are used for wavefield separation and enhancement of the signal-to-noise ratio (SNR) in the data set. The use of multi-linear methods such as the HOSVD is motivated by the natural modeling of a multi-dimensional data set using multi-way arrays. In particular, we present a multi-way model for signals recorded on arrays of vector-sensors acquiring seismic vibrations in different directions of the 3D space. Such acquisition schemes allow the recording of the polarization of waves and the proposed multi-way model ensures the effective use of polarization information in the processing. This leads to a substantial increase in the performances of the separation algorithms.

Before introducing the multi-way model and processing, we first describe the classical subspace method based on the SVD and ICA techniques for 2D (matrix) seismic data sets. Using a matrix model for these data sets, the SVD-based subspace method is presented and it is shown how an extra ICA step in the processing allows better wavefield separation. Then, considering signals recorded on vector-sensor arrays, the multi-way model is defined and discussed. The HOSVD is presented and some properties detailed. Based on this multi-linear decomposition, we propose a subspace method that allows separation of polarized waves under orthogonality constraints. We then introduce an ICA step in the process that is performed here uniquely on the temporal mode of the data set, leading to the so-called HOSVD–unimodal ICA subspace algorithm. Results on simulated and real polarized data sets show the ability of this algorithm to surpass a matrix-based algorithm and subspace method using only the HOSVD.

Section 5.2 presents matrix data sets and their associated model. In Section 5.3, the well-known SVD is detailed, as well as the matrix-based subspace method. Then, we present the ICA concept and its contribution to subspace formulation in Section 5.3.2. Applications of SVD–ICA to seismic wavefield separation are discussed by way of illustrations. Section 5.4 exposes how signal mixtures recorded on vector-sensor arrays can be described by a multi-way model. Then, in Section 5.5, we introduce the HOSVD and the associated subspace method for multi-way data processing. As in the matrix data set case, an extra ICA step is proposed leading to a HOSVD–unimodal ICA subspace method in Section 5.5.2. Finally, in Section 5.5.3 and Section 5.5.4, we illustrate the proposed algorithm on simulated and real multi-way polarized data sets. These examples emphasize the potential of using both HOSVD and ICA in multi-way data set processing.

5.2 Matrix Data Sets

In this section, we show how the signals recorded on scalar-sensor arrays can be modeled as a matrix data set having two *modes* or *diversities*: *time* and *distance*. Such a model allows the use of subspace-based processing using a SVD of the matrix data set. Also, an additional ICA step can be added to the processing to relax the unjustified orthogonality constraint for the propagation vectors by imposing a stronger constraint of (fourth-order) independence of the estimated waves. Illustrations of these matrix algebra techniques are presented on a simulated data set. Application to a real ocean bottom seismic (OBS) data set can be found in Refs. [1,2].

5.2.1 Acquisition

In geophysics, the most commonly used method to describe the structure of the earth is seismic reflection. This method provides images of the underground in 2D or 3D, depending on the geometry of the network of sensors used. Classical recorded data sets are usually gathered into a matrix having a time diversity describing the time or depth propagation through the medium at each sensor and a distance diversity related to the aperture of the array. Several methods exist to gather data sets and the most popular are *common shotpoint gather*, *common receiver gather*, or *common midpoint gather* [3]. Seismic processing consists in a series of elementary processing procedures used to transform field data, usually recorded in common shotpoint gather into a 2D or 3D common midpoint stacked 2D signals. Before stacking and interpretation, part of the processing is used to suppress unwanted coherent signals like multiple waves, ground-roll (surface waves), refracted waves, and also to cancel noise.

To achieve this goal, several filters are classically applied on seismic data sets. The SVD is a popular method to separate an initial data set into signal and noise subspaces. In some applications [4,5] when wavefield alignment is performed, the SVD method allows separation of the aligned wave from the other wavefields.

5.2.2 Matrix Model

Consider a uniform linear array composed of N_x omni-directional sensors recording the contributions of P waves, with $P < N_x$. Such a record can be written mathematically using a convolutive model for seismic signals first suggested by Robinson [6]. Using the superposition principle, the discrete-time signal $x_k(m)$ (m is the time index) recorded on sensor k is a linear combination of the P waves received on the array together with an additive noise $n_k(m)$:

$$x_k(m) = \sum_{i=1}^{P} a_{ki} s_i(m - m_{ki}) + n_k(m) \tag{5.1}$$

where $s_i(m)$ is the *i*th source waveform that has been propagated through the transfer function supposed here to consist in a delay m_{ki} and a factor attenuation a_{ki}. The noises on each sensor $n_k(m)$ are supposed centered, Gaussian, spatially white, and independent of the sources.

In the sequel, the use of the SVD to separate waves is only of significant interest if the subspace occupied by the *part of interest* contained in the mixture is of low rank. Ideally, the SVD performs well when the rank is 1. Thus, to ensure good results of the process, a preprocessing is applied on the data set. This consists of alignment (delay correction) of a chosen high amplitude wave. Denoting the aligned wave by $s_1(m)$, the model becomes after alignment:

$$y_k(m) = a_{k1} s_1(m) + \sum_{i=2}^{P} a_{ki} s_i(m - m'_{ki}) + n'_k(m) \tag{5.2}$$

where $y_k(m) = x_k(m + m_{k1})$, $m'_{ki} = m_{ki} - m_{k1}$ and $n'_k(m) = n_k(m + m_{k1})$.

In the following we assume that the wave $s_1(m)$ is independent from the others and therefore independent from $s_i(m - m'_{ki})$.

Considering the simplified model of the received signals (Equation 5.2) and supposing N_t time samples available, we define the matrix model of the recorded data set $\mathbf{Y} \in R^{N_x \times N_t}$ as

$$\mathbf{Y} = \{y_{km} = y_k(m) | \ 1 \leq k \leq N_x, \ 1 \leq m \leq N_t\} \tag{5.3}$$

That is, the data matrix $\mathbf{Y}$ has rows that are the N_x signals $y_k(m)$ given in Equation 5.2. Such a model allows the use of matrix decomposition, and especially the SVD, for its processing.

5.3 Matrix Processing

We now present the definition of the SVD of such a data matrix that will be of use for its decomposition into orthogonal subspaces and in the associated wave separation technique.

5.3.1 SVD

As the SVD is a widely used matrix algebra technique, we only recall here theoretical remarks and redirect readers interested in computational issues to the Golub and Van Loan book [7].

5.3.1.1 Definition

Any matrix $\mathbf{Y} \in R^{N_x \times N_t}$ can be decomposed into the product of three matrices as follows:

$$\mathbf{Y} = \mathbf{U}\Delta\mathbf{V}^T \tag{5.4}$$

where $\mathbf{U}$ is a $N_x \times N_x$ matrix, Δ is an $N_x \times N_t$ pseudo-diagonal matrix with singular values $\{\lambda_1, \lambda_2, \ldots, \lambda_N\}$ on its diagonal, satisfying $\lambda_1 \geq \lambda_2 \geq \ldots \geq \lambda_N \geq 0$, (with $N = \min(N_x, N_t)$), and $\mathbf{V}$ is an $N_t \times N_t$ matrix. The columns of $\mathbf{U}$ (respectively of $\mathbf{V}$) are called the *left* (respectively *right*) singular vectors, $\mathbf{u}_j$ (respectively $\mathbf{v}_j$), and form orthonormal bases. Thus $\mathbf{U}$ and $\mathbf{V}$ are orthogonal matrices. The rank r (with $r \leq N$) of the matrix $\mathbf{Y}$ is given by the number of nonvanishing singular values.

Such a decomposition can also be rewritten as

$$\mathbf{Y} = \sum_{j=1}^{r} \lambda_j \mathbf{u}_j \mathbf{v}_j^T \tag{5.5}$$

where $\mathbf{u}_j$ (respectively $\mathbf{v}_j$) are the columns of $\mathbf{U}$ (respectively $\mathbf{V}$). This notation shows that the SVD allows any matrix to be expressed as a sum of r rank-1 matrices[1].

5.3.1.2 Subspace Method

The SVD has been widely used in signal processing [8] because it gives the *best rank approximation* (in the least squares sense) of a given matrix [9]. This property allows denoising if the signal subspace is of relatively low rank. So, the subspace method consists of decomposing the data set into two orthogonal subspaces with the first one built from the p singular vectors related to the p highest singular values being the best rank approximation of the original data. This can be written as follows, using the SVD notation used in Equation 5.5, for a data matrix $\mathbf{Y}$ with rank r:

---

[1] Any matrix made up of the product of a column vector by a row vector is a matrix whose rank is equal to 1 [7].

$$\mathbf{Y} = \mathbf{Y}^{\text{Signal}} + \mathbf{Y}^{\text{Noise}} = \sum_{j=1}^{p} \lambda_j \mathbf{u}_j \mathbf{v}_j^T + \sum_{j=p+1}^{r} \lambda_j \mathbf{u}_j \mathbf{v}_j^T \qquad (5.6)$$

Orthogonality between the subspaces spanned by the two sets of singular vectors is ensured by the fact that left and right singular vectors form orthonormal bases.

From a practical point of view, the value of p is chosen by finding an abrupt change of slope in the curve of relative singular values (relative meaning percentile representation of) contained in the matrix Δ defined in Equation 5.4. For some cases where no "visible" change of slope can be found, the value of p can be fixed at 1 for a perfect alignment of waves, or at 2 for an imperfect alignment or for dispersive waves [10].

5.3.2 SVD and ICA

The motivation to relax the unjustified orthogonality constraint for the propagation vectors is now presented. ICA is the method used to achieve this by imposing a fourth-order independence on the estimated waves. This provides a new subspace method based on SVD–ICA.

5.3.2.1 Motivation

The SVD of the data matrix $\mathbf{Y}$ in Equation 5.4 provides two orthogonal matrices composed by the left $\mathbf{u}_j$ (respectively right $\mathbf{v}_j$) singular vectors. Note here that $\mathbf{v}_j$ are called *estimated waves* because they give the time dependence of received signals by the array sensor and $\mathbf{u}_j$ *propagation vectors* because they give the amplitude of $\mathbf{v}_j'$s on sensors [2].

As SVD provides orthogonal matrices, these vectors are also orthogonal. Orthogonality of the $\mathbf{v}_j'$s means that the estimated waves are decorrelated (second-order independence). Actually, this supports the usual cases in geophysical situations, in which recorded waves are supposed decorrelated. However, there is no physical reason to consider the orthogonality of propagation vectors $\mathbf{u}_j$. Why should we have different recorded waves with orthogonal propagation vectors? Furthermore, imposing the orthogonality of $\mathbf{u}_j'$s, the estimated waves $\mathbf{v}_j$ are forced to be a mixture of recorded waves [1].

One way to relax this limitation is to impose a stronger criterion for the estimated waves, that is, to be fourth-order statistically independent, and consequently to drop the unjustified orthogonality constraint for the propagation vectors. This step is motivated by cases encountered in geophysical situations, where the recorded signals can be approximated as an instantaneous linear mixture of unknown waves supposed to be mutually independent [11]. This can be done using ICA.

5.3.2.2 Independent Component Analysis

ICA is a blind decomposition of a multi-channel data set composed of an unknown linear mixture of unknown source signals, based on the assumption that these signals are mutually statistically independent. It is used in blind source separation (BSS) to recover independent sources (modeled as vectors) from a set of recordings containing linear combinations of these sources [12–15]. The statistical independence of sources means that the cross-cumulants of any order vanish. Generally, the third-order cumulants are discarded because they are generally close to zero. Therefore, here we will use fourth-order statistics, which have been found to be sufficient for instantaneous mixtures [12,13].

ICA is usually resolved by a two-step algorithm: prewhitening followed by high-order step. The first one consists in extracting decorrelated waves from the initial data set. The step is carried out directly by an SVD as the $\mathbf{v}'_j$s are orthogonal.

The second step consists in finding a rotation matrix $\mathbf{B}$, which leads to fourth-order independence of the estimated waves. We suppose here that the nonaligned waves in the data set $\mathbf{Y}$ are contained in a subspace of dimension $R-1$, smaller than the rank r of $\mathbf{Y}$. Assuming this, only the first R estimated waves $[\mathbf{v}_1, \ldots, \mathbf{v}_R]^{\text{notation}} = \mathbf{V}^R \in \mathbb{R}^{N_t \times R}$ are taken into account [2]. As the recorded waves are supposed mutually independent, this second step can be written as

$$\mathbf{V}^R \mathbf{B} = \tilde{\mathbf{V}}^R = [\tilde{\mathbf{v}}_1, \ldots, \tilde{\mathbf{v}}_R] \in \mathbb{R}^{N_t \times R} \qquad (5.7)$$

with $\mathbf{B} \in \mathbb{R}^{R \times R}$ the rotation (unitary) matrix having the property $\mathbf{BB}^T = \mathbf{B}^T \mathbf{B} = \mathbf{I}$. The new estimated waves $\tilde{\mathbf{v}}_j$ are now independent at the fourth order.

There are different methods of finding the rotation matrix: joint approximate diagonalization of eigenmatrices (JADE) [12], maximal diagonality (MD) [13], simultaneous third-order tensor diagonalization (STOTD) [14], fast and robust fixed-point algorithms for independent component analysis (FastICA) [15], and so on. To compare some cited ICA algorithms, Figure 5.1 shows the relative error (see Equation 5.12) of the estimated signal subspace versus the SNR (see Equation 5.11) for the data set presented in Section 5.3.3. For SNRs greater than -7.5 dB, FastICA using a "tanh" nonlinearity with the parameter equal to 1 in the fixed-point algorithm provides the smallest relative error, but with some erroneous points at different SNR. Note that the "tan h" nonlinearity is the one which gives the smallest error for this data set, compared with "pow3", "gauss" with the parameter equal to 1, or "skew" nonlinearities. MD and JADE algorithms are approximately equivalent according to the relative error. For SNRs smaller than -7.5 dB, MD provides the smallest relative error. Consequently, the MD algorithm was employed in the following.

Now, considering the SVD decomposition in Equation 5.5 and the ICA step in Equation 5.7, the subspace described by the first R estimated waves can be rewritten as

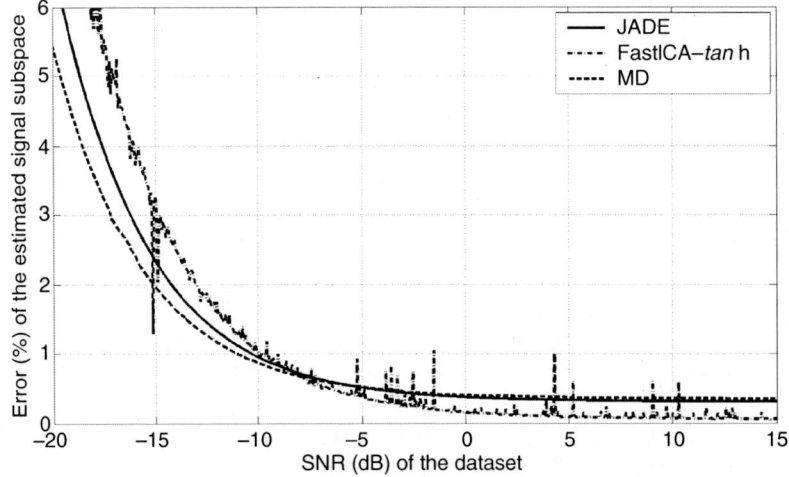

FIGURE 5.1
ICA algorithms—comparison.

$$\sum_{j=1}^{R}\lambda_j\mathbf{u}_j\mathbf{v}_j^T = \mathbf{U}^R\Delta^R(\mathbf{V}^R)^T = \mathbf{U}^R\Delta^R\mathbf{B}(\tilde{\mathbf{V}}^R)^T = \sum_{j=1}^{R}\beta_j\tilde{\mathbf{u}}_j\tilde{\mathbf{v}}_j^T = \sum_{i=1}^{R}\beta_i\tilde{\mathbf{u}}_i\tilde{\mathbf{v}}_i^T \quad (5.8)$$

where $\mathbf{U}^R = [\mathbf{u}_1, \ldots, \mathbf{u}_R]$ is made up of the first R vectors of the matrix $\mathbf{U}$ and $\Delta^R = \text{diag}(\lambda_1, \ldots, \lambda_R)$ is the $R \times R$ truncated version of Δ containing the greatest values of λ_j. The second equality is obtained using Equation 5.7. For the third equality, the $\tilde{\mathbf{u}}_j$ are the new propagation vectors obtained as the normalized[2] vectors (columns) of the matrix $\mathbf{U}^R\Delta^R\mathbf{B}$ and β_j are the "modified singular values" obtained as the ℓ^2-norm of the columns of the matrix $\mathbf{U}^R\Delta^R\mathbf{B}$.

The elements β_j are usually not ordered. For this reason, a permutation between the vectors $\tilde{\mathbf{u}}_j$ as well as between the vectors $\tilde{\mathbf{v}}_j$ is performed to order the modified singular values. Denoting with $\sigma(\cdot)$ this permutation and with $i = \sigma(j)$, the last equality of Equation 5.8 is obtained.

In this decomposition, which is similar to that given by Equation 5.5, a stronger criterion for the new estimated waves $\tilde{\mathbf{v}}_i$ has been imposed, that is, to be independent at the fourth order, and, at the same time, the condition of orthogonality for the new propagation vectors $\tilde{\mathbf{u}}_i$ has been relaxed.

In practical situations, the value of R becomes a parameter. Usually, it is chosen to completely describe the aligned wave by the first R estimated waves given by the SVD.

5.3.2.3 Subspace Method Using SVD–ICA

After the ICA and the permutation steps, the signal subspace is given by

$$\tilde{\mathbf{Y}}^{\text{Signal}} = \sum_{i=1}^{\tilde{p}}\beta_i\tilde{\mathbf{u}}_i\tilde{\mathbf{v}}_i^T \quad (5.9)$$

where $\tilde{p}$ is the number of the new estimated waves necessary to describe the aligned wave.

The noise subspace $\tilde{\mathbf{Y}}^{\text{Noise}}$ is obtained by subtraction of the signal subspace $\tilde{\mathbf{Y}}^{\text{Signal}}$ from the original data set $\mathbf{Y}$:

$$\tilde{\mathbf{Y}}^{\text{Noise}} = \mathbf{Y} - \tilde{\mathbf{Y}}^{\text{Signal}} \quad (5.10)$$

From a practical point of view, the value of $\tilde{p}$ is chosen by finding an abrupt change of slope in the curve of relative modified singular values. For cases with low SNR, no "visible" change of slope can be found and the value of $\tilde{p}$ can be fixed at 1 for a perfect alignment of waves, or at 2 for an imperfect alignment or for dispersive waves.

Note here that for very small SNR of the initial data set, (for example, smaller than −6.2 dB for the data set presented in Section 5.3.3, the aligned wave can be described by a less energetic estimated wave than by the first one (related to the highest singular value). For these extreme cases, a search must be done after the ICA and the permutation steps to identify the indexes for which the corresponding estimated waves $\tilde{\mathbf{v}}_i$ give the aligned wave. So the signal subspace $\tilde{\mathbf{Y}}^{\text{Signal}}$ in Equation 5.9 must be redefined by choosing the index values found in the search. For example, applying the MD algorithm to the data set presented in Section 5.3.3 for which the SNR was modified to −9 dB, the aligned wave is described by the third estimated wave $\tilde{\mathbf{v}}_3$. Note also that using SVD without ICA in the same conditions, the aligned wave is described by the eighth estimated wave $\mathbf{v}_8$.

---

[2] Vectors are normalized by their ℓ^2-norm.

5.3.3 Application

An application to a simulated data set is presented in this section to illustrate the behavior of the SVD–ICA versus the SVD subspace method. Application to a real data set obtained during an acquisition with OBS can be found in Refs. [1,2].

The preprocessed recorded signals $\mathbf{Y}$ on an 8-sensor array ($N_x = 8$) during $N_t = 512$ time samples are represented in Figure 5.2c. This synthetic data set was obtained by the addition of an original signal subspace $\mathbf{S}$ (Figure 5.2a) made up by a wavefront having *infinite celerity (velocity)*, consequently associated with the aligned wave $s_1(m)$, and an original noise subspace $\mathbf{N}$ (Figure 5.2b) made up by several nonaligned wavefronts. These nonaligned waves are contained in a subspace of dimension 7, smaller than the rank of $\mathbf{Y}$, which equals 8.

The SNR ratio of the presented data set is SNR = −3.9 dB. The SNR definition used here is[3]:

$$\text{SNR} = 20 \log_{10} \frac{\|\mathbf{S}\|}{\|\mathbf{N}\|} \tag{5.11}$$

Normalization to unit variance of each trace for each component was done before applying the described subspace methods. This ensures that even weak picked arrivals are well represented within the input data. After the computation of signal subspaces, a denormalization was applied to find the original signal subspace.

Firstly, the SVD subspace method was tested. The subspace method given by Equation 5.6 was employed, keeping only one singular vector (respectively one singular value). This choice was made by finding an abrupt change of slope after the first singular value (Figure 5.6) in the relative singular values for this data set. The obtained signal subspace $\mathbf{Y}^{\text{Signal}}$ and noise subspace $\mathbf{Y}^{\text{Noise}}$ are presented in Figure 5.3a and Figure 5.3b. It is clear

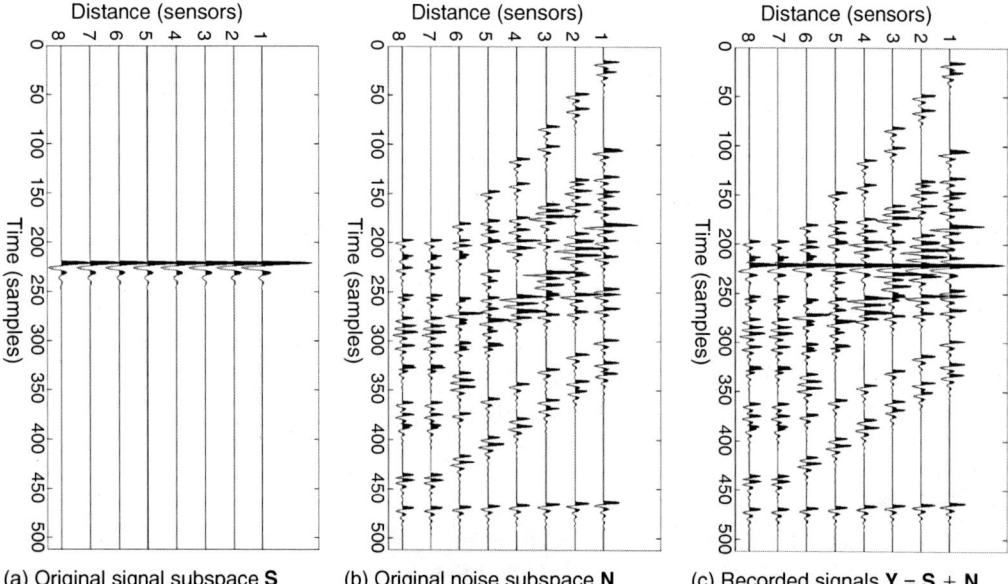

(a) Original signal subspace **S**   (b) Original noise subspace **N**   (c) Recorded signals **Y** = **S** + **N**

FIGURE 5.2
Synthetic data set.

---

[3] $\|\mathbf{A}\| = \sqrt{\sum_{i=1}^{I} \sum_{j=1}^{J} a_{ij}^2}$ is the Frobenius norm of the matrix $\mathbf{A} = \{a_{ij}\} \in \mathbb{R}^{I \times J}$

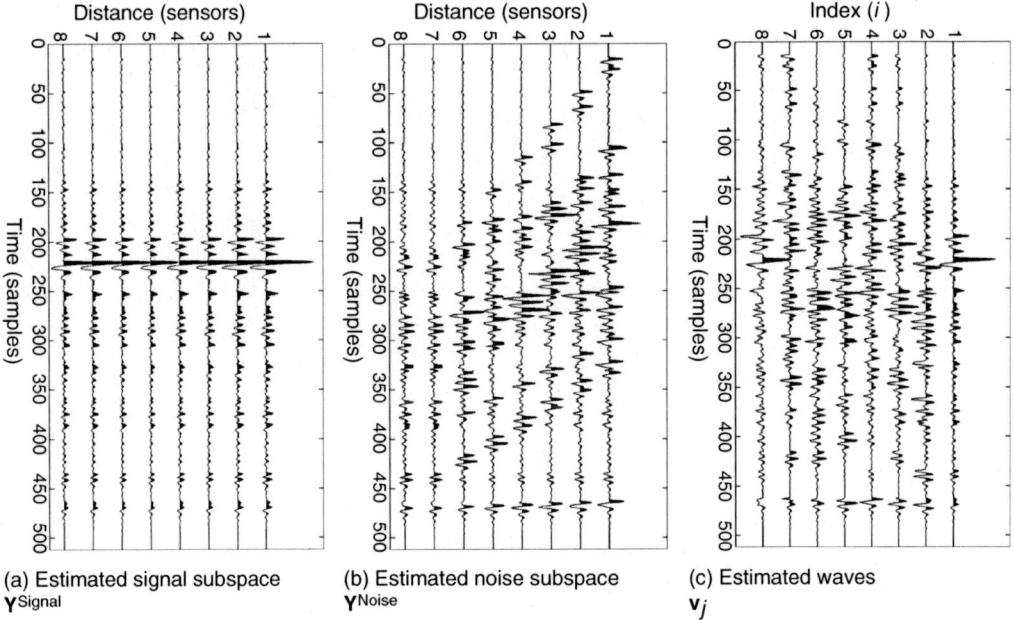

FIGURE 5.3
Results obtained using the SVD subspace method.

from these figures that the classical SVD implies artifacts in the two estimated subspaces for a wavefield separation objective. Moreover, the estimated waves v_j shown in Figure 5.3c are an instantaneous linear mixture of the recorded waves.

The signal subspace $\tilde{Y}^{Signal}$ and noise subspace $\tilde{Y}^{Noise}$ obtained using the SVD–ICA subspace method given by Equation 5.9 are presented in Figure 5.4a and Figure 5.4b. This improvement is due to the fact that using ICA we have imposed a fourth-order independence condition stronger than the decorrelation used in classical SVD. With this subspace method we have also relaxed the nonphysically justified orthogonality of the propagation vectors.

The dimension R of the rotation matrix $\mathbf{B}$ was chosen to be eight because the aligned wavelight is projected on all eight estimated waves v_j shown in Figure 5.3c. After the ICA and the permutation steps, the new estimated waves $\tilde{v}_i$ are presented in Figure 5.4c. As we can see, the first one describes the aligned wave "perfectly". As no visible change of slope can be found in the relative modified singular values shown in Figure 5.6, the value of $\tilde{p}$ was fixed at 1 because we are dealing with a perfectly aligned wave.

To compare the results qualitatively, the stack representation is usually employed [5]. Figure 5.5 shows, from left to right, the stacks on the initial data set $\mathbf{Y}$, the original signal subspace $\mathbf{S}$, and the estimated signal subspaces obtained with SVD and SVD–ICA subspace methods, respectively. As the stack on the estimated signal subspace $\tilde{Y}^{Signal}$ is very close to the stack on the original signal subspace $\mathbf{S}$, we can conclude that the SVD–ICA subspace method enhances the wave separation results.

To compare these methods quantitatively, we use the relative error ε of the estimated signal subspace defined as

$$\varepsilon = \frac{\|\mathbf{S} - \tilde{\mathbf{Y}}^{Signal}\|^2}{\|\mathbf{S}\|^2} \tag{5.12}$$

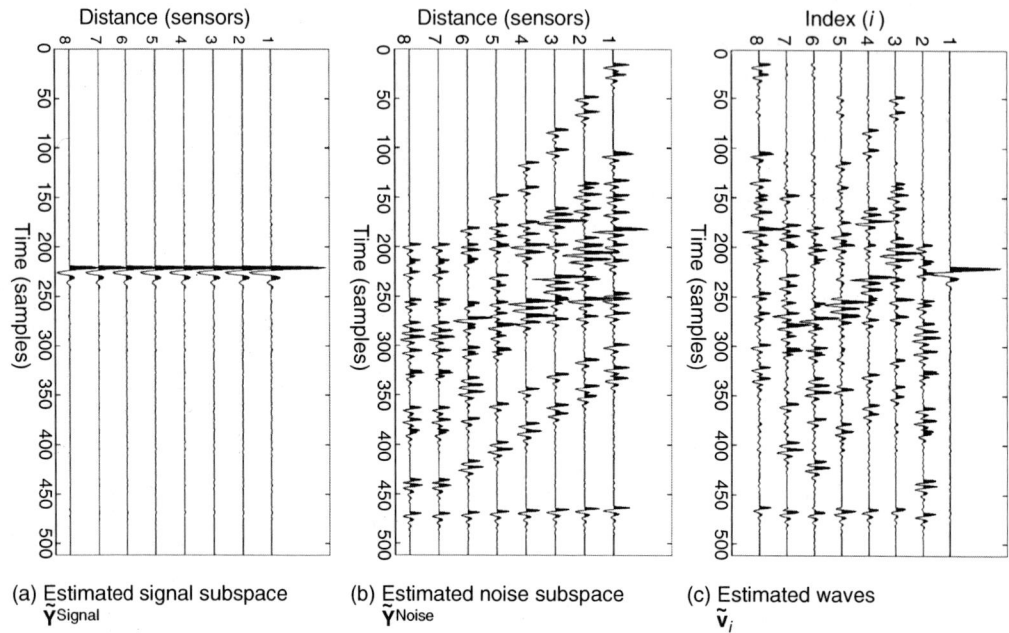

FIGURE 5.4
Results obtained using the SVD-ICA subspace method.

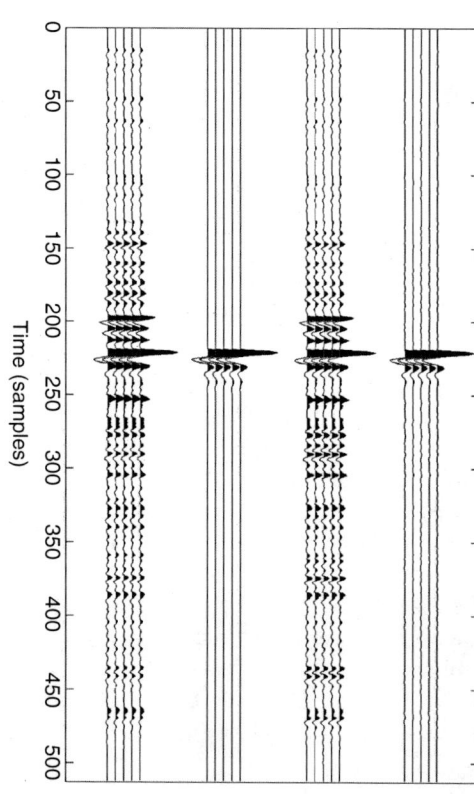

FIGURE 5.5
Stacks. From left to right: initial data set **Y**, original signal subspace **S**, SVD, and SVD–ICA estimated subspaces.

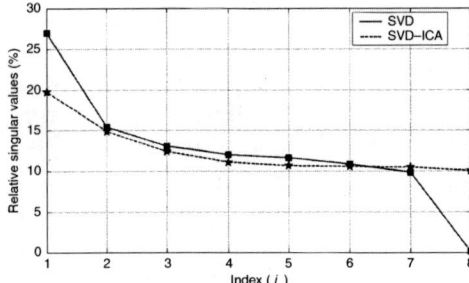

FIGURE 5.6
Relative singular values.

where $\|\cdot\|$ is the matrix Frobenius norm defined above, $\mathbf{S}$ is the original signal subspace and $\bar{\mathbf{Y}}^{Signal}$ represents either the estimated signal subspace $\mathbf{Y}^{Signal}$ obtained using SVD or the estimated signal subspace $\tilde{\mathbf{Y}}^{Signal}$ obtained using SVD–ICA. For the data set presented in Figure 5.2, we obtain $\varepsilon = 55.7\%$ for classical SVD and $\varepsilon = 0.5\%$ for SVD–ICA.

The SNR of this data set was modified by keeping the initial noise subspace constant and by adjusting the energy of the initial signal subspace. The relative errors of the estimated signal subspaces versus the SNR are plotted in Figure 5.7. For SNRs greater than 17 dB, the two methods are equivalent. For smaller SNR, the SVD–ICA subspace method is obviously better than the SVD subspace method. It provides a relative error lower than 1% for SNRs greater than −10 dB.

Note here that for other data sets, the SVD–ICA performance can be degraded by the unfulfilled independence assumption supposed for the aligned wave. However, for small SNR of the data set, the SVD–ICA usually gives better performances than SVD.

The ICA step leads to a fourth-order independence of the estimated waves and relaxes the unjustified orthogonality constraint for the propagation vectors. This step in the process enhances the wave separation results and minimizes the error on the estimated signal subspace, especially when the SNR ratio is low.

5.4 Multi-Way Array Data Sets

We now turn to the modelization and processing of data sets having more than two *modes* or *diversities*. Such data sets are recorded by arrays of vector-sensors (also called *multi-component* sensors) collecting, in addition to time and distance information, the *polarization* information. Note that there exist other acquisition schemes that output multi-way (or multi-dimensional, multi-modal) data sets, but they are not considered here.

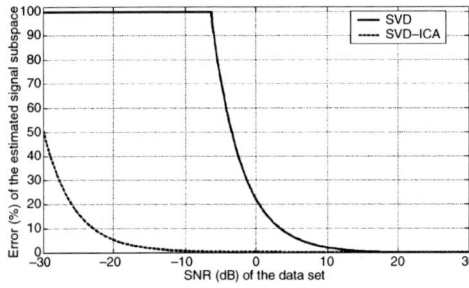

FIGURE 5.7
Relative error of the estimated subspaces.

5.4.1 Multi-Way Acquisition

In seismic acquisition campaigns, multi-component sensors have been used for more than ten years now. Such sensors allow the recording of the polarization of seismic waves. Thus, arrays of such sensors provide useful information about the nature of the propagated wavefields and allow a more complete description of the underground structures. The polarization information is very useful to differentiate and characterize waves in signal, but the specific (multi-component) nature of the data sets has to be taken into account in the processing. The use of vector-sensor arrays provides data sets with *time*, *distance*, and *polarization* modes, which are called *trimodal* or *three-mode* data sets. Here we propose to use *a multi-way model* to model and process them.

5.4.2 Multi-Way Model

To keep the *trimodal* (multi-dimensional) structure of data sets originated from vector-sensor arrays in their processing, we propose a multi-way model. This model is an extension of the one proposed in Section 5.2.2 Thus, a three-mode data set is modeled as a multi-way array of size $N_c \times N_x \times N_t$, where N_c is the number of components of each sensor used to recover the vibrations of the wavefield in the three directions of the 3D space, N_x is the number of sensors of the vector-sensor array, and N_t is the number of time samples.

Note that the number of components is defined by the vector-sensor configuration. As an example, for the illustration shown in Section 5.5.3, $N_c = 2$ because one geophone and one hydrophone were used, while $N_c = 3$ in Section 5.5.4 because three geophones were used.

Supposing that the propagation of waves only introduces delay and attenuation, the signal recorded on the cth component ($c = 1, \ldots, N_c$) of the kth sensor ($k = 1, \ldots, N_x$), using the superposition principle and assuming that P waves impinge on the array of vector-sensors, can be written as

$$x_{ck}(m) = \sum_{i=1}^{P} a_{cki} s_i(m - m_{ki}) + n_{ck}(m) \tag{5.13}$$

where a_{cki} represents the attenuation of the ith wave on the cth component of the kth sensor of the array. $s_i(m)$ is the ith wave and m_{ki} is the delay observed at sensor k. The time index is m.

$n_{ck}(m)$ is the noise, supposed Gaussian, centered, spatially white, and independent of the waves. As in the *matrix processing* approach, preprocessing is needed to ensure low rank of the signal subspace and to ensure good results for a subspace-based processing method.

Thus, a velocity correction applied on the dominant waveform (compensation of m_{k1}) leads for the signal recorded on component c of sensor k to:

$$y_{ck}(m) = a_{ck1} s_1(m) + \sum_{i=2}^{P} a_{cki} s_i(m - m'_{ki}) + n'_{ck}(m) \tag{5.14}$$

where $y_{ck}(m) = x_{ck}(m + m_{k1})$, $m'_{ki} = m_{ki} - m_{k1}$, and $n'_{ck}(m) = n_{ck}(m + m_{k1})$. In the sequel, the wave $s_1(m)$ is considered independent from other waves and from the noise. The subspace method developed thereafter will intend to isolate and estimate correctly this wave.

Thus, three-mode data sets recorded during N_t time samples on vector-sensor arrays made up by N_x sensors each one having N_c components can be modeled as multi-way arrays $Y \in \mathbb{R}^{N_c \times N_x \times N_t}$:

$$Y = \{y_{ckm} = y_{ck}(m) | 1 \leq c \leq N_c, 1 \leq k \leq N_x, 1 \leq m \leq N_t\} \tag{5.15}$$

This multi-way model can be used for extension of subspace method separation to multi-component data sets.

5.5 Multi-Way Array Processing

Multi-way data analysis arose firstly in the field of psychometrics with Tucker [16]. It is still an active field of research and it has found applications in many areas such as chemometrics, signal processing, communications, biological data analysis, food industry, etc.

It is an admitted fact that there exists no exact extension of the SVD for multi-way arrays of dimension greater than 2. Instead of such an extension, there exist mainly two decompositions: PARAFAC [17] and HOSVD [14,16]. The first one is also known as CANDECOMP as it gives a canonical decomposition of a multi-way array, that is, it expresses the multi-way array into a sum of rank-1 arrays. Note that the rank-1 arrays in the PARAFAC decomposition may not be orthogonal, unlike in the matrix case. The second multi-way array decomposition, HOSVD, gives orthogonal bases in the three ways of the array but is not a canonical decomposition as it does not express the original array into a sum of rank-1 arrays. However, in the sequel, we will make use of the HOSVD because of the orthogonal bases that allow extension of well-known subspace methods based on SVD to multi-way datasets.

5.5.1 HOSVD

We now introduce the HOSVD that was formulated and studied in detail in Ref. [14]. We give particular attention to the three-mode case because we will process such data in the sequel, but an extension to the multi-dimensional case exists [14]. One must notice that in the trimodal case the HOSVD is equivalent to the TUCKER3 model [16]; however, the HOSVD has a formulation that is more familiar in the signal processing community as its expression is given in terms of matrices of singular vectors just as in the SVD in the matrix case.

5.5.1.1 HOSVD Definition

Consider a multi-way array $Y \in \mathbb{R}^{N_c \times N_x \times N_t}$, the HOSVD of Y is given by

$$Y = C \times_1 \mathbf{V}_{(c)} \times_2 \mathbf{V}_{(x)} \times_3 \mathbf{V}_{(t)} \tag{5.16}$$

where $C \in \mathbb{R}^{N_c \times N_x \times N_t}$ is called the *core array* and $\mathbf{V}_{(i)} = [\mathbf{v}_{(i)1}, \ldots, \mathbf{v}_{(i)j}, \ldots, \mathbf{v}_{(i)r_i}] \in \mathbb{R}^{N_i \times r_i}$ are matrices containing the singular vectors $\mathbf{v}_{(i)j} \in \mathbb{R}^{N_i}$ of Y in the three modes ($i = c, x, t$). These matrices are orthogonal, $\mathbf{V}_{(i)} \mathbf{V}_{(i)}^T = \mathbf{I}$, just as in the matrix case. A schematic representation of the HOSVD is given in Figure 5.8.

The *core array* C is the counterpart of the diagonal matrix Δ in the SVD case in Equation 5.4, except that it is not hyperdiagonal but fulfils the less restrictive property of being *all-orthogonal*. All-orthogonality is defined as

$$\begin{aligned}&\langle C_{i=\alpha} C_{i=\beta}\rangle = 0 \text{ where } i = c, x, t \text{ and } \alpha \neq \beta \\ &\|C_{i=1}\| \geq \|C_{i=2}\| \geq \cdots \geq \|C_{i=r_i}\| \geq 0, \forall i\end{aligned} \tag{5.17}$$

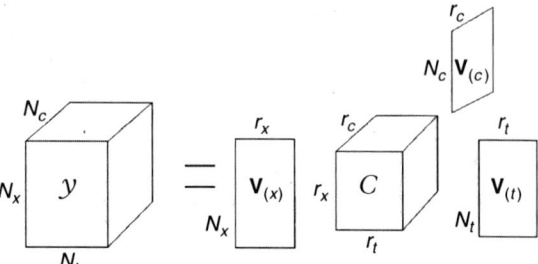

FIGURE 5.8
HOSVD of a three-mode data set **Y**.

where $\langle .,. \rangle$ is the classical scalar product between matrices[4] and $\|\cdot\|$ is the matrix Frobenius norm defined in Section 5.2 (because here we deal with three-mode data and the "slices" $C_{i\,=\,\alpha}$ define matrices). Thus, $\langle C_{i\,=\,\alpha}, C_{i\,-\,\beta} \rangle = 0$ corresponds to orthogonality between slices of the core array.

Clearly, the *all-orthogonality* property consists of orthogonality between two slices (matrices) of the core array cut in the same mode and ordering of the norm of these slices. This second property is the counterpart of the decreasing arrangement of the singular values in the SVD [7], with the special property of being valid here for norms of slices of C and in its three modes. As a consequence, the "energy" of the three-mode data set Y is concentrated at the (1,1,1) corner of the core array C.

The notation $\times_n$ in Equation 5.16 is called the *n-mode product* and there are three such products (namely $\times_1$, $\times_2$, and $\times_3$), which can be defined for the three-mode case. Given a multi-way array $A \in \mathbb{R}^{I_1 \times I_2 \times I_3}$, then the three possible *n*-mode products of A with matrices are:

$$(A \times_1 B)_{ji_2 i_3} = \sum_{i_1} a_{i_1 i_2 i_3} b_{j i_1}$$

$$(A \times_2 C)_{i_1 j i_3} = \sum_{i_2} a_{i_1 i_2 i_3} c_{j i_2}$$

$$(A \times_3 D)_{i_1 i_2 j} = \sum_{i_3} a_{i_1 i_2 i_3} d_{j i_3} \quad (5.18)$$

where $B \in \mathbb{R}^{J \times I_1}$, $C \in \mathbb{R}^{J \times I_2}$, and $D \in \mathbb{R}^{J \times I_3}$. This is a general notation in (multi-)linear algebra and even the SVD of a matrix can be expressed with such a product. For example, the SVD given in Equation 5.4 can be rewritten, using *n*-mode products, as $Y = \Delta \times_1 U \times_2 V$ [10].

5.5.1.2 Computation of the HOSVD

The problem of finding the elements of a three-mode decomposition was originally solved using alternate least square (ALS) techniques (see Ref. [18] for details). It was only in Ref. [14] that a technique based on unfolding matrix SVDs was proposed. We present briefly here a way to compute the HOSVD using this approach.

From a multi-way array $Y \in \mathbb{R}^{N_c \times N_x \times N_t}$, it is possible to build three unfolding matrices, with respect to the three modes c, x, and t, in the following way:

$$Y \in \mathbb{R}^{N_c \times N_x \times N_t} \Rightarrow \begin{cases} \mathbf{Y}_{(c)} \in \mathbb{R}^{N_c \times N_x N_t} \\ \mathbf{Y}_{(x)} \in \mathbb{R}^{N_x \times N_t N_c} \\ \mathbf{Y}_{(t)} \in \mathbb{R}^{N_t \times N_c N_x} \end{cases} \quad (5.19)$$

---

[4] $\langle A, B \rangle = \sum_{i=1}^{I} \sum_{j=1}^{J} a_{ij} b_{ij}$ is the scalar product between the matrices $A = \{a_{ij}\} \in \mathbb{R}^{I \times J}$ and $B = \{b_{ij}\} \in \mathbb{R}^{I \times J}$.

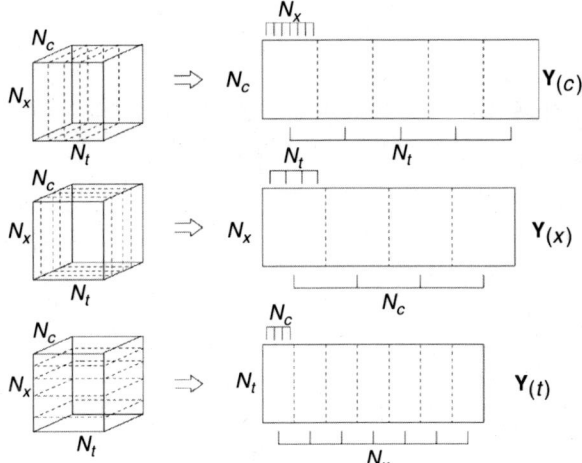

FIGURE 5.9
Schematic representation of the unfolding matrices.

A schematic representation of these unfolding matrices is presented in Figure 5.9.
Then, these three unfolding matrices admit the following SVD decompositions:

$$\mathbf{Y}_{(i)}^T \underset{\text{SVD}}{=} \mathbf{U}_{(i)} \Delta_{(i)} \mathbf{V}_{(i)}^T \tag{5.20}$$

where each matrix $\mathbf{V}_{(i)}$ ($i = c, x, t$) in Equation 5.16 is the right matrix given by the SVD of each transposed unfolding matrix $\mathbf{Y}_{(i)}^T$. The choice of the transpose was made to keep a homogeneous notation between matrix and multi-way processing. The matrices $\mathbf{V}_{(i)}$ define orthonormal bases in the three modes of the vector space $\mathbb{R}^{N_c \times N_x \times N_t}$. The core array is then directly obtained using the formula:

$$C = Y \times_1 \mathbf{V}_{(c)}^T \times_2 \mathbf{V}_{(x)}^T \times_3 \mathbf{V}_{(t)}^T \tag{5.21}$$

The singular values contained in the three matrices $\Delta_{(i)}$ ($i = c, x, t$) in Equation 5.20 are called *three-mode singular values*.

Thus, the HOSVD of a multi-way array can be easily obtained from the SVDs of the unfolding matrices, which makes computing of this decomposition easy using already existing algorithms.

5.5.1.3 The (r_c, r_x, r_t)-rank

Given a three-mode data $Y \in \mathbb{R}^{N_c \times N_x \times N_t}$, one gets three unfolded matrices $\mathbf{Y}_{(c)}$, $\mathbf{Y}_{(x)}$, and $\mathbf{Y}_{(t)}$, with respective ranks r_c, r_x, and r_t. That is, the r_j's are given as the number of nonvanishing singular values contained in the matrices $\Delta_{(i)}$ in Equation 5.20, with $i = c, x, t$.

As mentioned before, the HOSVD is not a canonical decomposition and so is not related to the generic rank (number of rank-1 arrays that lead to the original array by linear combination). Nevertheless, the HOSVD gives other information named, in the three-mode case, the *three-mode rank*. The three-mode rank consists of a triplet of ranks: the (r_c, r_x, r_t)-rank, which is made up of the ranks of matrices $\mathbf{Y}_{(c)}$, $\mathbf{Y}_{(x)}$, and $\mathbf{Y}_{(t)}$ in the HOSVD. In the sequel, the three-mode rank will be of use for determination of subspace dimensions, and so will be the counterpart of the classical rank used in matrix processing techniques.

5.5.1.4 Three-Mode Subspace Method

As in the matrix case, it is possible, using the HOSVD, to define a subspace method that decomposes the original three-mode data set into orthogonal subspaces. Such a technique was first proposed in Ref. [10] and can be stated as follows. Given a three-mode data set $Y \in \mathbb{R}^{N_c \times N_x \times N_t}$, it is possible to decompose it into the sum of a signal and a noise subspace as

$$Y = Y^{\text{Signal}} + Y^{\text{Noise}} \quad (5.22)$$

with a weaker constraint than in the matrix case (orthogonality), which is a *mode orthogonality*, that is, orthogonality in the three modes [10] (between subspaces defined using the unfolding matrices). Just as in the matrix case, the signal and noise subspaces are formed by different vectors obtained from the decomposition of the data set. In the three-mode case, the signal subspace Y^{Signal} is built using the first $p_c \leq r_c$ singular vectors in the first mode, $p_x \leq r_x$ in the second, and $p_t \leq r_t$ in the third:

$$Y^{\text{Signal}} = Y \times_1 P_{\mathbf{V}_{(c)}^{p_c}} \times_2 P_{\mathbf{V}_{(x)}^{p_x}} \times_3 P_{\mathbf{V}_{(t)}^{p_t}} \quad (5.23)$$

with $P_{\mathbf{V}_{(i)}^{p_i}}$ the projectors given by

$$P_{\mathbf{V}_{(i)}^{p_i}} = \mathbf{V}_{(i)}^{p_i} \mathbf{V}_{(i)}^{p_i^T} \quad (5.24)$$

where $\mathbf{V}_{(i)}^{p_i} = [\mathbf{v}_{(i)1}, \ldots, \mathbf{v}_{(i)p_i}]$ are the matrices containing the first p_i singular vectors ($i = c, x, t$). Then after estimation of the signal subspace, the noise subspace is simply obtained by subtraction, that is, $Y^{\text{Noise}} = Y - Y^{\text{Signal}}$.

The estimation of the signal subspace consists in finding a triplet of values p_c, p_x, p_t that allows recovery of the signal part by the (p_c, p_x, p_t)-rank truncation of the original data set Y. This truncation is obtained by classical matrix truncation of the three SVDs of the unfolding matrices. However, it is important to note that such a truncation is not the *best* (p_1, p_2, p_3)-*rank truncation* of the data [10]. Nevertheless, the decomposition of the original three-mode data set is possible and leads, under some assumptions, to the separation of the recorded wavefields.

From a practical point of view, the choice of p_c, p_x, p_t values is made by finding abrupt changes of the slope in the curves of relatives of three-mode singular values (the three sets of singular values contained in the matrices $\Delta_{(i)}$). For some special cases for which no "visible" change of slope can be found, the value of p_c can be fixed at 1 for a linear polarization of the aligned wavefield (denoted by $s_1(m)$), or at 2 for an elliptical polarization [10]. The value of p_x can be fixed at 1 and the value of p_t can be fixed at 1 for a perfect alignment of waves, or at 2 for not an imperfect alignment or for dispersive waves.

As in the matrix case, the HOSVD-based subspace method decomposes the original space of the data set into orthogonal subspaces, and so following the ideas developed in Section 5.3.2, it is possible to add an ICA step to modify the orthogonal constraint in the temporal mode.

5.5.2 HOSVD and Unimodal ICA

To enhance wavefield separation results, we now introduce a unimodal ICA step following the HOSVD-based subspace decomposition.

5.5.2.1 HOSVD and ICA

The SVD of the unfolded matrix $\mathbf{Y}_{(t)}$ in Equation 5.20 provides two orthogonal matrices $\mathbf{U}_{(t)}$ and $\mathbf{V}_{(t)}$ made up by the left and right singular vectors $\mathbf{u}_{(t)j}$ and $\mathbf{v}_{(t)j}$. As for the SVD, the $\mathbf{v}_{(t)j}s'$ are the estimated waves and $\mathbf{u}_{(t)j}$'s are the propagation vectors.

Based on the same motivations as in the SVD case, the unjustified orthogonality constraint for the propagation vectors can be relaxed by imposing a fourth-order independence for the estimated waves. Assuming the recorded waves are mutually independent, we can write:

$$\mathbf{V}_{(t)}^R \mathbf{B} = \tilde{\mathbf{V}}_{(t)}^R = [\tilde{\mathbf{v}}_{(t)1}, \ldots, \tilde{\mathbf{v}}_{(t)R}] \quad (5.25)$$

with $\mathbf{B} \in \mathbb{R}^{R \times R}$ the rotation (unitary) matrix given by one of the algorithms presented in Section 5.3.2 and $\mathbf{V}_{(t)}^R = [\mathbf{v}_{(t)1}, \ldots, \mathbf{v}_{(t)R}] \in \mathbb{R}^{N_t \times R}$ made up by the first R vectors of $\mathbf{V}_{(t)}$.

Here we also suppose that the nonaligned waves in the unfolded matrix $\mathbf{Y}_{(t)}$ are contained in a subspace of dimension $R - 1$, smaller than the rank r_t of $\mathbf{Y}_{(t)}$.

After the ICA step, a new matrix can be computed:

$$\tilde{\mathbf{V}}_{(t)} = \left[\tilde{\mathbf{V}}_{(t)}^R; \mathbf{V}_{(t)}^{N_t - R} \right] \quad (5.26)$$

This matrix is made up of the R vectors $\tilde{\mathbf{v}}_{(t)j}$ of the matrix $\tilde{\mathbf{V}}_{(t)}^R$, which are independent at the fourth order, and by the last $N_t - R$ vectors $\mathbf{v}_{(t)j}$ of the matrix $\mathbf{V}_{(t)}$, which are kept unchanged.

The HOSVD–unimodal ICA decomposition is defined as

$$Y = \tilde{C} \times_1 \mathbf{V}_{(c)} \times_2 \mathbf{V}_{(x)} \times_3 \tilde{\mathbf{V}}_{(t)} \quad (5.27)$$

with

$$\tilde{C} = Y \times_1 \mathbf{V}_{(c)}^T \times_2 \mathbf{V}_{(x)}^T \times_3 \tilde{\mathbf{V}}_{(t)}^T \quad (5.28)$$

Unimodal ICA implies here that ICA is only performed on one mode (the temporal mode). As in the SVD case, a permutation $\sigma(.)$ between the vectors of $\mathbf{V}_{(c)}, \mathbf{V}_{(x)}$, respectively, $\tilde{\mathbf{V}}_{(t)}$ must be performed for ordering the Frobenius norms of the subarrays (obtained by fixing one index) of the new core array $\tilde{C}$. Hence, we keep the same decomposition structure as in relations given in Equation 5.16 and Equation 5.21, the only difference is that we have modified the orthogonality into a fourth-order independence constraint for the first R estimated waves on the third mode. Note that the (r_c, r_x, r_t)-rank of the three-mode data set Y is unchanged.

5.5.2.2 Subspace Method Using HOSVD–Unimodal ICA

On the temporal mode, a new projector can be computed after the ICA and the permutation steps:

$$\tilde{P}_{\tilde{\mathbf{V}}_{(t)}^{\tilde{p}_t}} = \tilde{\mathbf{V}}_{(t)}^{\tilde{p}_t} \tilde{\mathbf{V}}_{(t)}^{\tilde{p}_t^T} \quad (5.29)$$

where $\tilde{\mathbf{V}}_{(t)}^{\tilde{p}_t} = [\tilde{\mathbf{v}}_{(t)1}, \ldots, \tilde{\mathbf{v}}_{(t)\tilde{p}_t}]$ is the matrix containing the first $\tilde{p}_t$ estimated waves, which are the columns of $\tilde{\mathbf{V}}_{(t)}$ defined in Equation 5.26. Note that the two projectors on the first two modes $P_{\mathbf{V}_{(c)}^{pc}}$ and $P_{\mathbf{V}_{(c)}^{px}}$, given by Equation 5.24, must be recomputed after the permutation step.

The signal subspace using the HOSVD–unimodal ICA is thus given as

$$\tilde{Y}^{\text{Signal}} = Y \times_1 P_{\mathbf{V}_{(c)}^{p_c}} \times_2 P_{\mathbf{V}_{(x)}^{p_x}} \times_3 \tilde{P}_{\tilde{\mathbf{V}}_{(t)}^{\tilde{p}_t}} \quad (5.30)$$

and the noise subspace $\tilde{Y}^{\text{Noise}}$ is obtained by the subtraction of the signal subspace from the original three-mode data:

$$\tilde{Y}^{\text{Noise}} = Y - \tilde{Y}^{\text{Signal}} \quad (5.31)$$

In practical situations, as in the SVD–ICA subspace case, the value of R becomes a parameter. It is chosen to fully describe the aligned wave by the first R estimated waves $\mathbf{v}_{(t)j}$ obtained while using the HOSVD.

The choice of the p_c, p_x, and $\tilde{p}_t$ values is made by finding abrupt changes of the slope in the curves of modified three-mode singular values, obtained after the ICA and the permutation steps. Note that in the HOSVD–unimodal ICA subspace method, the rank for the signal subspace in the third mode, $\tilde{p}_t$, is not necessarily equal to the rank p_t obtained using only the HOSVD.

For some special cases for which no "visible" change of slope can be found, the value of $\tilde{p}_t$ can be fixed at 1 for a perfect alignment of waves, or at 2 for an imperfect alignment or for dispersive waves. As in the HOSVD subspace method, p_x can be fixed at 1 and p_c can be fixed at 1 for a linear polarization of the aligned wavefield, or at 2 for an elliptical polarization.

Applications to simulated and real data are presented in the following sections to illustrate the behavior of the HOSVD–unimodal ICA method in comparison with component-wise SVD (SVD applied on each component of the multi-way data separately) and HOSVD subspace methods.

5.5.3 Application to Simulated Data

This simulation represents a multi-way data set $Y \in \mathbb{R}^{2 \times 18 \times 256}$ composed of $N_x = 18$ sensors each recording two directions ($N_c = 2$) in the 3D space for a duration of $N_t = 256$ time samples. The first component is related to a geophone $\mathbf{Z}$ and the second one to a hydrophone $\mathbf{Hy}$. The $\mathbf{Z}$ component was scaled by 5 to obtain the same amplitude range.

This data set shown in Figure 5.10c and Figure 5.11c has *polarization, distance,* and *time* as modes. It was obtained by the addition between an original signal subspace S with the two components shown in Figure 5.10a and Figure 5.11a, respectively, and an original noise subspace N (Figure 5.10b and Figure 5.11b respectively) obtained from a real geophysical acquisition after subtraction of aligned waves.

The original signal subspace is made of several wavefronts having *infinite apparent velocity*, associated with the aligned wave. The relation between $\mathbf{Z}$ and $\mathbf{Hy}$ is a linear relation, which is assimilated to the wave polarization (*polarization* mode) in the sense that it consists of phase and amplitude relations between the two components. Wave amplitudes vary along the sensors, simulating attenuation along the *distance* mode. The noise is uncorrelated from one sensor to the other (spatially white) and also unpolarized. The SNR ratio of this data set is SNR = −3 dB, where the SNR definition is:

$$\text{SNR} = 20 \log_{10} \frac{\|S\|}{\|N\|} \quad (5.32)$$

where $\|\cdot\|$ is the multi-way array Frobenius norm[5], S, and N the original signal and noise subspaces.

---

[5] For any multi-way array $X = \{x_{ijk}\} \in \mathbb{R}^{I \times J \times K}$, his Frobenius norm is $\|X\| = \sqrt{\sum_{i=1}^{i} \sum_{j=1}^{j} \sum_{k=1}^{K} x_{ijk}^2}$.

Multi-Dimensional Seismic Data Decomposition by Higher Order SVD and Unimodal ICA 91

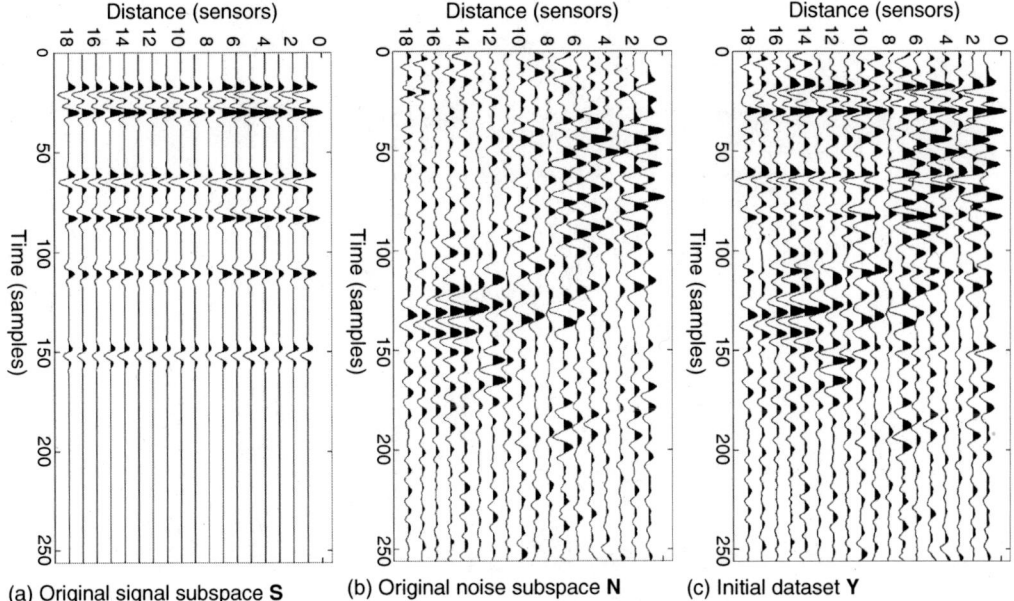

(a) Original signal subspace **S** (b) Original noise subspace **N** (c) Initial dataset **Y**

FIGURE 5.10
Simulated data: the **Z** component.

Our aim is to recover the original signal (Figure 5.10a and Figure 5.11a) from the mixture, which is, in practice, the only data available. Note that normalization to unit variance of each trace for each component was done before applying the described subspace methods. This ensures that even weak peaked arrivals are well represented

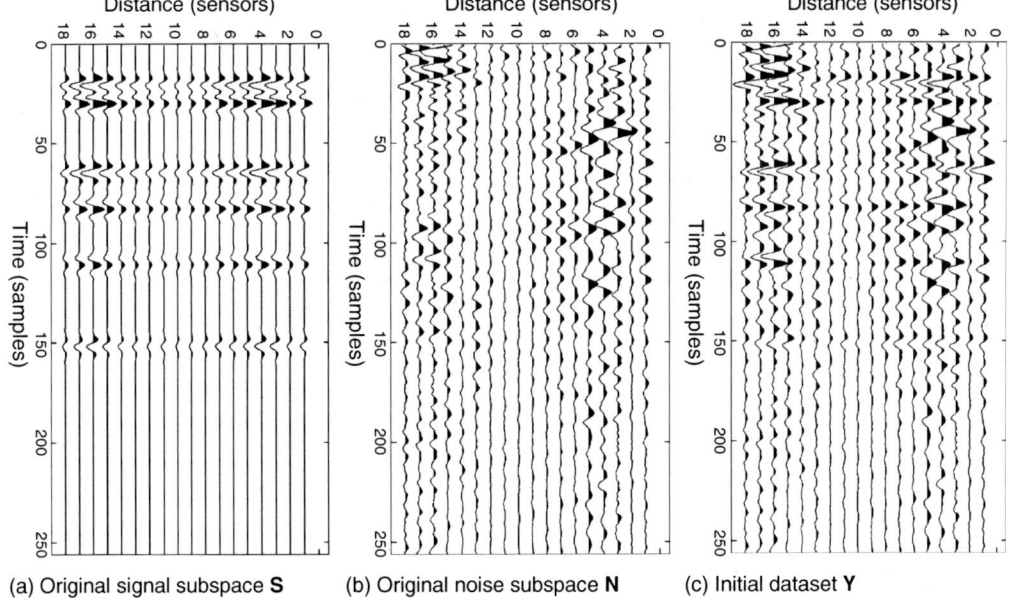

(a) Original signal subspace **S** (b) Original noise subspace **N** (c) Initial dataset **Y**

FIGURE 5.11
Simulated data: the **Hy** component.

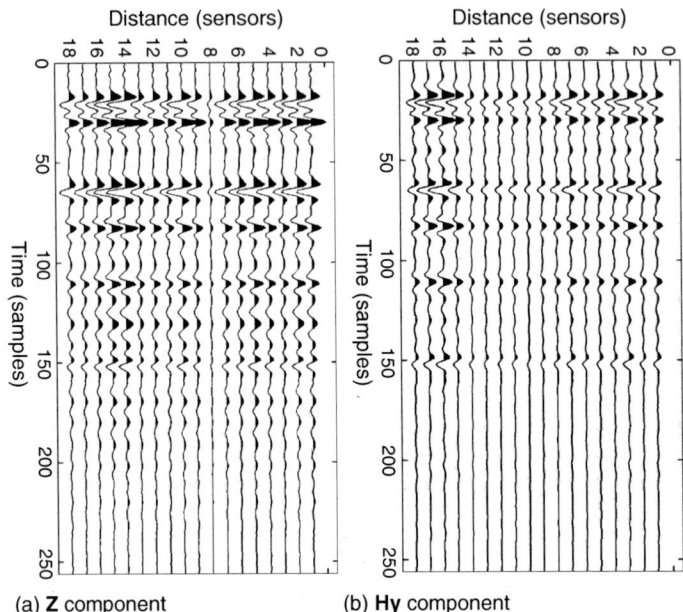

FIGURE 5.12
Signal subspace using component-wise SVD.

within the input data. After computation of signal subspaces, a denormalization was applied to find the original signal subspace.

Firstly, the SVD subspace method (described in Section 5.3) was applied separately on **Z** and **Hy** components of the mixture. The signal subspace components obtained keeping only one singular value for the two components are presented in Figure 5.12. This choice was made by finding an abrupt change of slope after the first singular value in the relative singular values shown in Figure 5.13 for each seismic 2D signal. The waveforms are not well recovered in respect of the original signal components (Figure 5.10a and Figure 5.11a). One can also see that the distinction between different wavefronts is not possible. Furthermore, no arrival time estimation is possible using this technique. Low signal level is a strong handicap for a component-wise process.

Applied on each component separately, the SVD subspace method does not find the same aligned polarized wave. The results depend therefore on the characteristics of each matrix signal. Using the SVD–ICA subspace method, the estimated aligned waves may be improved, but we can be confronted with the same problem.

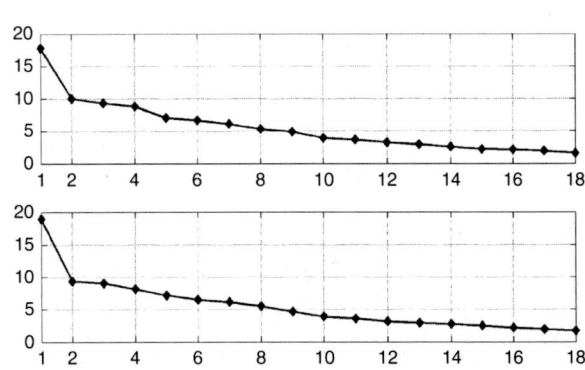

FIGURE 5.13
Relative singular values. Top: **Z** component.
Bottom: **Hy** component.

Multi-Dimensional Seismic Data Decomposition by Higher Order SVD and Unimodal ICA

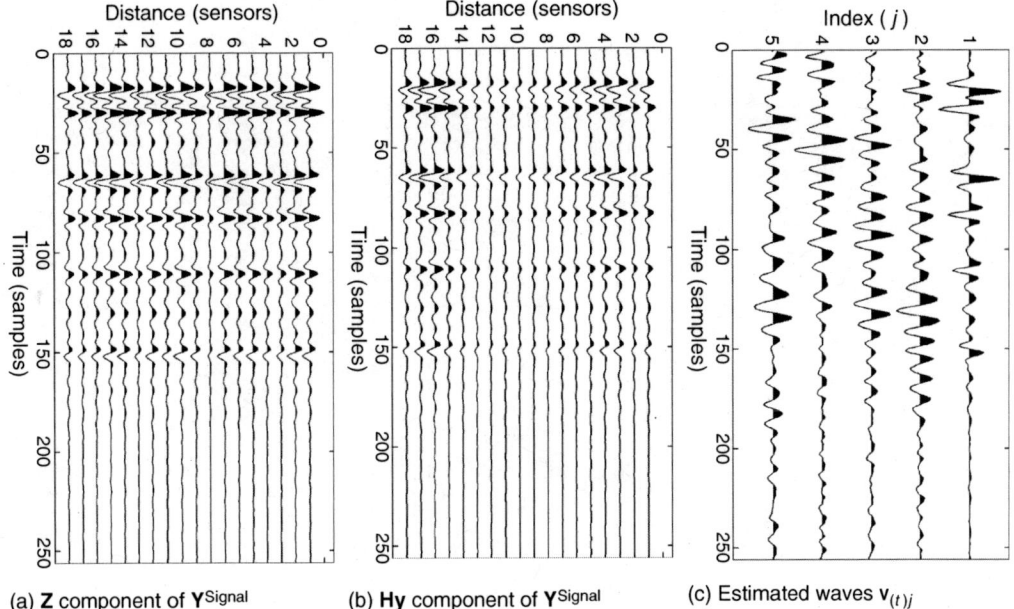

(a) **Z** component of **Y**^{Signal} (b) **Hy** component of **Y**^{Signal} (c) Estimated waves $v_{(t)j}$

FIGURE 5.14
Results obtained using the HOSVD subspace method.

Using the HOSVD subspace method, the components of the estimated signal subspace Y^{Signal} are presented in Figure 5.14a and Figure 5.14b. In this case, the number of singular vectors kept are: one on *polarization*, one on *distance*, and one on *time* mode, giving a (1,1,1)-rank truncation for the signal subspace. This choice is motivated here by the linear polarization of the aligned wave. For the other two modes the choice was made by finding an abrupt change of slope after the first singular value (Figure 5.17a).

There still remain some "oscillations" between the different wavefronts for the two components of the estimated signal subspace, that may induce some detection errors. An ICA step is required in this case to obtain a better signal separation and to cancel parasitic oscillations.

In Figure 5.15a and Figure 5.15b, the wavefronts of the estimated signal subspace $\tilde{Y}^{Signal}$ obtained with the HOSVD–unimodal ICA technique are very close to the original signal components.

Here, the ICA method was applied on the first $R = 5$ estimated waves shown in Figure 5.14c. These waves describe the aligned waves of the original signal subspace S. After the ICA step, the estimated waves $\tilde{v}_{(t)j}$ are shown in Figure 5.15c. As we can see, the first one $\tilde{v}_{(t)1}$ describes more precisely the aligned wave of the original subspace S than the first estimated wave $v_{(t)1}$ before the ICA step (Figure 5.14c). The estimation of signal subspace is more accurate and the aligned wavelet can be better estimated with our proposed procedure.

After the permutation step, the relative singular values on the three modes in the HOSVD–unimodal ICA case are shown in Figure 5.17b. This figure justifies the choice of a (1,1,1)-rank truncation for the signal subspace, due to the linear polarization and the abrupt changes of the slopes for the other two modes.

As for the bidimensional case, to compare these methods quantitatively, we use the relative error ε of the estimated signal subspace defined as

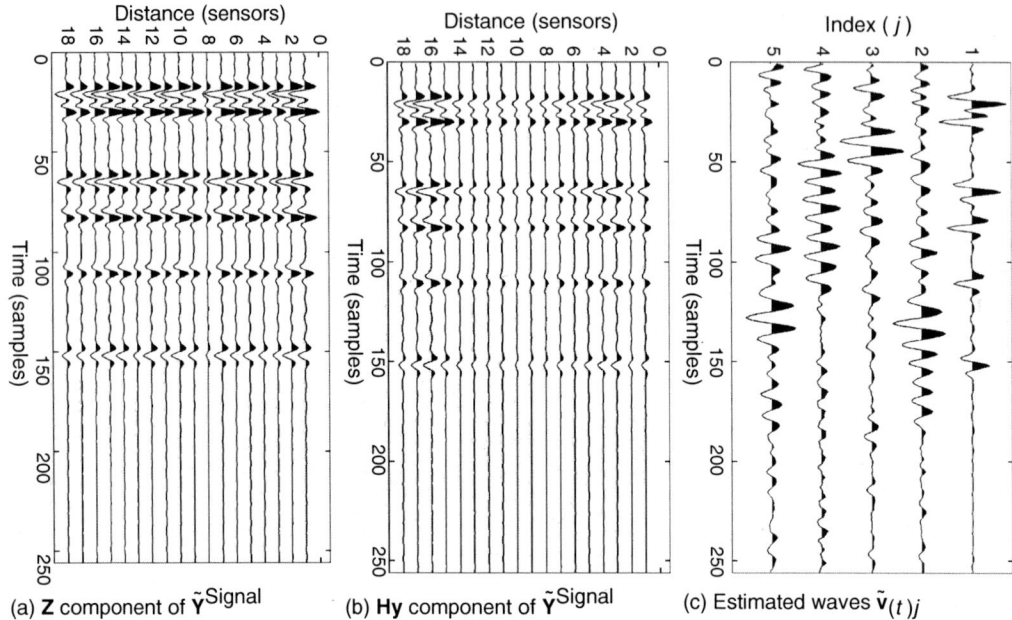

FIGURE 5.15
Results obtained using the HOSVD–unimodal ICA subspace method.

$$\varepsilon = \frac{\|S - \bar{Y}^{\text{Signal}}\|^2}{\|S\|^2} \qquad (5.33)$$

where $\|\cdot\|$ is the multi-way array Frobenius norm defined above, S is the original signal subspace, and $\bar{Y}^{\text{Signal}}$ represents the estimated signal subspaces obtained with the SVD, HOSVD, and HOSVD–unimodal ICA methods, respectively. For this data set we obtain $\varepsilon = 21.4\%$ for the component-wise SVD, $\varepsilon = 12.4\%$ for HOSVD, and $\varepsilon = 3.8\%$ for HOSVD–unimodal ICA. We conclude that the ICA step minimizes the error on the estimated signal subspace.

To compare the results qualitatively, the stack representation is employed. Figure 5.16 shows for each component, from left to right, the stacks on the initial data set Y, the original signal subspace S, and the estimated signal subspaces obtained with the component-wise SVD, HOSVD, and HOSVD–unimodal ICA subspace methods, respectively.

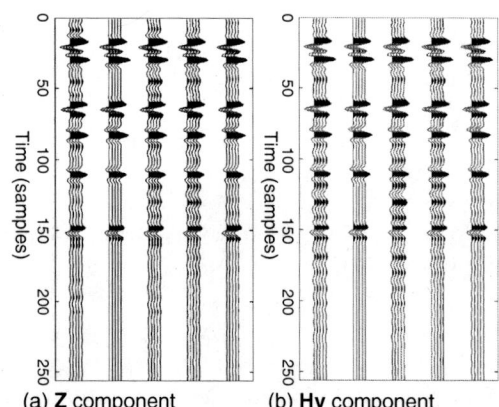

FIGURE 5.16
Stacks. From left to right: initial data set **Y**, original signal subspace S, SVD, HOSVD, and HOSVD–unimodal ICA estimated subspaces, respectively.

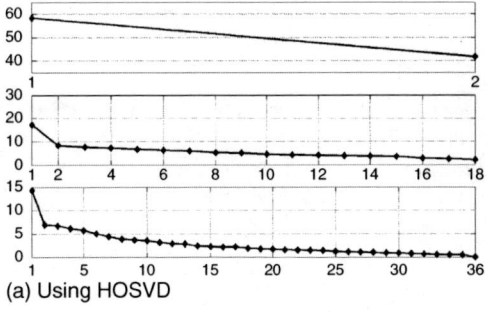
(a) Using HOSVD

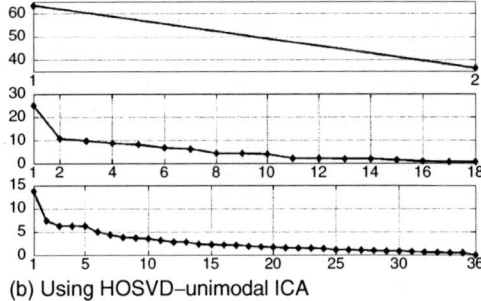
(b) Using HOSVD–unimodal ICA

FIGURE 5.17
Relative three-mode singular values.

As the stack on the estimated signal subspace $\tilde{Y}^{Signal}$ is very close to the stack on the original signal subspace S, we conclude that the HOSVD–unimodal ICA subspace method enhances the wave separation results.

5.5.4 Application to Real Data

We consider now a real vertical seismic profile (VSP) geophysical data set. This 3C data set was recorded by $N_x = 50$ sensors with the depth sampling 10 m, each one made up by $N_c = 3$ geophones recording three directions in the 3D space: **X**, **Y**, and **Z**, respectively. The recording time was 700 msec, corresponding to $N_t = 175$ time samples. The **Z** component was scaled seven times to obtain the same amplitude range.

After the preprocessing step (velocity correction based on the direct downgoing wave), the obtained data set $Y \in \mathbb{R}^{3 \times 50 \times 175}$ is shown in Figure 5.18.

As in the simulation case, normalization and denormalization of each trace for each component were done before and after applying the different subspace methods.

From the original data set we have constructed three seismic 2D matrix signals representing the three components of the data set Y. The SVD subspace method presented in Section 5.3.1 was applied on each matrix signal, keeping only one singular vector (respectively one singular value) for each one, due to an abrupt change of slope after the first singular value in the curves of relative singular values. As remarked in the simulation case, the SVD–ICA subspace method may improve the estimated aligned waves, but we will not find the same aligned polarized wave for all seismic matrix signals.

For the HOSVD subspace method, the estimated signal subspace Y^{Signal} can be defined as a (2,1,1)-rank truncation of the data set. This choice is motivated here by the elliptical polarization of the aligned wave. For the other two modes the choice was made by finding an abrupt change of slope after the first singular value (Figure 5.20a).

Using the HOSVD–unimodal ICA, the ICA step was applied here on the first $R = 9$ estimated waves $\mathbf{v}_{(t)j}$ shown in Figure 5.19a. As suggested, R becomes a parameter

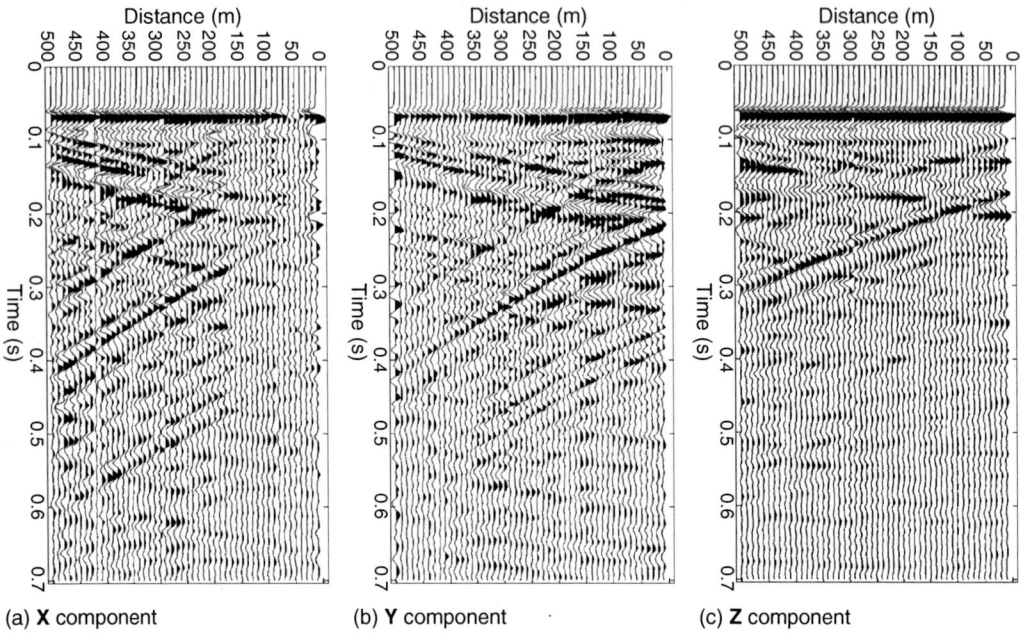

(a) **X** component (b) **Y** component (c) **Z** component

FIGURE 5.18
Real VSP geophysical data Y.

while using real data. However, the estimated waves $\tilde{\mathbf{v}}_{(t)j}$ shown in Figure 5.19b are more realistic (shorter wavelet and no side lobes) than those obtained without ICA. Due to the elliptical polarization and the abrupt change of slope after the first singular value for the other two modes (Figure 5.20b), the estimated signal subspace $\tilde{Y}^{\text{Signal}}$ is defined as a (2,1,1)-rank truncation of the data set. This step enhances the wave separation results, implying a minimization of the error on the estimated signal subspace.

When we deal with a real data set, only a qualitative comparison is possible. This is allowed by a stack representation. Figure 5.21 shows the stacks for the **X**, **Y**, and **Z** components, respectively, on the initial trimodal data Y and on the estimated signal

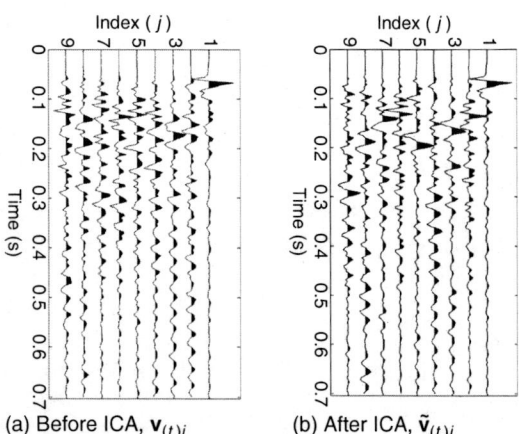

(a) Before ICA, $\mathbf{v}_{(t)j}$ (b) After ICA, $\tilde{\mathbf{v}}_{(t)j}$

FIGURE 5.19
The first 9 estimated waves.

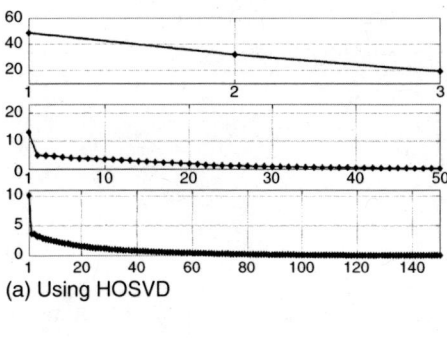

(a) Using HOSVD

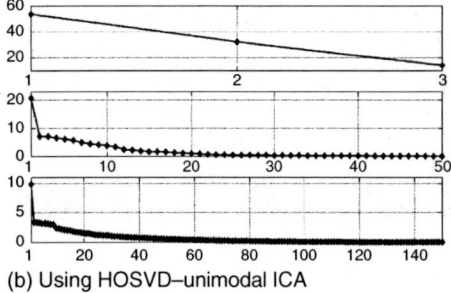

(b) Using HOSVD–unimodal ICA

FIGURE 5.20
Relative three-mode singular values.

subspaces given by the component-wise SVD, HOSVD, and HOSVD–unimodal ICA methods, respectively.

The results on simulated and real data suggest that the three-dimensional subspace methods are more robust than the component-wise techniques because they exploit the relationship between the components directly in the process. Also, the fourth-order independence constraint of the estimated waves enhances the wave separation results

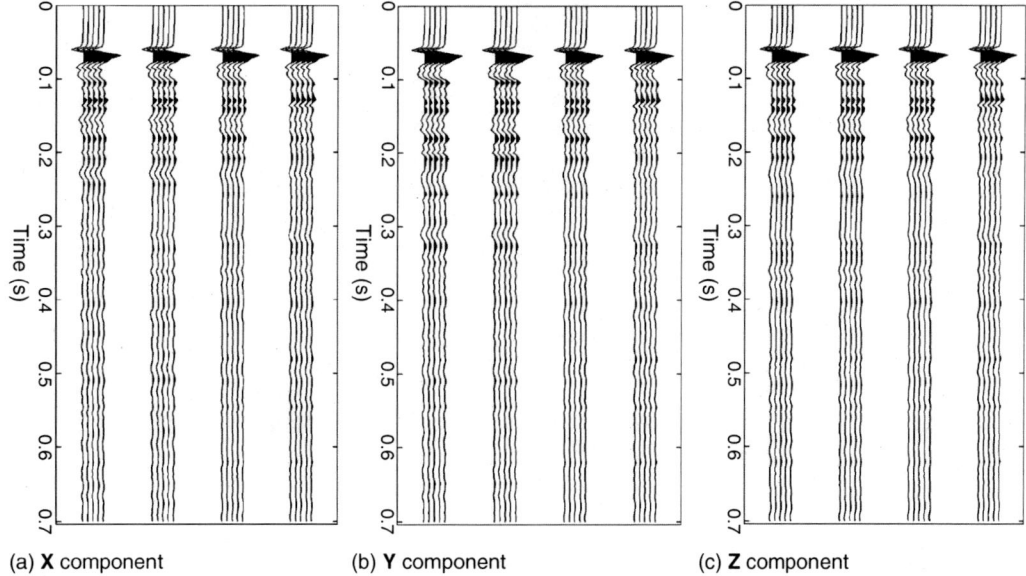

(a) **X** component   (b) **Y** component   (c) **Z** component

FIGURE 5.21
Stacks. From left to right: initial data set **Y**, SVD, HOSVD, and HOSVD–unimodal ICA estimated subspaces, respectively.

and minimizes the error on the estimated signal subspace. This emphasizes the potential of the HOSVD subspace method associated with a unimodal ICA step for vector-sensor array signal processing.

5.6 Conclusions

We have presented a subspace processing technique for multi-dimensional seismic data sets based on HOSVD and ICA. It is an extension of well-known subspace separation techniques for 2D (matrix) data sets based on SVD and more recently on SVD and ICA. The proposed multi-way technique can be used for the denoising and separation of polarized waves recorded on vector-sensor arrays. A multi-way (three-mode) model of polarized signals recorded on vector-sensor arrays allows us to take into account the additional polarization information in the processing and thus to enhance the separation results.

A decomposition of three-mode data sets into *all-orthogonal* subspaces has been proposed using HOSVD. An extra unimodal ICA step has been introduced to minimize the error on the estimated signal subspace and to improve the separation result. Also, we have shown on simulated and real data sets that the proposed approach gives better results than the component-wise SVD subspace method.

The use of multi-way array and associated decompositions for multi-dimensional data set processing is a powerful tool and ensures the extra dimension is fully taken into account in the process. This approach could be generalized to any multi-dimensional signal modelization and processing and could take advantage of recent work on tensors and multi-way array decomposition and analysis.

References

1. Vrabie, V.D., Statistiques d'ordre supérieur: applications en géophysique et électrotechnique, Ph.D. thesis, I.N.P. of Grenoble, France, 2003.
2. Vrabie, V.D., Mars, J.I., and Lacoume, J.-L., Modified singular value decomposition by means of independent component analysis, *Signal Processing*, 84(3), 645, 2004.
3. Sheriff, R.E. and Geldart, L.P., *Exploration Seismology*, Vol. 1 & 2, Cambridge University Press, 1982.
4. Freire, S. and Ulrich, T., Application of singular value decomposition to vertical seismic profiling, *Geophysics*, 53, 778–785, 1988.
5. Glangeaud, F., Mari, J.-L., and Coppens, F., *Signal processing for geologists and geophysicists*, Editions Technip, Paris, 1999.
6. Robinson, E.A. and Treitel, S., *Geophysical Signal Processing*, Prentice Hall, 1980.
7. Golub, G.H. and Van Loan, C.F., *Matrix Computation*, Johns Hopkins, 1989.
8. Moonen, M. and De Moor, B., SVD and signal processing III, in *Algorithms, Applications and Architecture*, Elsevier, Amsterdam, 1995.
9. Scharf, L.L., Statistical signal processing, detection, estimation and time series analysis, in *Electrical and Computer Engineering: Digital Signal Processing Series*, Addison Wesley, 1991.
10. Le Bihan, N. and Ginolhac, G., Three-mode dataset analysis using higher order subspace method: application to sonar and seismo-acoustic signal processing, *Signal Processing*, 84(5), 919–942, 2004.
11. Vrabie, V.D., Le Bihan, N., and Mars, J.I., 3D-SVD and partial ICA for 3D array sensors, *Proceedings of the 72nd Annual International Meeting of the Society of Exploration Geophysicists*, Salt Lake City, 2002, 1065.

12. Cardoso, J.-F. and Souloumiac, A., Blind beamforming for non-Gaussian signals, *IEE Proc.-F*, 140(6), 362, 1993.
13. Comon, P., Independent component analysis, a new concept? *Signal Processing*, 36(3), 287, 1994.
14. De Lathauwer, L., Signal processing based on multilinear algebra, Ph.D. thesis, Katholieke Universiteit Leuven, 1997.
15. Hyvärinen, A. and Oja, E., Independent component analysis: algorithms and applications, *Neural Networks*, 13(4–5), 411, 2000.
16. Tucker, L.R., The extension of factor analysis to three-dimensional matrices, H. Gulliksen and N. Frederiksen (Eds.), Contributions to mathematical psychology series, Holt, Rinehart and Winston, 109–127, 1964.
17. Caroll, J.D. and Chang, J.J., Analysis of individual differences in multidimensional scaling via n-way generalization of Eckart–Young decomposition, *Psychometrika*, 35(3), 283–319, 1970.
18. Kroonenberg, P.M., *Three-Mode Principal Component Analysis*, DSWO Press, Leiden, 1983.

6

Application of Factor Analysis in Seismic Profiling

Zhenhai Wang and Chi Hau Chen

CONTENTS

6.1 Introduction to Seismic Signal Processing 102
    6.1.1 Data Acquisition 102
    6.1.2 Data Processing 103
        6.1.2.1 Deconvolution 103
        6.1.2.2 Normal Moveout 103
        6.1.2.3 Velocity Analysis 104
        6.1.2.4 NMO Stretching 104
        6.1.2.5 Stacking 104
        6.1.2.6 Migration 104
    6.1.3 Interpretation 105
6.2 Factor Analysis Framework 105
    6.2.1 General Model 105
    6.2.2 Within the Framework 107
        6.2.2.1 Principal Component Analysis 107
        6.2.2.2 Independent Component Analysis 108
        6.2.2.3 Independent Factor Analysis 109
6.3 FA Application in Seismic Signal Processing 109
    6.3.1 Marmousi Data Set 109
    6.3.2 Velocity Analysis, NMO Correction, and Stacking 110
    6.3.3 The Advantage of Stacking 112
    6.3.4 Factor Analysis vs. Stacking 112
    6.3.5 Application of Factor Analysis 114
        6.3.5.1 Factor Analysis Scheme No. 1 114
        6.3.5.2 Factor Analysis Scheme No. 2 114
        6.3.5.3 Factor Analysis Scheme No. 3 116
        6.3.5.4 Factor Analysis Scheme No. 4 116
    6.3.6 Factor Analysis vs. PCA and ICA 118
6.4 Conclusions 120
References 120
Appendices 122
6.A Upper Bound of the Number of Common Factors 122
6.B Maximum Likelihood Algorithm 123

6.1 Introduction to Seismic Signal Processing

Formed millions of years ago from plants and animals that died and decomposed beneath soil and rock, fossil fuels, namely, coal and petroleum, due to their low cost availability, will remain the most important energy resource for at least another few decades. Ongoing petroleum research continues to focus on science and technology needs for increased petroleum exploration and production. The petroleum industry relies heavily on subsurface imaging techniques for the location of these hydrocarbons.

6.1.1 Data Acquisition

Many geophysical survey techniques exist, such as multichannel reflection seismic profiling, refraction seismic survey, gravity survey, and heat flow measurement. Among them, reflection seismic profiling method stands out because of its target-oriented capability, generally good imaging results, and computational efficiency. These reflectivity data resolve features such as faults, folds, and lithologic boundaries measured in 10s of meters, and image them laterally for 100s of kilometers and to depths of 50 kilometers or more. As a result, seismic reflection profiling becomes the principal method by which the petroleum industry explores for hydrocarbon-trapping structures.

The seismic reflection method works by processing echoes of seismic waves from boundaries between different Earth's subsurfaces that characterize different acoustic impedances. Depending on the geometry of surface observation points and source locations, the survey is called a 2D or a 3D seismic survey. Figure 6.1 shows a typical 2D seismic survey, during which, a cable with attached receivers at regular intervals is dragged by a boat. The source moves along the predesigned seismic lines and generates seismic waves at regular intervals such that points in the subsurfaces are sampled several times by the receivers, producing a series of seismic traces. These seismic traces are saved on magnetic tapes or hard disks in the recording boat for future processing.

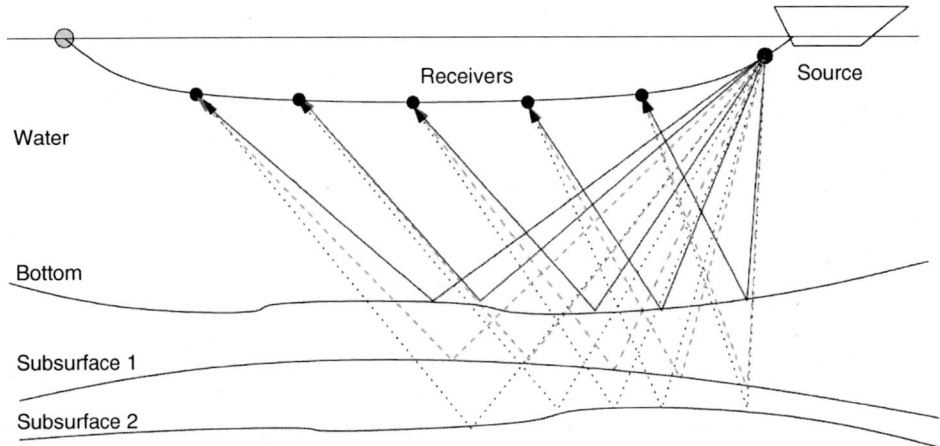

FIGURE 6.1
A typical 2D seismic survey.

6.1.2 Data Processing

Seismic data processing has been regarded as having a flavor of interpretive character; it is even considered as an art [1]. However, there is a well-established sequence for standard seismic data processing. Deconvolution, stacking, and migration are the three principal processes that make up the foundation. Besides, some auxiliary processes can also help improve the effectiveness of the principal processes. In the following subsections, we briefly discuss the principal processes and some auxiliary processes.

6.1.2.1 Deconvolution

Deconvolution can improve the temporal resolution of seismic data by compressing the basic seismic wavelet to approximately a spike and suppressing reverberations on the field data [2]. Deconvolution usually applied before stack is called prestack deconvolution. It is also a common practice to apply deconvolution to stacked data, which is named poststack deconvolution.

6.1.2.2 Normal Moveout

Consider the simplest case where the subsurfaces of the Earth are horizontal, and within this layer, the velocity is constant.

Here x is the distance (offset) between the source and the receiver positions, and v is the velocity of the medium above the reflecting interface. Given the midpoint location M, let $t(x)$ be the traveltime along the raypath from the shot position S to the depth point D, then back to the receiver position G. Let $t(0)$ be twice the traveltime along the vertical path MD. Utilizing the Pythagorean theorem, the traveltime equation as a function of offset is

$$t^2(x) = t^2(0) + x^2/v^2 \tag{6.1}$$

Note that the above equation describes a hyperbola in the plane of two-way time vs. offset. A common-midpoint (CMP) gather are the traces whose raypaths associated with each source–receiver pair reflect from the same subsurface depth point D. The difference between the two-way time at a given offset $t(x)$ and the two-way zero-offset time $t(0)$ is called NMO. From Equation 6.1, we see that velocity can be computed when offset x and the two-way times $t(x)$ and $t(0)$ are known. Once the NMO velocity is estimated, the travletimes can be corrected to remove the influence of offset.

$$\Delta t_{NMO} = t(x) - t(0)$$

Traces in the NMO-corrected gather are then summed to obtain a stack trace at the particular CMP location. The procedure is called stacking.

Now consider the horizontally stratified layers, with each layer's thickness defined in terms of two-way zero-offset time. Given the number of layers N, interval velocities are represented as $(v_1, v_2, \ldots, v_N)$. Considering the raypath from source S to depth D, back to receiver R, associated with offset x at midpoint location M, Equation 6.1 becomes

$$t^2(x) = t^2(0) + x^2/v_{rms}^2 \tag{6.2}$$

where the relation between the rms velocity and the interval velocity is represented by

$$v_{\text{rms}}^2 = \frac{1}{t(0)} \sum_{i=1}^{N} v_i^2 \Delta t_i(0)$$

where Δt_i is the vertical two-way time through the ith layer and $t(0) = \sum_{k=1}^{i} \Delta t_k$.

6.1.2.3 Velocity Analysis

Effective correction for normal moveout depends on the use of accurate velocities. In CMP surveys, the appropriate velocity is derived by computer analysis of the moveout in the CMP gathers. Dynamic corrections are implemented for a range of velocity values and the corrected traces are stacked. The stacking velocity is defined as the velocity value that produces the maximum amplitude of the reflection event in the stack of traces, which clearly represents the condition of successful removal of NMO.

In practice, NMO corrections are computed for narrow time windows down the entire trace, and for a range of velocities, to produce a velocity spectrum. The validity for each velocity value is assessed by calculating a form of multitrace correlation between the corrected traces of the CMP gathers. The values are shown contoured such that contour peaks occur at times corresponding to reflected wavelets and at velocities that produce an optimum stacked wavelet. By picking the location of the peaks on the velocity spectrum plot, a velocity function defining the increase of velocity with depth for that CMP gather can be derived.

6.1.2.4 NMO Stretching

After applying NMO correction, a frequency distortion appears, particularly for shallow events and at large offsets. This is called NMO stretching. The stretching is a frequency distortion where events are shifted to lower frequencies, which can be quantified as

$$\Delta f / f = \Delta t_{\text{NMO}} / t(0) \tag{6.3}$$

where f is the dominant frequency, Δf is change in frequency, and Δt_{NMO} is given by Equation 6.2. Because of the waveform distortion at large offsets, stacking the NMO-corrected CMP gather will severely damage the shallow events. Muting the stretched zones in the gather can solve this problem, which can be carried out by using the quantitative definition of stretching given in Equation 6.3. An alternative method for optimum selection of the mute zone is to progressively stack the data. By following the waveform along a certain event and observing where changes occur, the mute zone is derived. A trade-off exists between the signal-to-noise (SNR) ratio and mute, that is, when the SNR is high, more can be muted for less stretching; otherwise, when the SNR is low, a large amount of stretching is accepted to catch events on the stack.

6.1.2.5 Stacking

Among the three principal processes, CMP stacking is the most robust of all. Utilizing redundancy in CMP recording, stacking can significantly suppress uncorrelated noise, thereby increasing the SNR ratio. It also can attenuate a large part of the coherent noise in the data, such as guided waves and multiples.

6.1.2.6 Migration

On a seismic section such as that illustrated in Figure 6.2, each reflection event is mapped directly beneath the midpoint. However, the reflection point is located beneath the midpoint only if the reflector is horizontal. With a dip along the survey line the actual

Application of Factor Analysis in Seismic Profiling

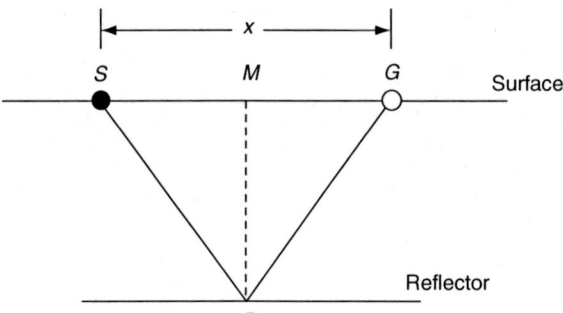

FIGURE 6.2
The NMO geometry of a single horizontal reflector.

reflection point is displaced in the up-dip direction; with a dip across the survey line the reflection point is displaced out of the plane of the section. Migration is a process that moves dipping reflectors into their true subsurface positions and collapses diffractions, thereby depicting detailed subsurface features. In this sense, migration can be viewed as a form of spatial deconvolution that increases spatial resolution.

6.1.3 Interpretation

The goal of seismic processing and imaging is to extract the reflectivity function of the subsurface from the seismic data. Once the reflectivity is obtained, it is the task of the seismic interpreter to infer the geological significance of a certain reflectivity pattern.

6.2 Factor Analysis Framework

Factor analysis (FA), a branch of multivariate analysis, is concerned with the internal relationships of a set of variates [3]. Widely used in psychology, biology, *chemometrics*[1] [4], and social science, the latent variable model provides an important tool for the analysis of multivariate data. It offers a conceptual framework within which many disparate methods can be unified and a base from which new methods can be developed.

6.2.1 General Model

In FA the basic model is

$$\mathbf{x} = A\mathbf{s} + \mathbf{n} \tag{6.4}$$

where $\mathbf{x} = (x_1, x_2, \ldots, x_p)^T$ is a vector of observable random variables (the test scores), $\mathbf{s} = (s_1, s_2, \ldots, s_r)^T$ is a vector $r < p$ unobserved or latent random variables (the common factor scores), A is a $(p \times r)$ matrix of fixed coefficients (factor loadings), $\mathbf{n} = (n_1, n_2, \ldots, n_p)^T$ is a vector of random error terms (unique factor scores of order p). The means are usually set to zero for convenience so that $E(\mathbf{x}) = E(\mathbf{s}) = E(\mathbf{n}) = 0$. The random error term consists

---

[1]Chemometrics is the use of mathematical and statistical methods for handling, interpreting, and predicting chemical data.

of errors of measurement and the unique individual effects associated with each variable x_j, $j = 1, 2, \ldots, p$. For the present model we assume that A is a matrix of constant parameters and **s** is a vector of random variables.

The following assumptions are usually made for the factor model [5]:

- rank $(A) = r < p$
- $E(\mathbf{x}|\mathbf{s}) = A\mathbf{s}$
- $E(\mathbf{xx}^T) = \Sigma$, $E(\mathbf{ss}^T) = \Omega$ and

$$\Psi = E(\mathbf{nn}^T) = \begin{bmatrix} \sigma_1^2 & & 0 \\ & \sigma_2^2 & \\ & & \ddots \\ 0 & & & \sigma_p^2 \end{bmatrix} \tag{6.5}$$

That is, the errors are assumed to be uncorrelated. The common factors however are generally correlated, and Ω is therefore not necessarily diagonal. For the sake of convenience and computational efficiency, the common factors are usually assumed to be uncorrelated and of unit variance, so that $\Omega = I$.

- $E(\mathbf{sn}^T) = 0$ so that the errors and common factors are uncorrelated.

From the above assumptions, we have

$$\begin{aligned} E(\mathbf{xx}^T) = \Sigma &= E\big[(A\mathbf{s} + \mathbf{n})(A\mathbf{s} + \mathbf{n})^T\big] \\ &= E(A\mathbf{ss}^T A^T + A\mathbf{sn}^T + \mathbf{ns}^T A^T + \mathbf{nn}^T) \\ &= AE(\mathbf{ss}^T)A^T + AE(\mathbf{sn}^T) + E(\mathbf{ns}^T)A^T + E(\mathbf{nn}^T) \\ &= A\Omega A^T + E(\mathbf{nn}^T) \\ &= \Gamma + \Psi \end{aligned} \tag{6.6}$$

where $\Gamma = A\Omega A^T$ and $\Psi = E(\mathbf{nn}^T)$ are the true and error covariance matrices, respectively.

In addition, postmultiplying Equation 6.4 by $\mathbf{s}^T$, considering the expectation, and using assumptions (6.3) and (6.4), we have

$$\begin{aligned} E(\mathbf{xs}^T) &= E(A\mathbf{ss}^T + \mathbf{ns}^T) \\ &= AE(\mathbf{ss}^T) + E(\mathbf{ns}^T) \\ &= A\Omega \end{aligned} \tag{6.7}$$

For the special case of $\Omega = I$, the covariance between the observation and the latent variables simplifies to $E(\mathbf{xs}^T) = A$.

A special case is found when **x** is a multivariate Gaussian; the second moments of Equation 6.6 will contain all the information concerning the factor model. The factor model Equation 6.4 will be linear, and given the factors **s** the variables **x** are conditionally independent. Let $\mathbf{s} \in N(0, I)$, the conditional distribution of **x** is

$$\mathbf{x}|\mathbf{s} \in N(A\mathbf{s}, \Psi) \tag{6.8}$$

or

$$p(\mathbf{x}|\mathbf{s}) = (2\pi)^{-p/2}|\Psi|^{-1/2}\exp\left\{-\frac{1}{2}(\mathbf{x}-A\mathbf{s})^T\Psi^{-1}(\mathbf{x}-A\mathbf{s})\right\} \quad (6.9)$$

with conditional independence following from the diagonality of Ψ. The common factors **s** therefore reproduce all covariances (or correlations) between the variables, but account for only a portion of the variance.

The marginal distribution for **x** is found by integrating the hidden variables **s**, or

$$p(\mathbf{x}) = \int p(\mathbf{x}|\mathbf{s})p(\mathbf{s})\,d\mathbf{s}$$
$$= (2\pi)^{-p/2}|\Psi + AA^T|^{-1/2}\exp\left\{-\frac{1}{2}\mathbf{x}^T(\Psi + AA^T)^{-1}\mathbf{x}\right\} \quad (6.10)$$

The calculation is straightforward because both $p(\mathbf{s})$ and $p(\mathbf{x}|\mathbf{s})$ are Gaussian.

6.2.2 Within the Framework

Many methods have been developed for estimating the model parameters for the special case of Equation 6.8. Unweighted least square (ULS) algorithm [6] is based on minimizing the sum of squared differences between the observed and estimated correlation matrices, not counting the diagonal. Generalized least square (GLS) [6] algorithm is adjusting ULS by weighting the correlations inversely according to their uniqueness. Another method, maximum likelihood (ML) algorithm [7], uses a linear combination of variables to form factors, where the parameter estimates are those most likely to have resulted in the observed correlation matrix. More details on the ML algorithm can be found in Appendix 6.B. These methods are all of second order, which find the representation using only the information contained in the covariance matrix of the test scores. In most cases, the mean is also used in the initial centering. The reason for the popularity of the second-order methods is that they are computationally simple, often requiring only classical matrix manipulations.

Second-order methods are in contrast to most higher order methods that try to find a meaningful representation. Higher order methods use information on the distribution of **x** that is not contained in the covariance matrix. The distribution of *f***x** must not be assumed to be Gaussian, because all the information of Gaussian variables is contained in the first two-order statistics from which all the high order statistics can be generated. However, for more general families of density functions, the representation problem has more degrees of freedom, and much more sophisticated techniques may be constructed for non-Gaussian random variables.

6.2.2.1 Principal Component Analysis

Principal component analysis (PCA) is also known as the Hotelling transform or the Karhunen–Loève transform. It is widely used in signal processing, statistics, and neural computing to find the most important directions in the data in the mean-square sense. It is the solution of the FA problem with minimum mean-square error and an orthogonal weight matrix.

The basic idea of PCA is to find the $r \leq p$ linearly transformed components that provide the maximum amount of variance possible. During the analysis, variables in **x** are transformed linearly and orthogonally into an equal number of uncorrelated new variables in **e**. The transformation is obtained by finding the latent roots and vectors of either the covariance or the correlation matrix. The latent roots, arranged in descending order of magnitude, are equal to the variances of the corresponding variables in **e**. Usually the first few components account for a large proportion of the total variance of **x**, accordingly, may then be used to reduce the dimensionality of the original data for further analysis. However, all components are needed to reproduce accurately the correlation coefficients within **x**.

Mathematically, the first principal component $\mathbf{e}_1$ corresponds to the line on which the projection of the data has the greatest variance

$$\mathbf{e}_1 = \arg\max_{\|\mathbf{a}\|=1} \sum_{t=1}^{T} \left(\mathbf{e}^T \mathbf{x}\right)^2 \tag{6.11}$$

The other components are found recursively by first removing the projections to the previous principal components:

$$\mathbf{e}_k = \arg\max_{\|\mathbf{e}\|=1} \sum \left[\mathbf{e}^T \left(\mathbf{x} - \sum_{i=1}^{k-1} \mathbf{e}_i \mathbf{e}_i^T \mathbf{x}\right)\right]^2 \tag{6.12}$$

In practice, the principal components are found by calculating the eigenvectors of the covariance matrix Σ of the data as in Equation 6.6. The eigenvalues are positive and they correspond to the variances of the projections of data on the eigenvectors.

The basic task in PCA is to reduce the dimension of the data. In fact, it can be proven that the representation given by PCA is an optimal linear dimension reduction technique in the mean-square sense [8,9]. The kind of reduction in dimension has important benefits [10]. First, the computational complexity of the further processing stages is reduced. Second, noise may be reduced, as the data not contained in the components may be mostly due to noise. Third, projecting into a subspace of low dimension is useful for visualizing the data.

6.2.2.2 Independent Component Analysis

The independent component analysis (ICA) model originates from the multi-input and multi-output (MIMO) channel equalization [11]. Its two most important applications are blind source separation (BSS) and feature extraction. The mixing model of ICA is similar to that of the FA, but in the basic case without the noise term. The data have been generated from the latent components **s** through a square mixing matrix A by

$$\mathbf{x} = A\mathbf{s} \tag{6.13}$$

In ICA, all the independent components, with the possible exception of one component, must be non-Gaussian. The number of components is typically the same as the number of observations. Such an A is searched for to enable the components $\mathbf{s} = A^{-1}\mathbf{x}$ to be as independent as possible.

In practice, the independence can be maximized, for example, by maximizing non-Gaussianity of the components or minimizing mutual information [12]. ICA can be approached from different starting points. In some extensions the number of independent components can exceed the number of dimensions of the observations making the basis overcomplete [12,13]. The noise term can be taken into the model. ICA can be viewed as a generative model when the 1D distributions for the components are modeled with, for example, mixtures of Gaussians (MoG).

The problem with ICA is that it has the ambiguities of scaling and permutation [12]; that is, the indetermination of the variances of the independent components and the order of the independent components.

6.2.2.3 Independent Factor Analysis

Independent factor analysis (IFA) is formulated by Attias [14]. It aims to describe p generally correlated observed variables **x** in terms of $r < p$ independent latent variables **s** and an additive noise term **n**. The proposed algorithm derives from the ML and more specifically from the expectation–maximization (EM) algorithm.

IFA model differs from the classic FA model in that the properties of the latent variables it involves are different. The noise variables **n** are assumed to be normally distributed, but not necessarily uncorrelated. The latent variables **s** are assumed to be mutually independent but not necessarily normally distributed; their densities are indeed modeled as mixtures of Gaussians. The independence assumption allows modeling the density of each s_i in the latent space separately.

There are some problems with the EM–MoG algorithm. First, approximating source densities with MoGs is not so straightforward because the number of Gaussians has to be adjusted. Second, EM–MoG is computationally demanding where the complexity of computation grows exponentially with the number of sources [14]. Given a small number of sources the EM algorithm is exact and all the required calculations can be done analytically, whereas it becomes intractable as the number of sources in the model increases.

6.3 FA Application in Seismic Signal Processing

6.3.1 Marmousi Data Set

Marmousi is a 2D synthetic data set generated at the Institut Françis du Pétrole (IFP). The geometry of this model is based on a profile through the North Quenguela trough in the Cuanza basin [15,16]. The geometry and velocity model was created to produce complex seismic data, which requires advanced processing techniques to obtain a correct Earth image. Figure 6.3 shows the velocity profile of the Marmousi model.

Based on the profile and the geologic history, a geometric model containing 160 layers was created. Velocity and density distributions were defined by introducing realistic horizontal and vertical velocities and density gradients. This resulted in a 2D density–velocity grid with dimensions of 3000 m in depth by 9200 m in offset.

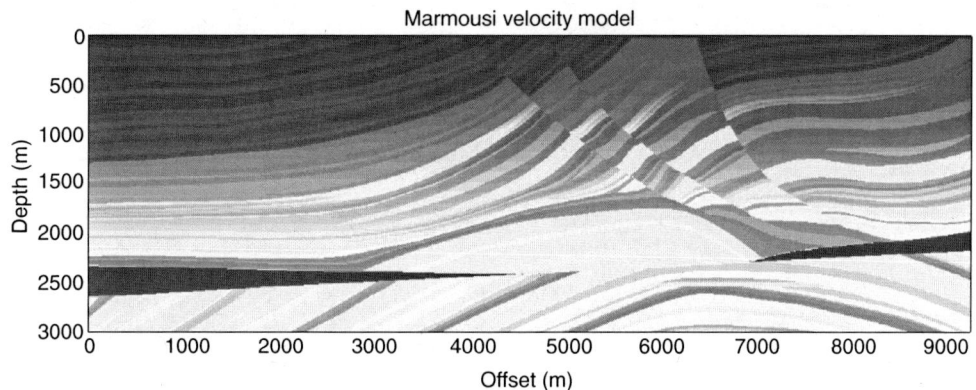

FIGURE 6.3
Marmousi velocity model.

Data were generated by a modeling package that can simulate a seismic line by computing successively the different shot records. The line was "shot" from west to east. The first and last shot points were, respectively, 3000 and 8975 m from the west edge of the model. Distance between shots was 25 m. Initial offset was 200 m and the maximum offset was 2575 m.

6.3.2 Velocity Analysis, NMO Correction, and Stacking

Given the Marmousi data set, after some conventional processing steps described in Section 6.2, the results of velocity analysis and normal moveout are shown in Figure 6.4.

The left-most plot is a CMP gather. There are totally 574 CMP gathers in the Marmousi data set; each includes 48 traces.

On the second plot, velocity spectrum is generated after the CMP gather is NMO-corrected and stacked using a range of constant velocity values, and the resultant stack traces for each velocity are placed side by side on a plane of velocity vs. two-way zero-offset time. By selecting the peaks on the velocity spectrum, an initial rms velocity can be defined, shown as a curve on the left of the second plot. The interval velocity can be calculated by using Dix formula [17] and shown on the right side of the plot.

Given the estimated velocity profile, the real moveout correction can be carried out, shown in the third plot. As compared with the first plot, we can see the hyperbolic curves are flattened out after NMO correction. Usually another procedure called muting will be carried out before stacking because as we can see in the middle of the third plot, there are

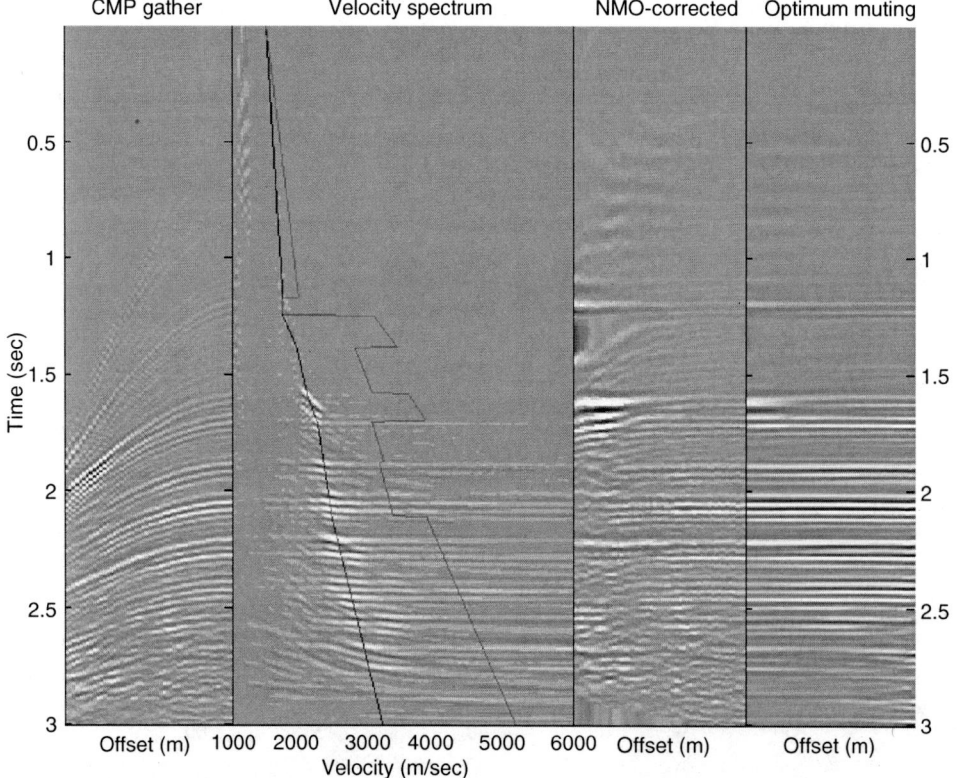

FIGURE 6.4
Velocity analysis and stacking of Marmousi data set.

Application of Factor Analysis in Seismic Profiling 111

great distortions because of the approximation. That part will be eliminated before stacking all the 48 traces together.

The fourth plot just shows a different way of highlighting the muting procedure. For details, see Ref. [1]. After we complete the velocity analysis, NMO correction, and stacking for the 56 of the CMPs, we get the following section of the subsurface image as on the left of Figure 6.5. There are two reasons that only 56 out of 574 of the CMPs are stacked. One reason is that the velocity analysis is too time consuming on a personal computer and the other is that although 56 CMPs are only one tenths of the 574 CMPs, it indeed covers nearly 700 m of the profile. It is enough to compare processing difference.

The right plot is the same image as the left one except that it is after the automatic amplitude adjustment, which is to stress the vague events so that both the vague events and strong events in the image are shown with approximately the same amplitude. The algorithm includes three easy steps:

1. Compute Hilbert envelope of a trace.
2. Convolve the envelope with a triangular smoother to produce the smoothed envelope.
3. Divide the trace by the smoothed envelope to produce the amplitude-adjusted trace.

By comparing the two plots, we can see that vague events at the top and bottom of the image are indeed stressed. In the following sections, we mainly use automatic amplitude-adjusted image to illustrate results.

It needs to be pointed out that due to NMO stretching and lack of data at small offset after muting, events before 0.2 sec in Figure 6.5 are shown as distorted and do not provide

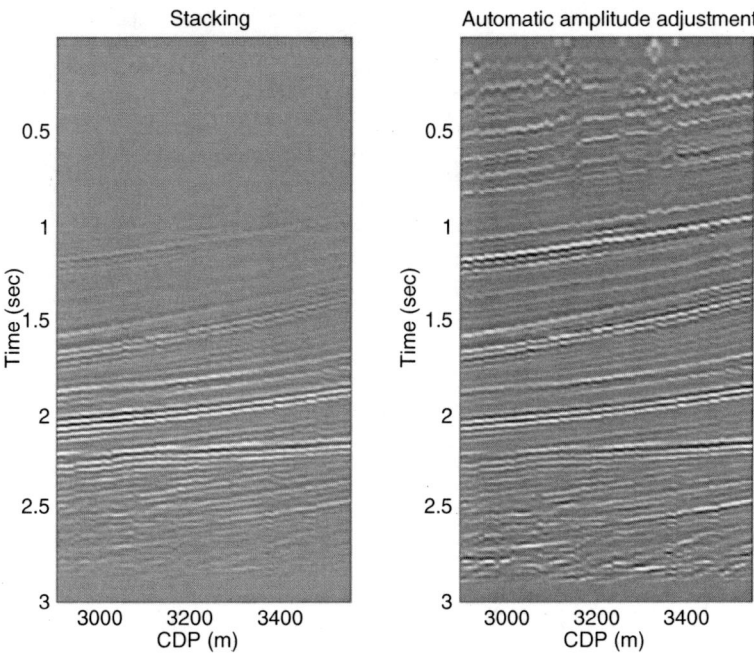

FIGURE 6.5
Stacking of 56 CMPs.

useful information. In the following sections, when we compare the result, we mainly consider events after 0.2 sec.

6.3.3 The Advantage of Stacking

Stacking is based on the assumption that all the traces in a CMP gather correspond to one single depth point. After they are NMO-corrected, the zero-offset traces should contain the same signal embedded in different random noises, which are caused by the different raypaths. The process of adding them together in this manner can increase the SNR ratio by adding up the signal components while canceling the noises among the traces. To see what stacking can do to improve the subsurface image quality, let us compare the image obtained from a single trace and that from stacking the 48 muted traces.

In Figure 6.6, the single trace result without stacking is shown in the right plot. For every CMP (or CDP) gather, only the trace of smallest offset is NMO-corrected and placed side by side together to produce the image, while in the stack result in the left plot, 48 NMO-corrected and muted traces are stacked and placed side by side. Clearly, after stacking, the main events at 0.5, 1.0, and 1.5 sec are stressed, and the noise in between is canceled out. Noise at 0.2 is effectively removed. Noise caused by multiples from 2.0 to 3.0 sec is significantly reduced. However, due to NMO stretching and muting, there are not enough data to depict events at 0 to 0.25 sec on both plots.

6.3.4 Factor Analysis vs. Stacking

Now we suggest an alternative way of obtaining the subsurface image by using FA instead of stacking. As presented in Appendix 6.A, FA can extract one unique common factor from the traces with maximum correlation among them. It fits well with what is

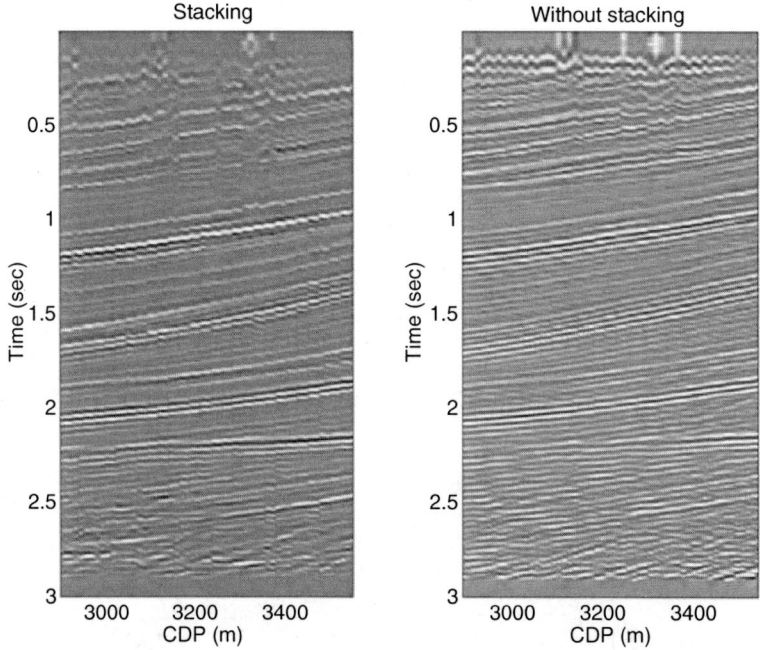

FIGURE 6.6
Comparison of stacking and single trace images.

Application of Factor Analysis in Seismic Profiling

expected of zero-offset traces in that after NMO correction they contain the same signal embedded in different random noises.

There are two reasons that FA works better than stacking. First, FA model considers scaling factor A as in Equation 6.14, while stacking assumes no scaling as in Equation 6.15.

$$\text{Factor analysis: } \mathbf{x} = A\mathbf{s} + \mathbf{n} \quad (6.14)$$

$$\text{Stacking: } \mathbf{x} = \mathbf{s} + \mathbf{n} \quad (6.15)$$

When the scaling information is lost, simple summation does not necessarily increase the SNR ratio. For example, if one scaling factor is 1 and the other is -1, summation will simply cancel out the signal component completely, leaving only the noise component. Second, FA makes use of the second-order statistics explicitly as the criterion to extract the signal while stacking does not. Therefore, SNR ratio will improve more in the case of FA than in the case of stacking.

To illustrate the idea, $\mathbf{x}(t)$ are generated using the following equation:

$$\begin{aligned} \mathbf{x}(t) &= A\mathbf{s}(t) + \mathbf{n}(t) \\ &= A\cos(2\pi t) + \mathbf{n}(t) \end{aligned}$$

where $s(t)$ is the sinusoidal signal, $\mathbf{n}(t)$ are 10 independent noise terms with Gaussian distribution. The matrix of factor loadings A is also generated randomly. Figure 6.7 shows the result of stacking and FA. The top plot is one of the ten observations $\mathbf{x}(t)$. The middle plot is the result of stacking and the bottom plot is the result of FA using ML algorithm as presented in Appendix 6.B. Comparing the two plots suggests that FA outperforms stacking in improving the SNR ratio.

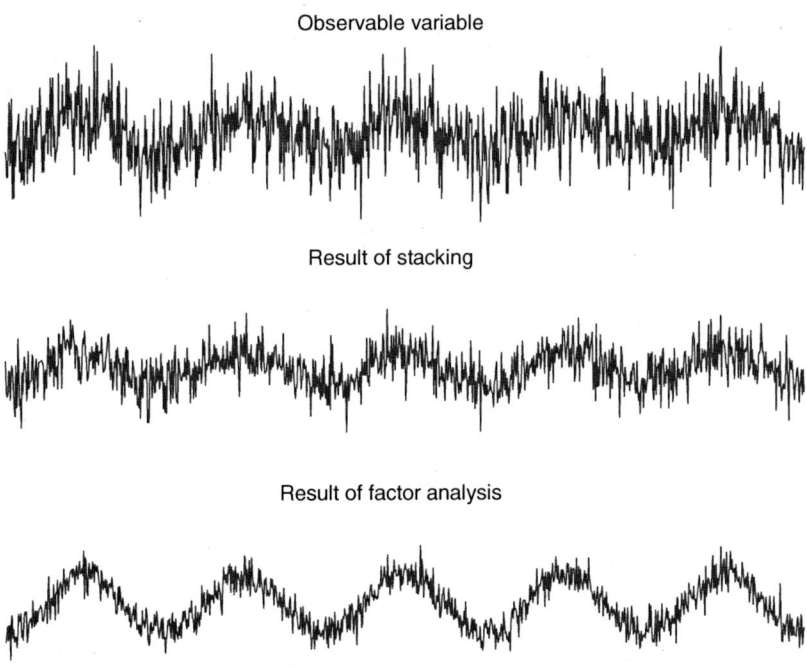

FIGURE 6.7
Comparison of stacking and FA.

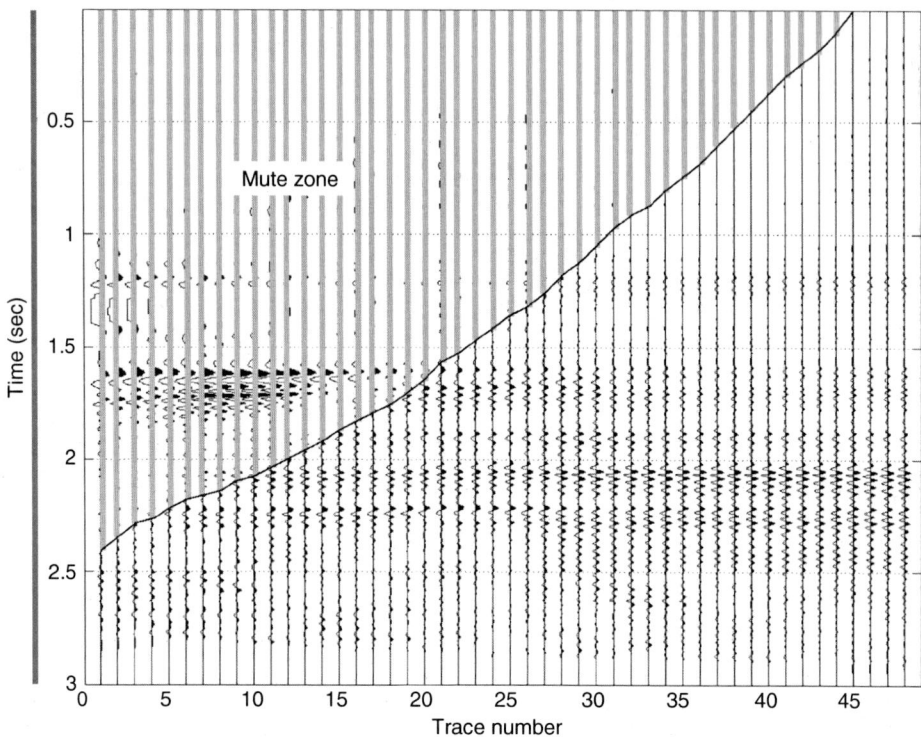

FIGURE 6.8
Factor analysis of Scheme no. 1.

6.3.5 Application of Factor Analysis

The simulation result in Section 6.3.4 suggests that FA can be applied to the NMO-corrected seismic data. One problem arises, however, when we inspect the zero-offset traces. They need to be muted because of the NMO stretching, which means almost all the traces will have a segment set to zero (mute zone), as is shown in Figure 6.8. Is it possible to just apply FA to the muted traces? Is it possible to have other schemes that make full use of the information at hand? In the following sections, we try to answer these questions by discussing different schemes to carry out FA.

6.3.5.1 Factor Analysis Scheme No. 1

Let us start with the easiest one. The scheme is illustrated in Figure 6.8. We will set the mute zone to zero and apply FA to a CMP gather using ML algorithm. Extracting one single common factor from the 48 traces, and placing all the resulting factors from 56 CMP gathers side by side, the right plot in Figure 6.9 is obtained.

Compared with the result of stacking shown on the left, events from 2.2 to 3.0 sec are more smoothly presented instead of the broken dashlike events after stacking. However, at near offset, from 0 to 0.7 sec, the image is contaminated with some vertical stripes.

6.3.5.2 Factor Analysis Scheme No. 2

In this scheme, the muted segments in each trace are replaced by segments of the nearest neighboring traces as is illustrated by Figure 6.10. Trace no. 44 borrows Segment 1

Application of Factor Analysis in Seismic Profiling

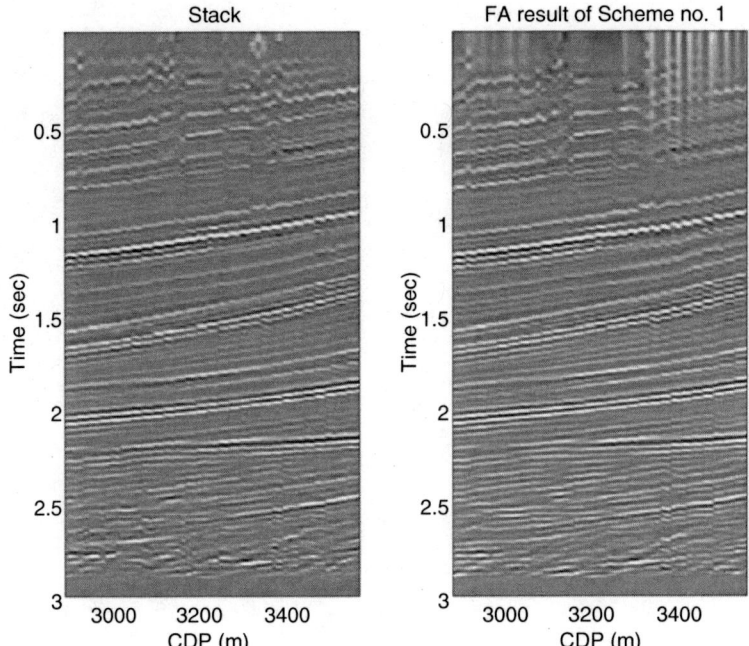

FIGURE 6.9
Comparison of stacking and FA result of Scheme no. 1.

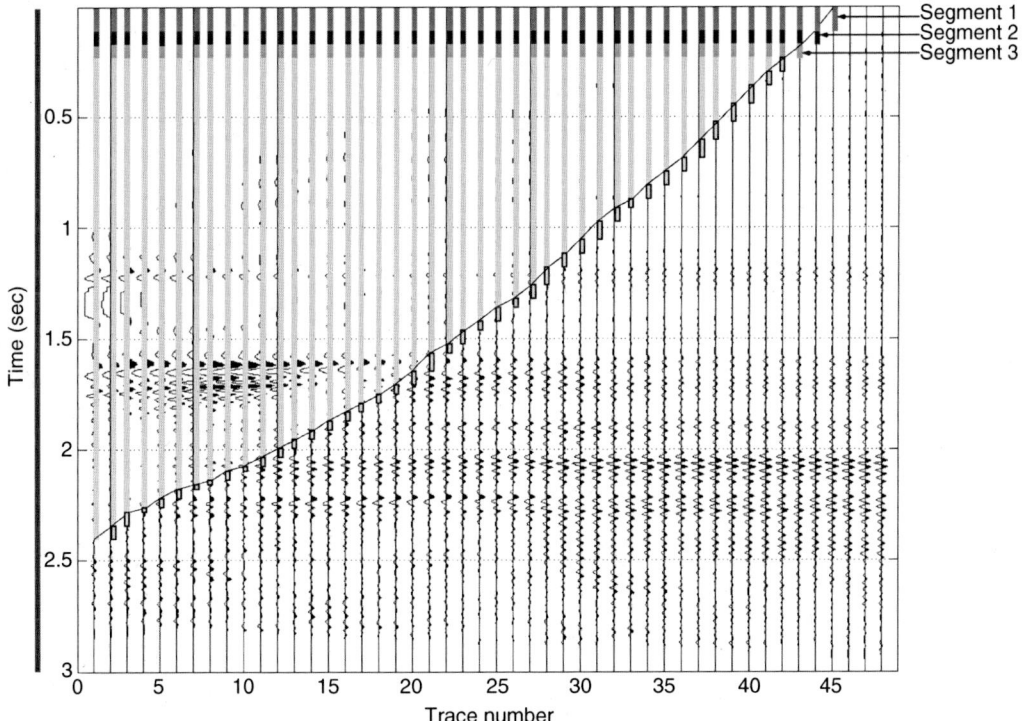

FIGURE 6.10
Factor analysis of Scheme no. 2.

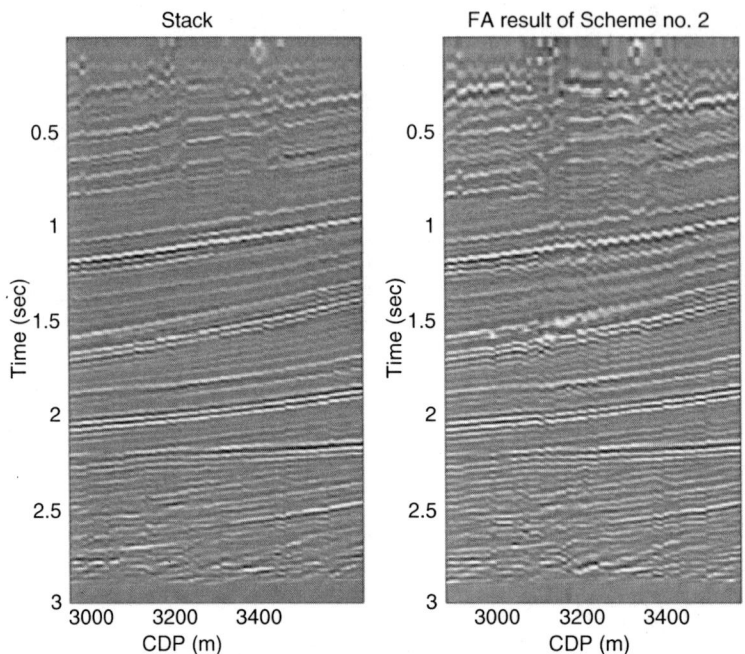

FIGURE 6.11
Comparison of stacking and FA result of Scheme no. 2.

from Trace no. 45 to fill out its muted segment. Trace no. 43 borrows Segments 1 and 2 from Trace no. 44 to fill out its muted segment and so on. As a result, Segment 1 from Trace no. 45 is copied to all the muted traces, from Trace no. 1 to 44. Segment 2 from Trace no. 44 is copied to traces from Trace no. 1 to 43. After the mute zone is filled out, FA is carried out to produce the result shown in Figure 6.11.

Compared to stacking shown on the left, there is no improvement in the obtained image. Actually, the result is worse. Some events are blurred. Therefore, Scheme no. 2 is not a good scheme.

6.3.5.3 Factor Analysis Scheme No. 3

In this scheme, instead of copying the neighboring segments to the mute zone, the segments obtained from applying FA to the traces included in the nearest neighboring box are copied. In Figure 6.12, we first apply FA to traces in Box 1 (Trace no. 45 to 48), then Segment 1 is extracted from the result and copied to Trace no. 44. Segment 2 obtained from applying FA to traces in Box 2 (Trace no. 44 to 48) will be copied to Trace no. 43. When done, the image obtained is shown in Figure 6.13.

Compared with Scheme no. 2, the result is better. But compared to stacking, there is still some contamination from 0 to 0.7 sec.

6.3.5.4 Factor Analysis Scheme No. 4

In the scheme, as is illustrated in Figure 6.14, Segment 1 will be extracted from applying FA to all the traces in Box 1 (traces from Trace no. 1 to 48), and Segment 2 will be extracted from applying FA to trace segments in Box 2 (traces from Trace no. 2 to 48). Note that the data are not muted before FA. In this manner, for every segment, all the data points available are fully utilized.

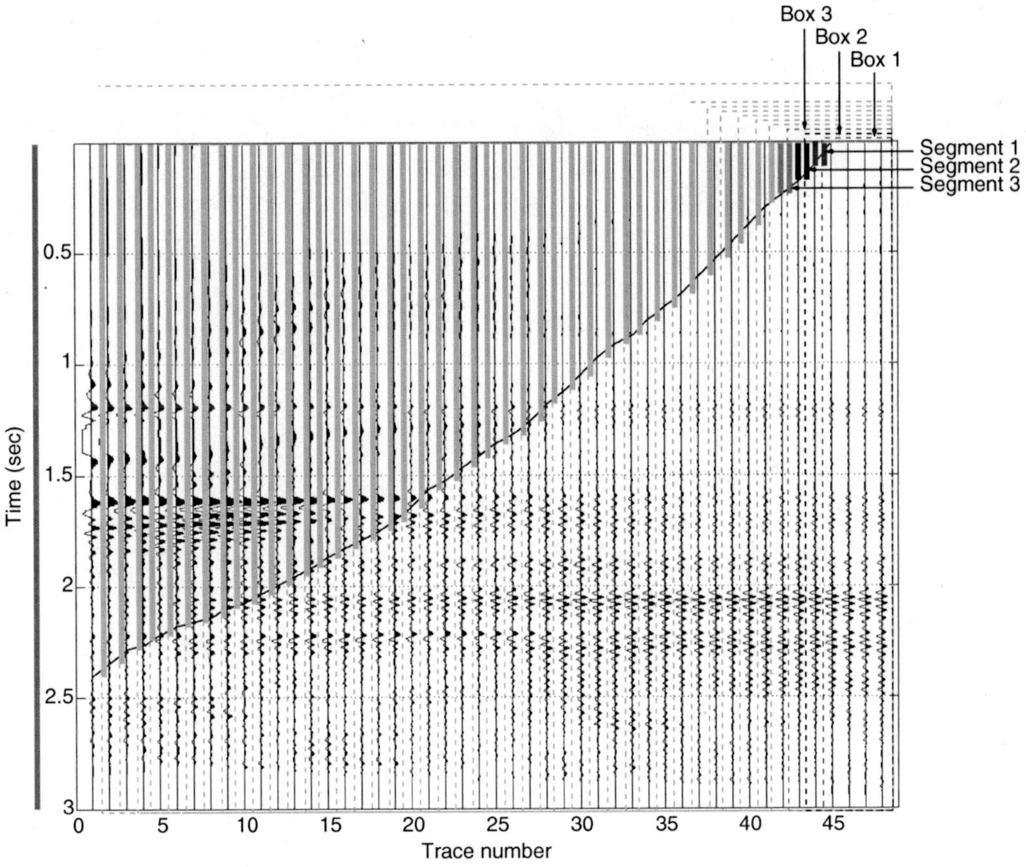

Application of Factor Analysis in Seismic Profiling 117

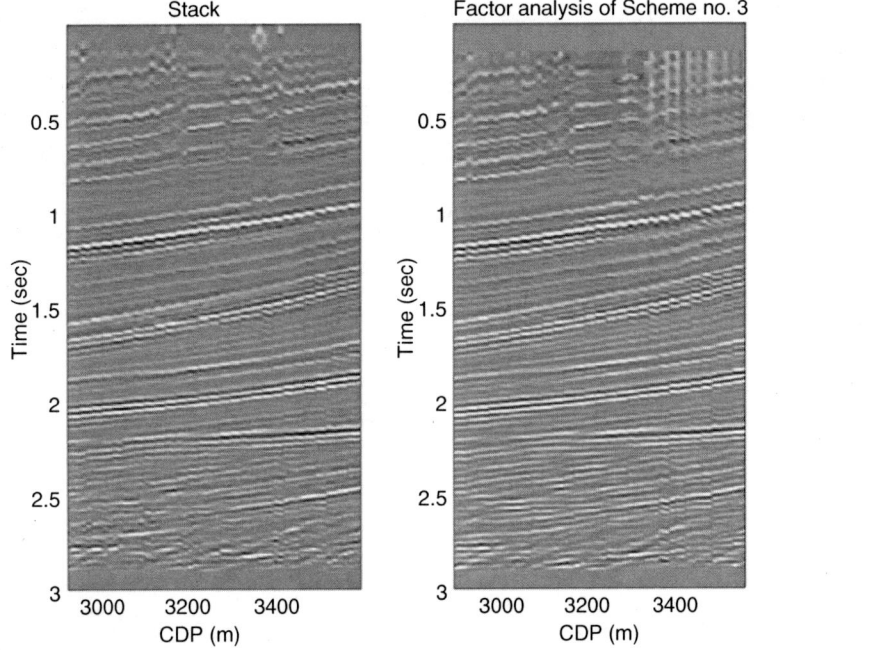

FIGURE 6.12
Factor analysis of Scheme no. 3.

FIGURE 6.13
Comparison of stacking and FA result of Scheme no. 3.

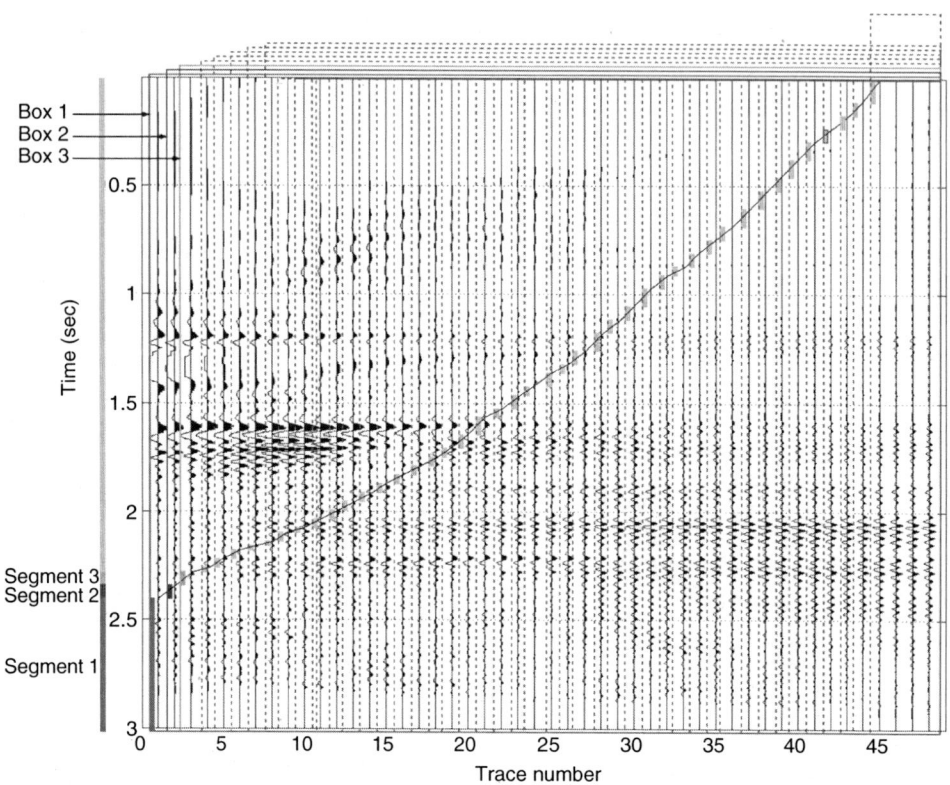

FIGURE 6.14
Factor analysis of Scheme no. 4.

In the result generated, we noticed that events from 0 to 0.16 sec are distorted. The amplitude is so large that it overshadows the other events. Comparing all the results obtained above, we conclude that both stacking and FA are unable to extract useful information from 0 to 0.16 sec. To better illustrate the FA result, we will mute the result from the distorted trace segments, and the final result is shown in Figure 6.15. Compare the results of FA and stacking; we can see that events at around 1 and 1.5 sec are strengthened. Events from 2.2 to 3.0 sec are more smoothly presented instead of the broken dashlike events in the stacked result. Overall, the SNR ratio of the image is improved.

6.3.6 Factor Analysis vs. PCA and ICA

The results of PCA and ICA (discussed in subsections 6.2.2.1 and 6.2.2.2) are placed side by side with the result of FA for comparison in Figure 6.16 and Figure 6.17. As we can see from both plots on the right side of the figures, important events are missing and the subsurface images are distorted.

The reason is that the criteria used in PCA and ICA to extract the signals are improper to this particular scenario. In PCA, traces are transformed linearly and orthogonally into an equal number of new traces that have the property of being uncorrelated, where the first component having the maximum variance is used to produce the image. In ICA, the algorithm tries to extract components that are as independent to each other as possible, where the obtained components suffer from the problems of scaling and permutation.

Application of Factor Analysis in Seismic Profiling 119

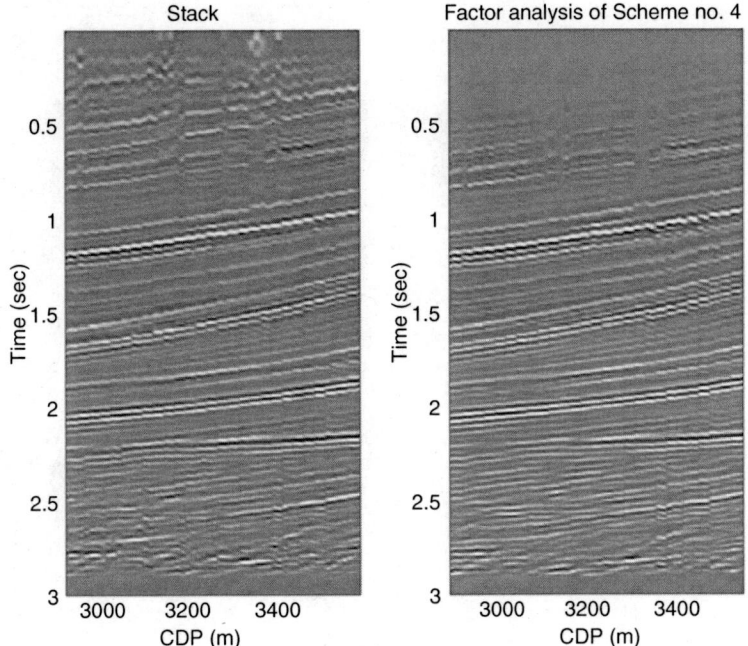

FIGURE 6.15
Comparison of stacking and FA result of Scheme no. 4.

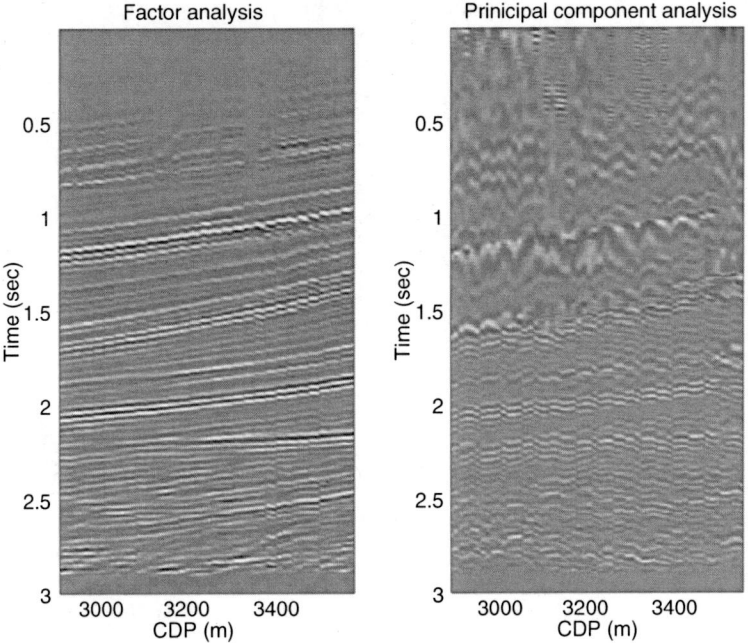

FIGURE 6.16
Comparison of FA and PCA results.

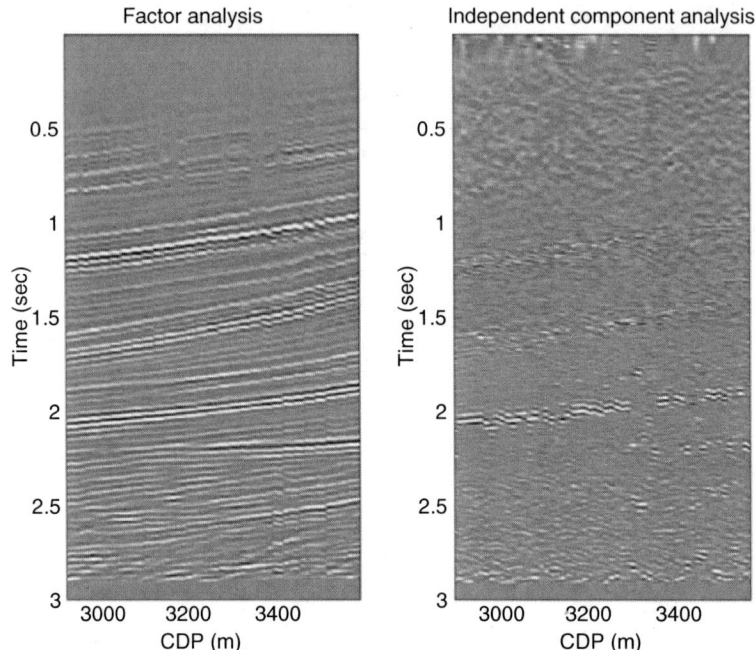

FIGURE 6.17
Comparison of FA and ICA results.

6.4 Conclusions

Stacking is one of the three most important and robust processing steps in seismic signal processing. By utilizing the redundancy of the CMP gathers, stacking can effectively remove noise and increase the SNR ratio. In this chapter we propose to use FA to replace stacking to obtain better subsurface images after applying FA algorithm to the synthetic Marmousi data set. Comparisons with PCA and ICA show that FA indeed has advantages over other techniques in this scenario.

It is noted that the conventional seismic processing steps adopted here are very basic and for illustrative purposes only. Better results may be obtained in velocity analysis and stacking if careful examination and iterative procedures are incorporated as is often the case in real situations.

References

1. Ö. Yilmaz, *Seismic Data Processing*, Society of Exploration Geophysicists, Tulsa, 1987.
2. E.A. Robinson, S. Treitei, R.A. Wiggins, and P.R. Gutowski, *Digital Seismic Inverse Methods*, International Human Resources Development Corporation, Boston, 1983.
3. D.N. Lawley and A.E. Maxwell, *Factor Analysis as a Statistical Method*, Butterworths, London, 1963.
4. E.R. Malinowski, *Factor Analysis in Chemistry*, 3rd ed., John Wiley & Sons, New York, 2002.
5. A.T. Basilevsky, *Statistical Factor Analysis and Related Methods: Theory and Applications*, 1st ed., Wiley-Interscience, New York, 1994.

6. H. Harman, *Modern Factor Analysis*, 2nd ed., University of Chicago Press, Chicago, 1967.
7. K.G. Jöreskog, Some contributions to maximum likelihood factor analysis, *Psychometrika*, 32:443–482, 1967.
8. I.T. Jolliffe, *Principal Component Analysis*, Springer-Verlag, Heidelberg, 1986.
9. M. Kendall, *Multivariate Analysis*, Charles Griffin, London, 1975.
10. A. Hyvärinen, Survey on independent component analysis, *Neural Computing Surveys*, 2:94–128, 1999.
11. P. Comon, Independent component analysis, a new concept? *Signal Processing*, 36:287–314, 1994.
12. J. Karhunen, A. Hyvärinen, and E. Oja, *Independent Component Analysis*, John Wiley & Sons, New York, 2001.
13. T. Lee, M. Girolami, M. Lewicki, and T. Sejnowski, Blind source separation of more sources than mixtures using overcomplete representations, *IEEE Signal Processing Letters*, 6:87–90, 1999.
14. H. Attias, Independent factor analysis, *Neural Computation*, 11:803–851, 1998.
15. R.J. Versteeg, Sensitivity of prestack depth migration to the velocity model, *Geophysics*, 58(6):873–882, 1993.
16. R.J. Versteeg, The Marmousi experience: velocity model determination on a synthetic complex data set, *The Leading Edge*, 13:927–936, 1994.
17. C.H. Dix, Seismic velocities from surface measurements, *Geophysics*, 20:68–86, 1955.
18. R. Bellman, *Introduction to Matrix Analysis*, McGraw-Hill, New York, 1960.

Appendices

6.A Upper Bound of the Number of Common Factors

Suppose that there is a unique Ψ, matrix $\Sigma - \Psi$ must be of rank r. This is the covariance matrix for x where each diagonal element represents the part of the variance that is due to the r common factors instead of the total variance of the corresponding variate. This is known as communality of the variate.

When $r = 1$, A reduces to a column vector of p elements. It is unique, apart from a possible change of sign of all its elements.

With $1 < r < p$ common factors, it is not generally possible to determine A and $\mathbf{s}$ uniquely, even in the case of a normal distribution. Although every factor model specified by Equation 6.8 leads to a multivariate normal, the converse is not necessarily true when $1 < r < p$. The difficulty is known as the factor identification or factor rotation problem.

Let H be any $(r \times r)$ orthogonal matrix, so that $HH^T = H^T H = I$, then

$$\mathbf{x} = AHH^T \mathbf{s} + \mathbf{n}$$
$$= A^{\mathring{a}} \mathbf{s}^{\mathring{a}} + \mathbf{n}$$

Thus, $\mathbf{s}$ and $\mathbf{s}^{\mathring{a}}$ have the same statistical properties since

$$E\left(\mathbf{s}^{\mathring{a}}\right) = H^T E(\mathbf{s})$$

$$\operatorname{cov}\left(\mathbf{s}^{\mathring{a}}\right) = H^T \operatorname{cov}(\mathbf{s})H = H^T H = I$$

Assume there exist $1 < r < p$ common factors such that $\Gamma = A \Omega A^T$ and Ψ is Grammian and diagonal. The covariance matrix Σ has

$$C\binom{p}{2} + p = \frac{1}{2}p(p+1)$$

distinct elements, which equals the total number of normal equations to be solved. However, the number of solutions is infinite, as can be seen from the following derivation. Since Ω is Grammian, its Cholesky decomposition exists. That is, there exists a nonsingular $(r \times r)$ matrix U, such that $\Omega = U^T U$ and

$$\Sigma = A \Omega A^T + \Psi$$
$$= A U^T U A^T + \Psi$$
$$= (AU^T)(AU^T)^T + \Psi$$
$$= A^{\mathring{a}} A^{\mathring{a}T} + \Psi \tag{6A.1}$$

Apparently both factorization Equation 6.6 and Equation 6A.1 leave the same residual error Ψ and therefore must represent equally valid factor solutions. Also, we can substitute $A^{\aa} = AB$ and $\Omega^{\aa} = B^{-1}\Omega (B^{T})^{-1}$, which again yields a factor model that is indistinguishable from Equation 6.6. Therefore, no sample estimator can distinguish between such an infinite number of transformations. The coefficients A and $A^{\aa}$ are thus statistically equivalent and cannot be distinguished from each other or identified uniquely; that is, both the transformed and untransformed coefficients, together with Ψ, generate Σ in exactly the same way and cannot be differentiated by any estimation procedure without the introduction of additional restrictions.

To solve the rotational indeterminacy of the factor model we require restrictions on Ω, the covariance matrix of the factors. The most straightforward and common restriction is to set $\Omega = I$. The number m of free parameters implied by the equation

$$\Sigma = AA^T + \Psi \qquad (6A.2)$$

is then equal to the total number $pr + p$ for unknown parameters in A and Ψ, minus the number of zero restrictions placed on the off-diagonal elements of Ω, which is equal to $1/2(r^2 - r)$ since Ω is symmetric. We then have

$$\begin{aligned} m &= (pr + p) - 1/2(r^2 - r) \\ &= p(r + 1) - 1/2(r^2 - r) \end{aligned} \qquad (6A.3)$$

where the columns of A are assumed to be orthogonal. The number of degrees of freedom d is then given by the number of equations implied by Equation 6A.2, that is, the number of distinct elements in Σ minus the number of free parameters m. We have

$$\begin{aligned} d &= 1/2p(p+1) - \left[p(r+1) - 1/2(r^2 - r) \right] \\ &= 1/2 \left[(p-r)^2 - (p-r) \right] \end{aligned} \qquad (6A.4)$$

which for a meaningful (i.e., nontrivial) empirical application must be strictly positive. This places an upper bound on the number of common factors r, which may be obtained in practice, a number which is generally somewhat smaller than the number of variables p.

6.B Maximum Likelihood Algorithm

The maximum likelihood (ML) algorithm presented here is proposed by Jöreskog [7]. The algorithm uses an iterative procedure to compute a linear combination of variables to form factors. Assume that the random vector x has a multivariate normal distribution as defined in Equation 6.9. The elements of A, Ω, and Ψ are the parameters of the model to be estimated from the data. From a random sample of N observations of x we can find the mean vector and the estimated covariance matrix Σ, whose elements are the usual estimates of variances and covariances of the components of x.

$$\mathbf{m_x} = \frac{1}{N} \sum_{i=1}^{N} \mathbf{x}_i$$

$$\Sigma = \frac{1}{N-1} \sum_{i=1}^{N} (\mathbf{x} - \mathbf{m_x})(\mathbf{x} - \mathbf{m_x})^T$$

$$= \frac{1}{N-1} \left(\sum_{i=1}^{N} \mathbf{x}\mathbf{x}^T - N\mathbf{m_x}\mathbf{m_x}^T \right). \tag{6B.1}$$

The distribution of Σ is the Wishart distribution [3]. The log-likelihood function is given by

$$\log L = -\frac{1}{2}(N-1)\left[\log|\hat{\Sigma}| + \operatorname{tr}\left(\Sigma\hat{\Sigma}^{-1}\right)\right]$$

However, it is more convenient to minimize

$$F(A, \Omega, \Psi) = \log|\hat{\Sigma}| + \operatorname{tr}\left(\Sigma\hat{\Sigma}^{-1}\right) - \log|\Sigma| - p$$

instead of maximizing $\log L$ [7]. They are equivalent because $\log L$ is a constant minus $\frac{1}{2}(N-1)$ times F. The function F is regarded as a function of A and Ψ. Note that if H is any nonsingular ($k \times k$) matrix, then

$$F(AH^{-1}, H\Omega H^T, \Psi) = F(A, \Omega, \Psi)$$

which means that the parameters in A and Ω are not independent of one another, and to make the ML estimates of A and Ω unique, k^2 independent restrictions must be imposed on A and Ω.

To find the minimum of F we shall first find the conditional minimum for a given Ψ and then find the overall minimum. The partial derivative of F with respect to A is

$$\frac{\partial F}{\partial A} = 2\hat{\Sigma}^{-1}(\hat{\Sigma} - \Sigma)\hat{\Sigma}^{-1}A$$

See details in Ref. [7]. For a given Ψ, the minimization of A is to be found in the solution of

$$\hat{\Sigma}^{-1}(\hat{\Sigma} - \Sigma)\hat{\Sigma}^{-1}A = 0$$

Premultiplying with $\hat{\Sigma}$ gives

$$(\hat{\Sigma} - \Sigma)\hat{\Sigma}^{-1}A = 0$$

Using the following expression for the inverse $\hat{\Sigma}^{-1}$ [3]

$$\hat{\Sigma}^{-1} = \Psi^{-1} - \Psi^{-1}A(I + A^T\Psi^{-1}A)^{-1}A^T\Psi^{-1} \tag{6B.2}$$

whose left side may be further simplified [7] so that

$$(\hat{\Sigma} - \Sigma)\Psi^{-1}A(I + A^T\Psi^{-1}A)^{-1} = 0$$

Postmultiplying by $I + A^T\Psi^{-1}A$ gives

$$(\tilde{\Sigma} - \Sigma)\Psi^{-1}A = 0 \tag{6B.3}$$

which after substitution of Σ from Equation 6A.2 and rearrangement of terms gives

$$\tilde{\Sigma}\Psi^{-1}A = A(I + A^T\Psi^{-1}A)$$

Premultiplying by $\Psi^{-1/2}$ finally gives

$$(\Psi^{-1/2}\tilde{\Sigma}\Psi^{-1/2})(\Psi^{-1/2}A) = (\Psi^{-1/2}A)(I + A^T\Psi^{-1}A) \tag{6B.4}$$

From Equation 6B.4, we can see that it is convenient to take $A^T\Psi^{-1}A$ to be diagonal, since F is unaffected by postmultiplication of A by an orthogonal matrix and $A^T\Psi^{-1}A$ can be reduced to diagonal form by orthogonal transformations [18]. In this case, Equation 6B.4 is a standard eigen decomposition form. The columns of $\Psi^{-1/2}A$ are latent vectors of $\Psi^{-1/2}\tilde{\Sigma}\Psi^{-1/2}$, and the diagonal elements of $I + A^T\Psi^{-1}A$ are the corresponding latent roots. Let $\tilde{\lambda}_1 \geq \tilde{\lambda}_2 \geq \cdots \geq \tilde{\lambda}_p$ be the latent roots of $\Psi^{-1/2}\tilde{\Sigma}\Psi^{-1/2}$ and let $\tilde{e}_1, \tilde{e}_2, \cdots, \tilde{e}_k$ be a set of latent vectors corresponding to the k largest roots. Let $\tilde{\Lambda}_k$ be the diagonal matrix with $\tilde{\lambda}_1, \tilde{\lambda}_2, \ldots, \tilde{\lambda}_k$ as diagonal elements and let E_k be the matrix with $\tilde{e}_1, \tilde{e}_2, \ldots, \tilde{e}_k$ as columns. Then

$$\Psi^{-1/2}\tilde{A} = E_k(\tilde{\Lambda}_k - I)^{1/2}$$

Premultiplying by $\Psi^{1/2}$ gives the conditional ML estimate of A as

$$\tilde{A} = \Psi^{1/2}E_k(\tilde{\Lambda}_k - I)^{1/2} \tag{6B.5}$$

Up to now, we have considered the minimization of F with respect to A for a given Ψ. Now let us examine the partial derivative of F with respect to Ψ [3],

$$\frac{\partial F}{\partial \Psi} = \text{diag}\left[\Sigma^{-1}(\Sigma - \tilde{\Sigma})\right]\Sigma^{-1}$$

Substituting $\tilde{\Sigma}^{-1}$ with Equation 6B.2 and using Equation 6B.3 gives

$$\frac{\partial F}{\partial \Psi} = \text{diag}\left[\Psi^{-1}(\Sigma - \tilde{\Sigma})\right]\Psi^{-1}$$

which by Equation 6.6 becomes

$$\frac{\partial F}{\partial \Psi} = \text{diag}\left[\Psi^{-1}(\tilde{A}\tilde{A}^T + \Psi - \tilde{\Sigma})\right]\Psi^{-1}$$

Minimizing it, we will get,

$$\Psi = \text{diag}(\tilde{\Sigma} - \tilde{A}\tilde{A}^T) \tag{6B.6}$$

By iterating Equation 6B.5 and Equation 6B.6, the ML estimation of the FA model of Equation 6.4 can be obtained.

7

Kalman Filtering for Weak Signal Detection in Remote Sensing

Stacy L. Tantum, Yingyi Tan, and Leslie M. Collins

CONTENTS
7.1 Signal Models 129
    7.1.1 Harmonic Signal Model 129
    7.1.2 Interference Signal Model 129
7.2 Interference Mitigation 130
7.3 Postmitigation Signal Models 131
    7.3.1 Harmonic Signal Model 132
    7.3.2 Interference Signal Model 132
7.4 Kalman Filters for Weak Signal Estimation 133
    7.4.1 Direct Signal Estimation 134
        7.4.1.1 Conventional Kalman Filter 134
        7.4.1.2 Kalman Filter with an AR Model for Colored Noise 135
        7.4.1.3 Kalman Filter for Colored Noise 137
    7.4.2 Indirect Signal Estimation 138
7.5 Application to Landmine Detection via Quadrupole Resonance 140
    7.5.1 Quadrupole Resonance 141
    7.5.2 Radio-Frequency Interference 141
    7.5.3 Postmitigation Signals 142
        7.5.3.1 Postmitigation Quadrupole Resonance Signal 143
        7.5.3.2 Postmitigation Background Noise 143
    7.5.4 Kalman Filters for Quadrupole Resonance Detection 143
        7.5.4.1 Conventional Kalman Filter 143
        7.5.4.2 Kalman Filter with an Autoregressive Model
                for Colored Noise 144
        7.5.4.3 Kalman Filter for Colored Noise 144
        7.5.4.4 Indirect Signal Estimation 144
7.6 Performance Evaluation 145
    7.6.1 Detection Algorithms 145
    7.6.2 Synthetic Quadrupole Resonance Data 145
    7.6.3 Measured Quadrupole Resonance Data 146
7.7 Summary 148
References 149

Remote sensing often involves probing a region of interest with a transmitted electromagnetic signal, and then analyzing the returned signal to infer characteristics of the investigated region. It is not uncommon for the measured signal to be relatively weak or for ambient noise to interfere with the sensor's ability to isolate and measure only the desired return signal. Although there are potential hardware solutions to these obstacles, such as increasing the power in the transmit signal to strengthen the return signal, or altering the transmit frequency, or shielding the system to eliminate the interfering ambient noise, these solutions are not always viable. For example, regulatory constraints on the amount of power that may be radiated by the sensor or the trade-off between the transmit power and the battery life for a portable sensor may limit the power in the transmitted signal, and effectively shielding a system in the field from ambient electromagnetic signals is very difficult. Thus, signal processing is often utilized to improve signal detectability in situations such as these where a hardware solution is not sufficient.

Adaptive filtering is an approach that is frequently employed to mitigate interference. This approach, however, relies on the ability to measure the interference on auxiliary reference sensors. The signals measured on the reference sensors are utilized to estimate the interference, and then this estimate is subtracted from the signal measured by the primary sensor, which consists of the signal of interest and the interference. When the interference measured by the reference sensors is completely correlated with the interference measured by the primary sensor, the adaptive filtering can completely remove the interference from the primary signal. When there are limitations in the ability to measure the interference, that is, the signals from the reference sensors are not completely correlated with the interference measured by the primary sensor, this approach is not completely effective. Since some residual interference remains after the adaptive interference cancellation, signal detection performance is adversely affected. This is particularly true when the signal of interest is weak. Thus, methods to improve signal detection when there is residual interference would be useful.

The Kalman filter (KF) is an important development in linear estimation theory. It is the statistically optimal estimator when the noise is Gaussian-distributed. In addition, the Kalman filter is still the optimal linear estimator in the minimum mean square error (MMSE) sense even when the Gaussian assumption is dropped [1]. Here, Kalman filters are applied to improve detection of weak harmonic signals. The emphasis in this chapter is not on developing new Kalman filters but, rather, on applying them in novel ways for improved weak harmonic signal detection. Both direct estimation and indirect estimation of the harmonic signal of interest are considered. Direct estimation is achieved by applying Kalman filters in the conventional manner; the state of the system is equal to the signal to be estimated. Indirect estimation of the harmonic signal of interest is achieved by reversing the usual application of the Kalman filter so the background noise is the system state to be estimated, and the signal of interest is the observation noise in the Kalman filter problem statement.

This approach to weak signal estimation is evaluated through application to quadrupole resonance (QR) signal estimation for landmine detection. Mine detection technologies and systems that are in use or have been proposed include electromagnetic induction (EMI) [2], ground penetrating radar (GPR) [3], and QR [4,5]. Regardless of the technology utilized, the goal is to achieve a high probability of detection, P_D, while maintaining a low probability of false alarm, P_{FA}. This is of particular importance for landmine detection since the nearly perfect P_D required to comply with safety requirements often comes at the expense of a high P_{FA}, and the time and cost required to remediate contaminated areas is directly proportional to P_{FA}. In areas such as a former battlefield, the average ratio of real mines to suspect objects can be as low as 1:100, thus the process of clearing the area often proceeds very slowly.

QR technology for explosive detection is of crucial importance in an increasing number of applications. Most explosives, such as RDX, TNT, PETN, etc., contain nitrogen (N). Some of its isotopes, such as ^{14}N, possess electric quadrupole moments. When compounds with such moments are probed with radio-frequency (RF) signals, they emit unique signals defined by the specific nucleus and its chemical environment. The QR frequencies for explosives are quite specific and are not shared by other nitrogenous materials. Since the detection process is specific to the chemistry of the explosive and therefore is less susceptible to the types of false alarms experienced by sensors typically used for landmine detection, such as EMI or GPR sensors, the pure QR of ^{14}N nuclei supports a promising method for detecting explosives in the quantities encountered in landmines. Unfortunately, QR signals are weak, and thus vulnerable to both the thermal noise inherent in the sensor coil and external radio-frequency interference (RFI). The performance of the Kalman filter approach is evaluated on both simulated data and measured field data collected by Quantum Magnetics, Inc. (QM). The results show that the proposed algorithm improves the performance of landmine detection.

7.1 Signal Models

In this chapter, it is assumed that the sensor operates by repeatedly transmitting excitation pulses to investigate the potential target and acquires the sensor response after each pulse. The data acquired after each excitation pulse are termed a segment, and a group of segments constitutes a measurement. In general, for each potential target there are multiple measurements with each measurement containing many segments.

7.1.1 Harmonic Signal Model

The discrete-time harmonic signal of interest, at frequency f_0, in a single segment can be represented by

$$s(n) = A_0 \cos(2\pi f_0 n + \phi_0), \quad n = 0, 1, \ldots, N-1 \tag{7.1}$$

The measured signal may be demodulated at the frequency of the desired harmonic signal, f_0, to produce a baseband signal, $\tilde{s}(n)$,

$$\tilde{s}(n) = A_0 e^{j\phi_0}, \quad n = 0, 1, \ldots, N-1 \tag{7.2}$$

Assuming the frequency of the harmonic signal of interest is precisely known, the signal of interest after demodulation and subsequent low-pass filtering to remove any aliasing introduced by the demodulation is a DC constant.

7.1.2 Interference Signal Model

A source of interference for this type of signal detection problem is ambient harmonic signals. For example, sensors operating in the RF band could experience interference due to other transmitters operating in the same band, such as radio stations. Since there may be many sources transmitting harmonic signals operating simultaneously, the demodulated interference signal measured in each segment may be modeled as the sum of the contribution from M different sources, each operating at its own frequency f_m,

$$\tilde{I} = \sum_{m=1}^{M} \tilde{A}_m(n) e^{j(2\pi f_m n + \phi_m)}, \quad n = 0, 1, \ldots, N-1 \tag{7.3}$$

where the superscript $\sim$ denotes a complex value and we assume the frequencies are distinct, meaning $f_i \neq f_j$ for $i \neq j$. The amplitudes $\tilde{A}_m(n)$ from a discrete time series. Although the amplitudes are not restricted to constant values in general, they are assumed to remain essentially constant over the short time intervals during which each data segment is collected. For time intervals on this order, it is reasonable to assume $\tilde{A}_m(n)$ is constant for each data segment, but may change from segment to segment. Therefore, the interference signal model may be expressed as

$$\tilde{I} = \sum_{m=1}^{M} \tilde{A}_m e^{j(2\pi f_m n + \phi_m)}, \quad n = 0, \ldots, N-1 \tag{7.4}$$

This model represents all frequencies even though the harmonic signal of interest exists in a very narrow band. In practice, only the frequency corresponding to the harmonic signal of interest needs to be considered.

7.2 Interference Mitigation

Adaptive filtering is a widely applied approach for noise cancellation [6]. The basic approach is illustrated in Figure 7.1. The primary signal consists of both the harmonic signal of interest and the interference. In contrast, the signal measured on each auxiliary antenna or sensor consists of only the interference. Adaptive noise cancellation utilizes the measured reference signals to estimate the noise present in the measured primary signal. The noise estimate is then subtracted from the primary signal to find the signal of interest.

Adaptive noise cancellation, such as the least mean square (LMS) algorithm, is well suited for those applications in which one or more reference signals are available [6].

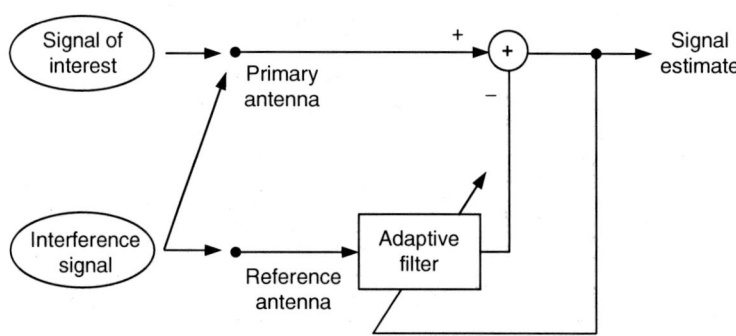

FIGURE 7.1
Interference mitigation based on adaptive noise cancellation.

In this application, the adaptive noise cancellation is performed in the frequency domain by applying the normalized LMS algorithm to each frequency component of the frequency domain representation of the measured primary signal. The primary measured signal at time n may be denoted by $d(n)$ and the measured reference signal with $u(n)$. The tap input vector may be represented by $\mathbf{u}(n) = [u(n)\ u(n-1) \cdots u(n-M+1)]^T$ and the tap weight vector may be given by $\hat{\mathbf{w}}(n)$. Both the tap input and tap weight vectors are of length M. Given these definitions, the filter output at time n, $e(n)$, is

$$e(n) = d(n) - \hat{\mathbf{w}}^H(n)\mathbf{u}(n) \tag{7.5}$$

The quantity $\hat{\mathbf{w}}^H(n)\mathbf{u}(n)$ represents the interference estimate. The tap weights are updated according to

$$\hat{\mathbf{w}}(n+1) = \hat{\mathbf{w}}(n) + \mu_0 \mathbf{P}^{-1}(n)\mathbf{u}(n)e^*(n) \tag{7.6}$$

where the parameter μ_0 is an adaptation constant that controls the convergence rate and $\mathbf{P}(n)$ is given by

$$\mathbf{P}(n) = \beta \mathbf{P}(n-1) + (1-\beta)|\mathbf{u}(n)|^2 \tag{7.7}$$

with $0 < \beta < 1$ [7]. The extension of this approach to utilize multiple reference signals is straightforward.

7.3 Postmitigation Signal Models

Under perfect circumstances the interference present in the primary signal is completely correlated with the reference signals and all interference can be removed by the adaptive noise cancellation, leaving only Gaussian noise associated with the sensor system. Since the interference often travels over multiple paths and the sensing systems are not perfect, however, the adaptive interference mitigation rarely removes all the interference. Thus, there is residual interference that remains after the adaptive noise cancellation. In addition, the adaptive interference cancellation process alters the characteristics of the signal of interest.

The real-valued observed data prior to interference mitigation may be represented by

$$x(n) = s(n) + I(n) + w(n), \quad n = 0, 1, \ldots, N-1 \tag{7.8}$$

where $s(n)$ is the signal of interest, $I(n)$ is the interference, and $w(n)$ is Gaussian noise associated with the sensor system. The baseband signal after demodulation becomes

$$\tilde{x}(n) = \tilde{s}(n) + \tilde{I}(n) + \tilde{w}(n) \tag{7.9}$$

where $\tilde{I}(n)$ is the interference, which is reduced but not completely eliminated by adaptive filtering.

The signal remaining after interference mitigation is

$$\tilde{y}(n) = \tilde{s}(n) + \tilde{v}(n), \quad n = 0, 1, \ldots, N-1 \tag{7.10}$$

where $\tilde{s}(n)$ is now the altered signal of interest and $\tilde{v}(n)$ is the background noise remaining after interference mitigation, both of which are described in the following sections.

7.3.1 Harmonic Signal Model

Due to the nonlinear phase effects of the adaptive filter, the demodulated harmonic signal of interest is altered by the interference mitigation; it is no longer guaranteed to be a DC constant. The postmitigation harmonic signal is modeled with a first-order Gaussian–Markov model,

$$\tilde{s}(n+1) = \tilde{s}(n) + \tilde{\varepsilon}(n) \tag{7.11}$$

where $\tilde{\varepsilon}(n)$ is zero mean white Gaussian noise with variance $\sigma_{\tilde{\varepsilon}}^2$.

7.3.2 Interference Signal Model

Although interference mitigation can remove most of the interference, some residual interference remains in the postmitigation signal. The background noise remaining after interference mitigation, $\tilde{v}(n)$, consists of the residual interference and the remaining sensor noise, which is altered by the mitigation. Either an autoregressive (AR) model or an autoregressive moving average (ARMA) model is appropriate for representing the sharp spectral peaks, valleys, and roll-offs in the power spectrum of $\tilde{v}(n)$. An AR model is a causal, linear, time-invariant discrete-time system with a transfer function containing only poles, whereas an ARMA model is a causal, linear, time-invariant discrete-time system with a transfer function containing both zeros and poles. The AR model has a computational advantage over the ARMA model in the coefficient computation. Specifically, the AR coefficient computation involves solving a system of linear equations known as Yule–Walker equations, whereas the ARMA coefficient computation is significantly more complicated because it requires solving systems of nonlinear equations.

Estimating and identifying an AR model for real-valued time series is well understood [8]. However, for complex-valued AR models, few theoretical and practical identification and estimation methods could be found in the literature. The most common approach is to adapt methods originally developed for real-valued data to complex-valued data. This strategy works well only when the the complex-valued process is the output of a linear system driven by white circular noise whose real and imaginary parts are uncorrelated and white [9]. Unfortunately, circularity is not always guaranteed for most complex-valued processes in practical situations. Analysis of the postmitigation background noise for the specific QR signal detection problem considered here showed that it is not a pure circular complex process. However, since the cross-correlation between the real and imaginary parts is small compared to the autocorrelation, we assume that the noise is a circular complex process and can be modeled as a P-th order complex AR process. Thus,

$$\tilde{v}(n) = -\sum_{p=1}^{P} \tilde{a}_p \tilde{v}(n-p) + \tilde{\varepsilon}(n) \tag{7.12}$$

where the driving noise $\tilde{\varepsilon}(n)$ is white and complex-valued.

Conventional modeling methods extended from real-valued time series are used for estimating the complex AR parameters. The Burg algorithm, which estimates the AR parameters by determining reflection coefficients that minimize the sum of forward and backward residuals, is the preferred estimator among various estimators for AR parameters [10]. Furthermore, the Burg algorithm has been reformulated so that it can be used to analyze data containing several separate segments [11]. The usual way of combining the information across segments is to take the average of the models estimated from the individual segments [12], such as average AR parameters (AVA) and average reflection coefficient (AVK). We found that for this application, the model coefficients are stable enough to be averaged. Although this will reduce the estimate variance, it still contains a bias that is proportional to $1/N$, where N is the number of observations [13]. Another novel extension of the Burg algorithm to segments, the segment Burg algorithm (SBurg), proposed in [14], estimates the reflection coefficients by minimizing the sum of the forward and backward residuals of all the segments taken together. This means a single model is fit to all the segments simultaneously.

An important aspect of AR modeling is choosing the appropriate order P. Although there are several criteria to determine the order for a real AR model, no formal criterion exists for a complex AR model. We adopt the Akaike information criterion (AIC) [15] that is a common choice for real AR models. The AIC is defined as

$$\text{AIC}(p) = \ln(\hat{\sigma}_\varepsilon^2(p)) + \frac{2p}{T}, \quad p = 1, 2, 3, \ldots \quad (7.13)$$

where $\hat{\sigma}_\varepsilon^2(p)$ is the prediction error power at P-th order and T is the total number of samples. The prediction error power for a given value of P is simply the variance in the difference between the true signal and the estimate for the signal using a model of order P. For signal measurements consisting of multiple segments, i.e. S segments and N samples/segment, $T = N \times S$. The model order for which the AIC has the smallest value is chosen. The AIC method tends to overestimate the order [16] and is good for short data records.

7.4 Kalman Filters for Weak Signal Estimation

Kalman filters are appropriate for discrete-time, linear, and dynamic systems whose output can be characterized by the system state. To establish notation and terminology, this section provides a brief overview consisting of the problem statement and the recursive solution of the Kalman filter variants applied for weak harmonic signal detection.

Two approaches to estimate a weak signal in nonstationary noise, both employing Kalman filter approaches, are proposed [17]. The first approach directly estimates the signal of interest. In this approach, the Kalman filter is utilized in a traditional manner, in that the signal of interest is the state to be estimated. The second approach indirectly estimates the signal of interest. This approach utilizes the Kalman filter in an unconventional way because the noise is the state to be estimated, and then the noise estimate is subtracted from the measured signal to obtain an estimate of the signal of interest. The problem statements and recursive Kalman filter solutions for each of the Kalman filters considered are provided with their application to weak signal estimation.

These approaches have an advantage over the more intuitive approach of simply subtracting a measurement of the noise recorded in the field because the measured background noise is nonstationary, and the residual background noise after interference mitigation is also nonstationary. Due to the nonstationarity of the background noise, it must be measured and subtracted in real time. This is exactly what the adaptive frequency-domain interference mitigation attempts to do, and the interference mitigation does significantly improve harmonic signal detection. There are, however, limitations to the accuracy with which the reference background noise can be measured, and these limitations make the interference mitigation insufficient for noise reduction. Thus, further processing, such as the Kalman filtering proposed here, is necessary to improve detection performance.

7.4.1 Direct Signal Estimation

The first approach assumes the demodulated harmonic signal is the system state to be estimated. The conventional Kalman filter was designed for white noise. Since the measurement noise in this application is colored, it is necessary to investigate modified Kalman filters. Two modifications are considered: utilizing an autoregressive (AR) model for the colored noise and designing a Kalman filter for colored noise, which can be modeled as a Markov process.

In applying each of the following Kalman filter variants to directly estimate the demodulated harmonic signal, the system is assumed to be described by

$$\text{state equation: } \tilde{s}_k = \tilde{s}_{k-1} + \tilde{\varepsilon}_k \tag{7.14}$$

$$\text{observation equation: } \tilde{y}_k = \tilde{s}_k + \tilde{v}_k \tag{7.15}$$

where $\tilde{s}_k$ is the postmitigation harmonic signal, modeled using a first-order Gaussian–Markov model, and $\tilde{v}_k$ is the postmitigation background noise.

7.4.1.1 Conventional Kalman Filter

The system model for the conventional Kalman filter is described by two equations: a state equation and an observation equation. The state equation relates the current system state to the previous system state, while the observation equation relates the observed data to the current system state. Thus, the system model is described by

$$\text{state equation: } \mathbf{x}_k = \mathbf{F}_k \mathbf{x}_{k-1} + \mathbf{G}_k \mathbf{w}_k \tag{7.16}$$

$$\text{observation equation: } \mathbf{z}_k = \mathbf{H}_k^H \mathbf{x}_k + \mathbf{u}_k \tag{7.17}$$

where the M-dimensional parameter vector $\mathbf{x}_k$ represents the state of the system at time k, the $M \times M$ matrix $\mathbf{F}_k$ is the known state transition matrix relating the states of the system at time k and $k-1$, and the N-dimensional parameter vector $\mathbf{z}_k$ represents the measured data at time k. The $M \times 1$ vector $\mathbf{w}_k$ represents process noise, and the $N \times 1$ vector $\mathbf{u}_k$ is measurement noise. The $M \times M$ diagonal coefficient matrix $\mathbf{G}_k$ modifies the variances of the process noise. If both $\mathbf{u}_k$ and $\mathbf{w}_k$ are independent, zero mean, white noise processes with $E\{\mathbf{u}_k \mathbf{u}_k^H\} = \mathbf{R}_k$ and $E\{\mathbf{w}_k \mathbf{w}_k^H\} = \mathbf{Q}_k$, then the initial system state $\mathbf{x}_0$ is a random vector, with mean $\bar{\mathbf{x}}_0$ and covariance Σ_0, independent of $\mathbf{u}_k$ and $\mathbf{w}_k$.

The Kalman filter determines the estimates of the system state, $\hat{\mathbf{x}}_{k|k-1} = E\{\mathbf{x}_k | \mathbf{z}_{k-1}\}$ and $\hat{\mathbf{x}}_{k|k} = E\{\mathbf{x}_k | \mathbf{z}_k\}$, and the associated error covariance matrices $\Sigma_{k|k-1}$ and $\Sigma_{k|k}$. The recursive solution is achieved in two steps. The first step predicts the current state and the error covariance matrix using the previous data,

$$\hat{\mathbf{x}}_{k|k-1} = \mathbf{F}_{k-1} \hat{\mathbf{x}}_{k-1|k-1} \tag{7.18}$$

$$\Sigma_{k|k-1} = \mathbf{F}_{k-1} \Sigma_{k-1|k-1} \mathbf{F}_{k-1}^H + \mathbf{G}_{k-1} \mathbf{Q}_{k-1} \mathbf{G}_{k-1}^H \tag{7.19}$$

The second step first determines the Kalman gain, $\mathbf{K}_k$, and then updates the state and the error covariance prediction using the current data,

$$\mathbf{K}_k = \Sigma_{k|k-1}\mathbf{H}_k(\mathbf{H}_k^H\Sigma_{k|k-1} + \mathbf{R}_k)^{-1} \quad (7.20)$$

$$\hat{\mathbf{x}}_{k|k} = \hat{\mathbf{x}}_{k|k-1} + \mathbf{K}_k(\mathbf{z}_k - \mathbf{H}_k^H\hat{\mathbf{x}}_{k|k-1}) \quad (7.21)$$

$$\Sigma_{k|k} = (\mathbf{I} - \mathbf{K}_k\mathbf{H}_k^H)\Sigma_{k|k-1} \quad (7.22)$$

The initial conditions on the state estimate and error covariance matrix are $\hat{\mathbf{x}}_{1|0} = \bar{\mathbf{x}}_0$ and $\Sigma_{1|0} = \Sigma_0$.

In general, the system parameters $\mathbf{F}$, $\mathbf{G}$, and $\mathbf{H}$ are time-varying, which is denoted in the preceding discussion by the subscript k. In the discussions of the Kalman filters implemented for harmonic signal estimation, the system is assumed to be time-invariant. Thus, the subscript k is omitted and the system model becomes

$$\text{state equation: } \mathbf{x}_k = \mathbf{F}\mathbf{x}_{k-1} + \mathbf{G}\mathbf{w}_k \quad (7.23)$$

$$\text{observation equation: } \mathbf{z}_k = \mathbf{H}^H\mathbf{x}_k + \mathbf{u}_k \quad (7.24)$$

7.4.1.2 Kalman Filter with an AR Model for Colored Noise

A Kalman filter for colored measurement noise, based on the fact that colored noise can often be simulated with sufficient accuracy by a linear dynamic system driven by white noise, is proposed in [18]. In this approach, the colored noise vector is included in an augmented state variable vector, and the observations now contain only linear combinations of the augmented state variables.

The state equation is unchanged from Equation 7.23; however, the observation equation is modified so the measurement noise is colored, thus the system model is

$$\text{state equation: } \mathbf{x}_k = \mathbf{F}\mathbf{x}_{k-1} + \mathbf{G}\mathbf{w}_k \quad (7.25)$$

$$\text{observation equation: } \mathbf{z}_k = \mathbf{H}^H\mathbf{x}_k + \tilde{v}_k \quad (7.26)$$

where $\tilde{v}_k$ is colored noise modeled by the complex-valued AR process in Equation 7.12. Expressing the P-th order AR process $\tilde{v}_k$ in state space notion yields

$$\text{state equation: } \mathbf{v}_k = \mathbf{F}_v\mathbf{v}_{k-1} + \mathbf{G}_v\tilde{\varepsilon}_k \quad (7.27)$$

$$\text{observation equation: } \tilde{v}_k = \mathbf{H}_v^H\mathbf{v}_k \quad (7.28)$$

where

$$\mathbf{v}_k = \begin{bmatrix} \tilde{v}(k-P+1) \\ \tilde{v}(k-P+2) \\ \vdots \\ \tilde{v}(k) \end{bmatrix}_{P \times 1} \quad (7.29)$$

$$\mathbf{F}_v = \begin{bmatrix} 0 & 1 & 0 & \cdots & 0 & 0 \\ 0 & 0 & 1 & \cdots & 0 & 0 \\ \vdots & \vdots & \vdots & \ddots & \vdots & \vdots \\ 0 & 0 & 0 & \cdots & 0 & 1 \\ -\tilde{a}_P & -\tilde{a}_{P-1} & -\tilde{a}_{P-2} & \cdots & -\tilde{a}_2 & -\tilde{a}_1 \end{bmatrix}_{P \times P} \quad (7.30)$$

$$\mathbf{G}_v = \mathbf{H}_v = \begin{bmatrix} 0 \\ 0 \\ \vdots \\ 0 \\ 1 \end{bmatrix}_{P \times 1} \tag{7.31}$$

and $\tilde{\varepsilon}_k$ is the white driving process in Equation 7.12 with zero mean and variance $\sigma_{\tilde{\varepsilon}}^2$. Combining Equation 7.25, Equation 7.26, and Equation 7.28, yields a new Kalman filter expression whose dimensions have been extended,

$$\text{state equation: } \bar{\mathbf{x}}_k = \overline{\mathbf{F}} \bar{\mathbf{x}}_{k-1} + \overline{\mathbf{G}} \bar{\mathbf{w}}_k \tag{7.32}$$

$$\text{observation equation: } \bar{\mathbf{z}}_k = \overline{\mathbf{H}}^H \bar{\mathbf{x}}_k \tag{7.33}$$

where

$$\bar{\mathbf{x}}_k = \begin{bmatrix} \mathbf{x}_k \\ \mathbf{v}_k \end{bmatrix}, \quad \bar{\mathbf{w}}_k = \begin{bmatrix} \mathbf{w}_k \\ \tilde{\varepsilon}_k \end{bmatrix}, \quad \overline{\mathbf{H}}^H = \begin{bmatrix} \mathbf{H}^H & \mathbf{H}_v^H \end{bmatrix},$$

$$\overline{\mathbf{F}} = \begin{bmatrix} \mathbf{F} & 0 \\ 0 & \mathbf{F}_v \end{bmatrix}, \quad \text{and} \quad \overline{\mathbf{G}} = \begin{bmatrix} \mathbf{G} & 0 \\ 0 & \mathbf{G}_v \end{bmatrix} \tag{7.34}$$

In estimation literature, this is termed the noise-free [19] or perfect measurement [20] problem. The process noise, $\mathbf{w}_k$, and colored noise state process noise, $\tilde{\varepsilon}_k$, are assumed to be uncorrelated, so

$$\overline{\mathbf{Q}} = E\{\bar{\mathbf{w}}_k \bar{\mathbf{w}}_k^H\} = \begin{bmatrix} \mathbf{Q} & 0 \\ 0 & \sigma_{\tilde{\varepsilon}}^2 \end{bmatrix} \tag{7.35}$$

Since there is no noise in Equation 7.33, the covariance matrix of the observation noise $\overline{\mathbf{R}} = 0$. The recursive solution for this problem, defined in Equation 7.32 and Equation 7.33, is the same as for the conventional Kalman filter given in Equation 7.18 through Equation 7.22.

When estimating the harmonic signal with the traditional Kalman filter with an AR model for the colored noise, the coefficient matrices are

$$\bar{\mathbf{x}}_k = \begin{bmatrix} \tilde{x}(k) \\ \tilde{v}(k-P+1) \\ \tilde{v}(k-P+2) \\ \vdots \\ \tilde{v}(k) \end{bmatrix}, \quad \overline{\mathbf{H}} = \begin{bmatrix} 1 \\ 0 \\ \vdots \\ 1 \end{bmatrix}_{P+1} \tag{7.36}$$

$$\overline{\mathbf{F}} = \begin{bmatrix} 1 & 0 & 0 & \cdots & 0 & 0 \\ 0 & 1 & 0 & \cdots & 0 & 0 \\ \vdots & \vdots & \vdots & \ddots & \vdots & \vdots \\ 0 & 0 & 0 & \cdots & 0 & 1 \\ 0 & -a_P & -a_{P-1} & \cdots & -a_2 & -a_1 \end{bmatrix} \tag{7.37}$$

$$\bar{\mathbf{w}}_k = \begin{bmatrix} \tilde{w}_k \\ \tilde{\varepsilon}_k \end{bmatrix}, \quad \overline{\mathbf{G}} = \begin{bmatrix} 1 & 0 \\ 0 & 0 \\ \vdots & \vdots \\ 0 & 1 \end{bmatrix}_{(P+1) \times 2} \tag{7.38}$$

and

$$\overline{\mathbf{Q}} = E\{\overline{\mathbf{w}}_k \overline{\mathbf{w}}_k^H\} = \begin{bmatrix} \sigma_{\tilde{w}}^2 & 0 \\ 0 & \sigma_{\tilde{\varepsilon}}^2 \end{bmatrix} \quad (7.39)$$

Since the theoretical value $\overline{\mathbf{R}} = 0$ turns off the Kalman filter it does not track the signal, $\overline{\mathbf{R}}$ should be set to a positive value.

7.4.1.3 Kalman Filter for Colored Noise

The Kalman filter has been generalized to systems for which both the process noise and measurement noise are colored and can be modeled as Markov processes [21]. A Markov process is a process for which the probability density function of the current sample depends only on the previous sample, not the entire process history. Although the Markov assumption may not be accurate in practice, this approach remains applicable. The system model is

$$\text{state equation: } \mathbf{x}_k = \mathbf{F}\mathbf{x}_{k-1} + \mathbf{G}w_k \quad (7.40)$$

$$\text{observation equation: } \mathbf{z}_k = \mathbf{H}_k^H \mathbf{x} + u_k \quad (7.41)$$

where the process noise w_k and the measurement noise u_k are both zero mean and colored with arbitrary covariance matrices at time k, $\mathbf{Q}$, and $\mathbf{R}$, so

$$\mathbf{Q}_{ij} = \text{cov}(w_i, w_j), i, j = 0, 1, 2, \ldots$$

$$\mathbf{R}_{ij} = \text{cov}(u_i, u_j), i, j = 0, 1, 2, \ldots$$

The initial state $\mathbf{x}_0$ is a random vector with mean $\bar{\mathbf{x}}_0$ and covariance matrix $\mathbf{P}_0$, and $\mathbf{x}_0$, w_k, and u_k are independent.

The prediction step of the Kalman filter solution is

$$\hat{\mathbf{x}}_{k|k-1} = \mathbf{F}\hat{\mathbf{x}}_{k-1|k-1} \quad (7.42)$$

$$\Sigma_{k|k-1} = \mathbf{F}\Sigma_{k-1|k-1}\mathbf{F}^H + \mathbf{G}\mathbf{Q}_{k-1}\mathbf{G}^H + \mathbf{F}\Psi_{k-1} + (\mathbf{F}\Psi_{k-1})^H \quad (7.43)$$

where $\Psi_{k-1} = \Psi_{k-1|k-1}^{k-1}$ and $\Psi_{k-1|k-1}^{k-1}$ is given recursively by

$$\Psi_{i|i-1}^{k-1} = \mathbf{F}\Psi_{i-1|i-1}^{k-1} + \mathbf{G}\mathbf{Q}_{i-1,k-1}\mathbf{G}^H \quad (7.44)$$

$$\Psi_{i|i}^{k-1} = \Psi_{i|i-1}^{k-1} - \mathbf{K}_i \mathbf{H}^H \Psi_{i|i-1}^{k-1} \quad (7.45)$$

with the initial value $\Psi_{0|0}^{k} = 0$ and $\mathbf{Q}_0 = 0$. The k in the superscript denotes time k.

The subsequent update step is

$$\hat{\mathbf{x}}_{k|k} = \hat{\mathbf{x}}_{k|k-1} + \mathbf{K}_k(\mathbf{z}_k - \mathbf{H}^H \hat{\mathbf{x}}_{k|k-1}) \quad (7.46)$$

$$\Sigma_{k|k} = \Sigma_{k|k-1} - \mathbf{K}_k \mathbf{S}_k \mathbf{K}_k^H \quad (7.47)$$

where $\mathbf{K}_k$ and $\mathbf{S}_k$ are given by

$$\mathbf{K}_k = (\Sigma_{k|k-1}\mathbf{H} + \Omega_k)\mathbf{S}_k^{-1} \quad (7.48)$$

$$\mathbf{S}_k = \mathbf{H}^H \Sigma_{k|k-1} \mathbf{H} + \mathbf{H}^H \Omega_k + (\mathbf{H}^H \Omega_k)^H + \mathbf{R}_k \tag{7.49}$$

$\Omega_k = \Omega_{k|k-1}^k$ is given recursively by

$$\Omega_{i|i-1}^k = \mathbf{F}\Omega_{i-1|i-1}^k \Omega_{i|i}^k = \Omega_{i|i-1}^k + \mathbf{K}_i(-\mathbf{R}_{i,k} - \mathbf{H}^H \Omega_{i|i-1}^k) \tag{7.50}$$

with $-\mathbf{R}_k$ being the new "data" and the initial value $\Omega_{0|0}^k = 0$.

When the noise is white $\Omega = 0$ and $\Psi = 0$, and this Kalman filter reduces to the conventional Kalman filter. It is worth noting that this filter is optimal only when $E\{w_k\}$ and $E\{u_k\}$ are known.

7.4.2 Indirect Signal Estimation

A central premise in Kalman filter theory is that the underlying state-space model is accurate. When this assumption is violated, the performance of the filter can deteriorate appreciably. The sensitivity of the Kalman filter to signal modeling errors has led to the development of robust Kalman filtering techniques based on modeling the noise.

The conventional point of view in applying Kalman filters to a signal detection problem is to assign the system state, x_k, to the signal to be detected. Under this paradigm, the hypotheses "signal absent" (H_0) and "signal present" (H_1) are represented by the state itself. When the conventional point of view is applied for this particular application, the demodulated harmonic signal (a DC constant) is the system state to be estimated. Although a model for the desired signal can be developed from a relatively pure harmonic signal obtained in a shielded lab environment, that signal model often differs from the harmonic signal measured in the field, sometimes substantially, because the measured signal may be a function of system and environmental parameters, such as temperature. Therefore the harmonic signal measured in the field may deviate from the assumed signal model and the Kalman filter may produce a poor state estimate due to inaccuracies in the state equation. However, the background noise in the field can be measured, from which a reliable noise model can be developed, even though the measured noise is not exactly the same as the noise corrupting the measurements.

The background noise in the postmitigation signal (Equation 7.10), $\tilde{v}(n)$, can be modeled as a complex-valued AR process as previously described. The Kalman filter is applied to estimate the background noise in the postmitigation signal and then the background noise estimate is subtracted from the postmitigation signal to indirectly estimate the postmitigation harmonic signal (a DC constant). Thus, the state in the Kalman filter, x_k, is the background noise $\tilde{v}(n)$,

$$\mathbf{x}_k = \begin{bmatrix} \tilde{v}(k-P+1) \\ \tilde{v}(k-P+2) \\ \vdots \\ \tilde{v}(k) \end{bmatrix} \tag{7.51}$$

and the measurement noise in the observation, $\mathbf{u}_k$, is the demodulated harmonic signal $\tilde{s}(n)$,

$$\mathbf{u}_k = \tilde{s}_k \tag{7.52}$$

The other system parameters are

$$F = \begin{bmatrix} 0 & 1 & 0 & \cdots & 0 & 0 \\ 0 & 0 & 1 & \cdots & 0 & 0 \\ \vdots & \vdots & \vdots & \ddots & \vdots & \vdots \\ 0 & 0 & 0 & \cdots & 0 & 1 \\ -\tilde{a}_P & -\tilde{a}_{P-1} & -\tilde{a}_{P-2} & \cdots & -\tilde{a}_2 & -\tilde{a}_1 \end{bmatrix} \qquad (7.53)$$

$$G = H = \begin{bmatrix} 0 \\ 0 \\ \vdots \\ 0 \\ 1 \end{bmatrix}_{P \times 1} \qquad (7.54)$$

and $w_k = \tilde{\varepsilon}_k$. Therefore, the measurement equation becomes

$$z_k = \tilde{v}_k + \tilde{s}_k \qquad (7.55)$$

where z_k is the measured data corresponding to $\tilde{y}(k)$ in Equation 7.10.

A Kalman filter assumes complete *a priori* knowledge of the process and measurement noise statistics **Q** and **R**. These statistics, however, are inexactly known in most practical situations. The use of incorrect *a priori* statistics in the design of a Kalman filter can lead to large estimation errors, or even to a divergence of errors. To reduce or bound these errors, an adaptive filter is employed by modifying or adapting the Kalman filter to the real data. The approaches to adaptive filtering are divided into four categories: Bayesian, maximum likelihood, correlation, and covariance matching [22]. The last technique has been suggested for the situations when **Q** is known but **R** is unknown. The covariance matching algorithm ensures that the residuals remain consistent with the theoretical covariance. The residual, or innovation, is defined by

$$v_k = z_k - \mathbf{H}^H \hat{\mathbf{x}}_{k|k-1} \qquad (7.56)$$

which has a theoretical covariance of

$$E\{v_k v_k^H\} = \mathbf{H}^H \Sigma_{k|k-1} \mathbf{H} + \mathbf{R} \qquad (7.57)$$

If the actual covariance of v_k is much larger than the covariance obtained from the Kalman filter, **R** should be increased to prevent divergence. This has the effect of increasing $\Sigma_{k|k-1}$, thus bringing the actual covariance of v_k closer to that given in Equation 7.57. In this case, **R** is estimated as

$$\hat{\mathbf{R}}_k = \frac{1}{m} \sum_{j=1}^{m} v_{k-j} v_{k-j}^H - \mathbf{H}^H \Sigma_{k|k-1} \mathbf{H} \qquad (7.58)$$

Here, a two-step adaptive Kalman filter using the covariance matching method is proposed. First, the covariance matching method is applied to estimate $\hat{\mathbf{R}}$. Then, the conventional Kalman filter is implemented with $\mathbf{R} = \hat{\mathbf{R}}$ to estimate the background noise. In this application, there are several measurements of data, with each measurement containing tens to hundreds of segments. For each segment, the covariance matching method is

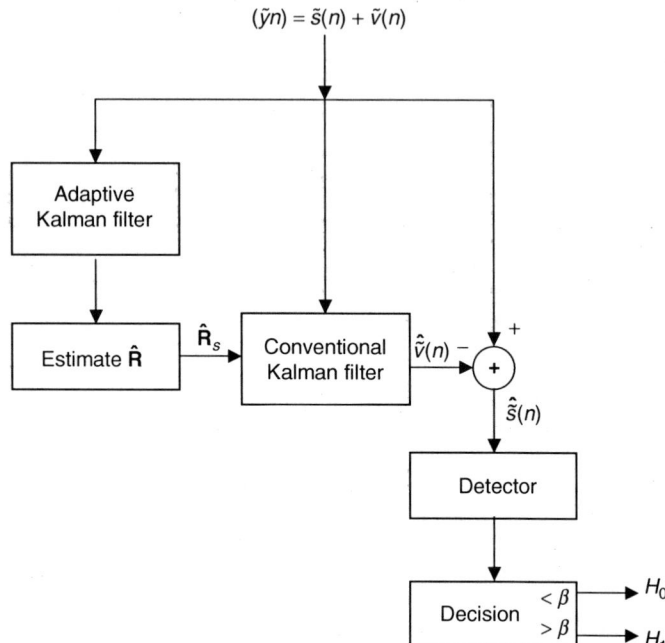

FIGURE 7.2
Block diagram of the two-step adaptive Kalman filter strategy. (From Tan et al., *IEEE Transactions on Geoscience and Remote Sensing* 43(7), 1507–1516, 2005. With permission.)

employed to estimate $\hat{\mathbf{R}}_k$, where the subscript k denotes the sample index. Since it is an adaptive procedure, only the steady-state values of $\hat{\mathbf{R}}_k$ ($k \geq m$) are retained and averaged to find the value of $\hat{\mathbf{R}}_s$ for each segment,

$$\hat{\mathbf{R}}_s = \frac{1}{N-m} \sum_{k=m}^{N-1} \hat{\mathbf{R}}_k \tag{7.59}$$

where the subscript s denotes the segment index. Then, the average is taken over all the segments in each measurement. Thus,

$$\hat{\mathbf{R}} = \frac{1}{L} \sum_{s=1}^{L} \hat{\mathbf{R}}_s \tag{7.60}$$

is used in the conventional Kalman filter in this two-step process. A block diagram depicting the two-step adaptive Kalman filter strategy is shown in Figure 7.2.

7.5 Application to Landmine Detection via Quadrupole Resonance

Landmines are a form of unexploded ordnance, usually emplaced on or just under the ground, which are designed to explode in the presence of a triggering stimulus such as pressure from a foot or vehicle. Generally, landmines are divided into two categories: antipersonnel mines and antitank mines. Antipersonnel (AP) landmines are devices usually designed to be triggered by a relatively small amount of pressure, typically 40 lbs, and generally contain a small amount of explosive so that the explosion aims or kills the person who triggers the device. In contrast, antitank (AT) landmines are specifically designed to destroy tanks and vehicles. They explode only if compressed by an object weighing hundreds of pounds. AP landmines are generally small (less than 10 cm in diameter) and are usually more difficult to detect than the larger AT landmines.

7.5.1 Quadrupole Resonance

When compounds with quadrupole moments are excited by a properly designed EMI system, they emit unique signals characteristic of the compound's chemical structure. The signal for a compound consists of a set of spectral lines, where the spectral lines correspond to the QR frequencies for that compound, and every compound has its own set of resonant frequencies. This phenomenon is similar to nuclear magnetic resonance (NMR). Although there are several subtle distinctions between QR and NMR, in this context it is sufficient to view QR as NMR without the external magnetic field [4].

The QR phenomenon is applicable for landmine detection because many explosives, such as RDX, TNT, and PETN, contain nitrogen, and some of nitrogen's isotopes, namely ^{14}N, have electric quadrupole moments. Because of the chemical specificity of QR, the QR frequencies for explosives are unique and are not shared with other nitrogenous materials. In summary, landmine detection using QR is achieved by observing the presence, or absence, of a QR signal after applying a sequence of RF pulses designed to excite the resonant frequency of frequencies for the explosive of interest [4].

RFI presents a problem since the frequencies of the QR response fall within the commercial AM radio band. After the QR response is measured, additional processing is often utilized to reduce the RFI, which is usually a non-Gaussian colored noise process. Adaptive filtering is a common method for cancelling RFI when RFI reference signals are available. The frequency-domain LMS algorithm is an efficient method for extracting the QR signal from the background RFI. Under perfect circumstances, when the RFI measured on the main antenna is completely correlated with the signals measured on the reference antennas, all RFI can be removed by RFI mitigation, leaving only Gaussian noise associated with the QR system. Since the RFI travels over multiple paths and the antenna system is not perfect, however, the RFI mitigation cannot remove all of the non-Gaussian noise. Consequently, more sophisticated signal processing methods must be employed to estimate the QR signal after the RFI mitigation, and thus improve the QR signal detection. Figure 7.3 shows the basic block diagram for QR signal detection.

The data acquired during each excitation pulse are termed a segment, and a group of segments constitutes a measurement. In general, for each potential target there are multiple measurements, with each measurement containing several hundred segments. The measurements are demodulated at the expected QR resonant frequency. Thus, if the demodulated frequency equals the QR resonant frequency, the QR signal after demodulation is a DC constant.

7.5.2 Radio-Frequency Interference

Although QR is a promising technology due to its chemical specificity, it is limited by the inherently weak QR signal and susceptibility to RFI. TNT is one of the most prevalent explosives in landmines, and also one of the most difficult explosives to detect. TNT possesses 18 resonant frequencies, 12 of which are clustered in the range of 700–900 kHz. Consequently, AM radio transmitters strongly interfere with TNT-QR detection in the

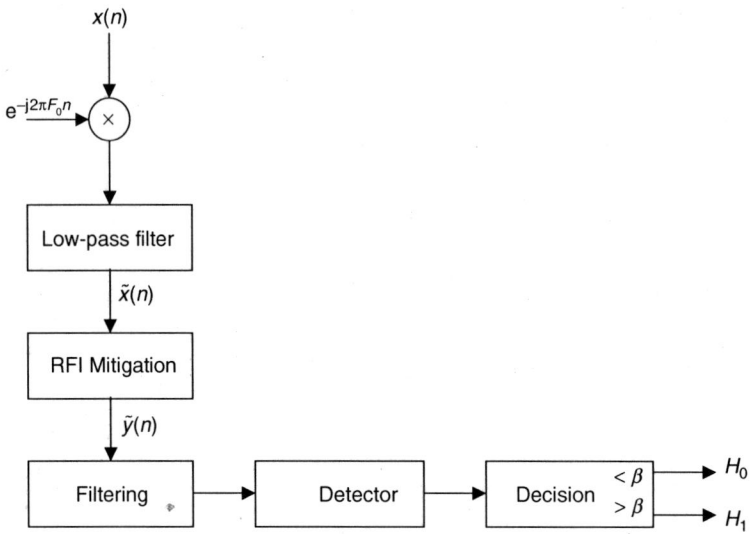

FIGURE 7.3
Signal processing block diagram for QR signal detection. F_0 is the resonant QR frequency for the explosive of interest and $\sim$ denotes a complex-valued signal. (From Tan et al., *IEEE Transactions on Geoscience and Remote Sensing* 43(7), 1507–1516, 2005. With permission.)

field, and are the primary source of RFI. Since there may be several AM radio transmitters operating simultaneously, the baseband (after demodulation) RFI signal measured by the QR system in each segment may be modeled as

$$\tilde{I}(n) = \sum_{m=1}^{M} \tilde{A}_m(n) e^{j(2\pi f_m n + \phi_m)}, \quad n = 0, \ldots, N-1 \qquad (7.61)$$

where the superscript $\sim$ denotes a complex value and we assume the frequencies are distinct, meaning $f_i \neq f_j$ for $i \neq j$. For the RFI, $A_m(n)$ is the discrete time series of the message signal from an AM transmitter, which may be a nonstationary speech or music signal from a commercial AM radio station. The statistics of this signal, however, can be assumed to remain essentially constant over the short time intervals during which data are collected. For time intervals of this order, it is reasonable to assume $A_m(n)$ is constant for each data segment, but may change from segment to segment. Therefore, Equation 7.61 may be expressed as

$$\tilde{I}(n) = \sum_{m=1}^{M} \tilde{A}_m e^{j(2\pi f_m n + \phi_m)}, \quad n = 0, \ldots, N-1 \qquad (7.62)$$

This model represents all frequencies even though each of the QR signals exists in a very narrow band. In practice, only the frequencies corresponding to the QR signals need be considered.

7.5.3 Postmitigation Signals

The applicability of the postmitigation signal models described previously is demonstrated by examining measured QR and RFI signals.

7.5.3.1 Postmitigation Quadrupole Resonance Signal

An example simulated QR signal before and after RFI mitigation is shown in Figure 7.4. Although the QR signal is a DC constant prior to RFI mitigation, that is no longer true after mitigation. The first-order Gaussian–Markov model introduced previously is an appropriate model for postmitigation QR signal. The value of $\sigma_\varepsilon^2 = 0.1$ is estimated from the data.

7.5.3.2 Postmitigation Background Noise

Although RFI mitigation can remove most of the RFI, some residual RFI remains in the postmitigation signal. The background noise remaining after RFI mitigation, $\tilde{v}(n)$, consists of the residual RFI and the remaining sensor noise, which is altered by the mitigation. An example power spectrum of $\tilde{v}(n)$ derived from experimental data is shown in Figure 7.5. The power spectrum contains numerous peaks and valleys. Thus, the AR and ARMA models discussed previously are appropriate for modeling the residual background noise.

7.5.4 Kalman Filters for Quadrupole Resonance Detection

7.5.4.1 Conventional Kalman Filter

For the system described by Equation 7.14 and Equation 7.15, the variables in the conventional Kalman filter are $\mathbf{x}_k = \tilde{s}(k)$, $\mathbf{w}_k = \tilde{w}(k)$, $\mathbf{u}_k = \tilde{v}(k)$, and $\mathbf{z}_k = \tilde{y}(k)$, and the coefficient and covariance matrices in the Kalman filter are $\mathbf{F} = [1]$, $\mathbf{G} = [1]$, $\mathbf{H} = [1]$, and $\mathbf{Q} = [\sigma_w^2]$. The covariance $\mathbf{R}$ is estimated from the data, and the off-diagonal elements are set to 0.

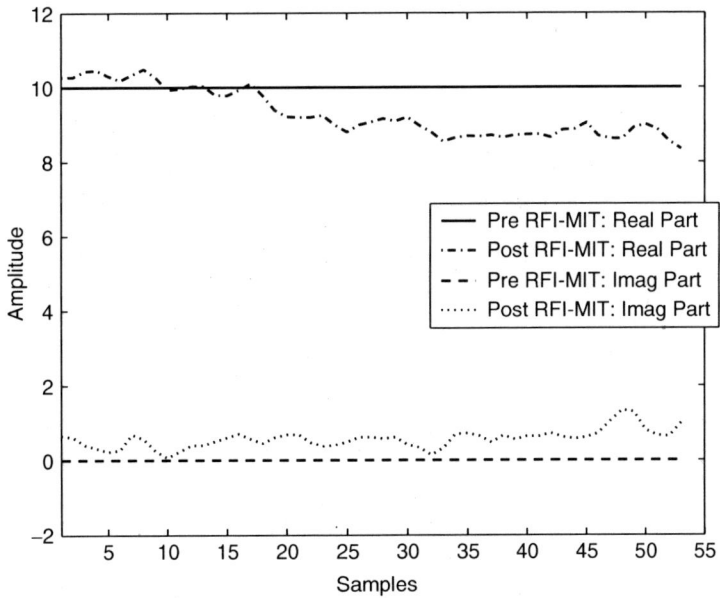

FIGURE 7.4
Example realization of the the simulated QR signal before and after RFI mitigation. (From Tan et al., *IEEE Transactions on Geoscience and Remote Sensing* 43(7), 1507–1516, 2005. With permission.)

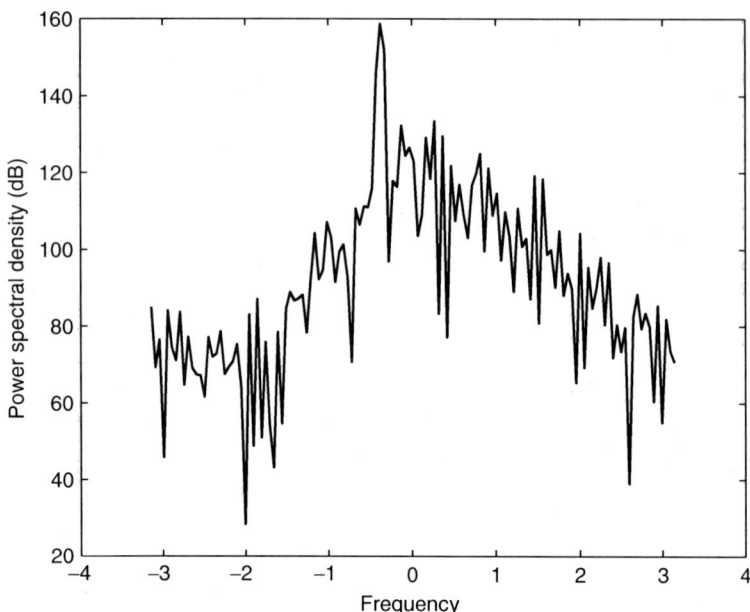

FIGURE 7.5
Power spectrum of the remaining background noise $\tilde{\nu}(n)$ after RFI mitigation. The x-axis units are normalized frequency ($[-\pi,\pi]$). (From Tan et al., *IEEE Transactions on Geoscience and Remote Sensing* 43(7), 1507–1516, 2005. With permission.)

7.5.4.2 Kalman Filter with an Autoregressive Model for Colored Noise

The multi-segment Burg algorithms are used to estimate the AR parameters, and the simulation results do not show significant differences between AVA, AVK, and SBurg algorithms. The AVA algorithm was chosen to estimate the AR coefficients $\tilde{a}_p$ and error power $\sigma_{\tilde{\varepsilon}}^2$ describing the background noise for each measurement. The optimal order, as determined by the AIC, is 6.

7.5.4.3 Kalman Filter for Colored Noise

For this Kalman filter, the coefficient and covariance matrices are the same as for the conventional Kalman filter, with the exception of the observation noise covariance **R**. In this filter, all elements of **R** are retained, as opposed to the conventional Kalman filter in which only the diagonal elements are retained.

7.5.4.4 Indirect Signal Estimation

For the adaptive Kalman filter in the first step, all coefficient and covariance matrices, **F, G, H, Q, R**, and Σ_0, are the same under both H_1 and H_0. The practical value of $\mathbf{x}_0$ is given by

$$[\tilde{y}(0)\ \tilde{y}(1) \cdots \tilde{y}(P-1)]^{\mathrm{T}} \tag{7.63}$$

where P is the order of the AR model representing $\tilde{v}(n)$. For the conventional Kalman filter in the second step, **F, G, H, Q**, and Σ_0 are the same under both H_1 and H_0; however, the estimated observation noise covariance, $\hat{\mathbf{R}}$, depends on the postmitigation signal and therefore is different under the two hypotheses.

7.6 Performance Evaluation

The proposed Kalman filter methods are evaluated on experimental data collected in the field by QM. A typical QR signal consists of multiple sinusoids where the amplitude and frequency of each resonant line are the parameters of interest. Usually, the amplitude of only one resonant line is estimated at a time, and the frequency is known *a priori*. For the baseband signal demodulated at the desired resonant frequency, adaptive RFI mitigation is employed. A 1-tap LMS mitigation algorithm is applied to each frequency component of the frequency-domain representation of the experimental data [23].

First, performance is evaluated for synthetic QR data. In this case, the RFI data are measured experimentally in the absence of an explosive material, and an ideal QR signal (complex-valued DC signal) is injected into the measured RFI data. Second, performance is evaluated for experimental QR data. In this case, the data are measured for both explosive and nonexplosive samples in the presence of RFI. The matched filter is employed for the synthetic QR data, while the energy detector is utilized for the measured QR data.

7.6.1 Detection Algorithms

After an estimate of the QR signal has been obtained, a detection algorithm must be applied to determine whether or not the QR signal is present. For this binary decision problem, both an energy detector and a matched filter are applied.

The energy detector simply computes the energy in the QR signal estimate, $\hat{s}(n)$.

$$E_s = \sum_{n=0}^{N-1} |\hat{s}(n)|^2 \tag{7.64}$$

As it is a simple detection algorithm that does not incorporate any prior knowledge of the QR signal characteristics, it is easy to compute.

The matched filter computes the detection statistic

$$\lambda = R_e \left\{ \sum_{n=0}^{N-1} \hat{s}(n) \tilde{s}^* \right\} \tag{7.65}$$

where $\tilde{S}$ is the reference signal, which, in this application, is the known QR signal. The matched filter is optimal only if the reference signal is precisely known *a priori*. Thus, if there is uncertainty regarding the resonant frequency of the QR signal, the matched filter will no longer be optimal. QR resonant frequency uncertainty may arise due to variations in environmental parameters, such as temperature.

7.6.2 Synthetic Quadrupole Resonance Data

Prior to detection using the matched filter, the QR signal is estimated directly. Three Kalman filter approaches to estimate the QR signal are considered: the conventional Kalman filter, the extended Kalman filter, and the Kalman filter for arbitrary colored noise. Each of these filters requires the initial value of the system state, x_0, and the selection of the initial value may affect the estimate of x. The sample mean of the observation $\tilde{y}(n)$ is used to set x_0. Since only the steady-state output of the Kalman filter is reliable, the first 39 points are removed from the data used for detection to ensure the system has reached steady state.

Although the measurement noise is known to be colored, the conventional Kalman filter (conKF) designed for white noise is applied to estimate the QR signal. This algorithm has the benefits of being simpler with lower computational burden than either of the other two Kalman filters proposed for direct estimation of the QR signal. Although its assumptions regarding the noise structure are recognized as inaccurate, its performance can be used as a benchmark for comparison. For a given process noise covariance $\mathbf{Q}$, the covariance of the intial state $\mathbf{x}_0$, Σ_0, affects the speed with which Σ reaches steady state in the conventional Kalman filter. As $\mathbf{Q}$ increases, it takes less time for Σ to converge. However, the steady-state value of Σ also increases. In these simulations, $\Sigma_0 = 10$ in the conventional Kalman filter.

The second approach applied to estimate the QR signal is the Kalman filter with an AR model for the colored noise (extKF). Since $\overline{\mathbf{R}} = 0$ shuts down the Kalman filter it does not track the signal, we set $\overline{\mathbf{R}} = 10$. It is shown that the detection performance does not decrease when $\mathbf{Q}$ increases.

Finally, the Kalman filter for colored noise (arbKF) is applied to the synthetic data. Compared to the conventional Kalman filter, the Kalman filter for arbitrary noise has a smaller Kalman gain, and therefore, slower convergence speed. Consequently, the error covariances for the Kalman filter for arbitrary noise are larger. Since only the steady state is used for detection, $\Sigma_0 = 50$ is chosen for $\mathbf{Q} = 0.1$ and $\Sigma_0 = 100$ is chosen for $\mathbf{Q} = 1$ and $\mathbf{Q} = 10$.

Results for each of these Kalman filtering methods for different values of $\mathbf{Q}$ are presented in Figure 7.6. When $\mathbf{Q}$ is small ($\mathbf{Q} = 0.1$ and $\mathbf{Q} = 1$), all three methods have similar detection performance. However, when greater model error is introduced in the state equation ($\mathbf{Q} = 10$) both the conventional Kalman filter and the Kalman filter for colored noise have poorer detection performance than the Kalman filter with an AR model for the noise. Thus, the Kalman filter with an AR model for the noise shows robust performance. Considering the computational efficiency, the Kalman filter for colored noise is the poorest because it recursively estimates Ψ and Ω for each k. In this application, although the measurement noise $\tilde{v}_k$ is colored, the diagonal elements of the covariance matrix dominate. Therefore, the conventional Kalman filter is preferable to the Kalman filter for colored noise.

7.6.3 Measured Quadrupole Resonance Data

Indirect estimation of the QR signal is validated using two groups of real data collected both with and without an explosive present. The two explosives for which measured QR data are collected, denoted Type A and Type B, are among the more common explosives found in landmines. It is well known that the Type B explosive is the more challenging of the two explosives to detect. The first group, denoted Data II, has Type A explosive, and the second group, denoted Data III, has both Type A and Type B explosives.

For each data group, a 50–50% training–testing strategy is employed. Thus, 50% of the data are used to estimate the coefficient and covariance matrices, and the remaining 50% of the data are used to test the algorithm. The AVA algorithm is utilized to estimate the AR parameters from the training data. Table 7.1 lists the four training–testing strategies considered. For example, if there are 10 measurements and measurements 1–5 are used for training and measurements 6–10 are used for testing, then this is termed "first 50% training, last 50% testing." If the training–testing strategy measurements 1,3,5,7,9 are used for training, and the other measurements for testing, this is termed "odd 50% training, even 50% testing."

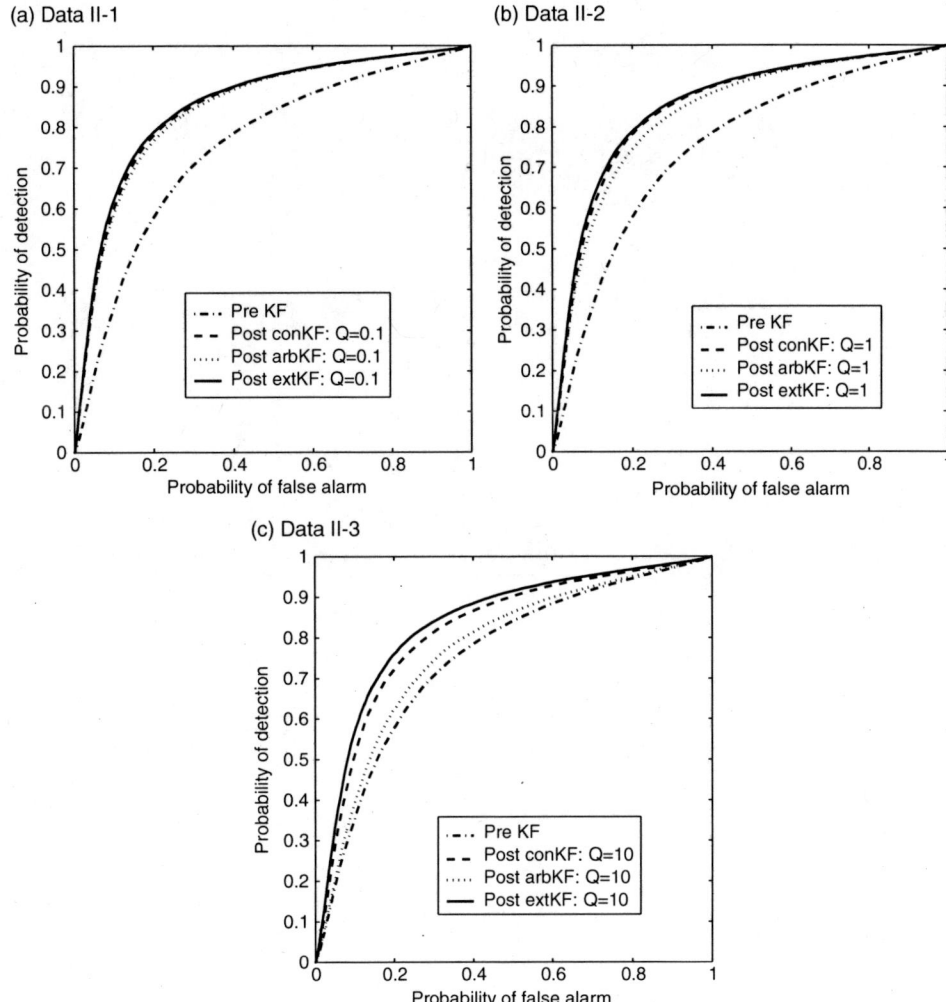

FIGURE 7.6
Direct estimation of QR signal tested on Data I (synthetic data) with different noise model variance **Q**. (From Tan et al., *IEEE Transactions on Geoscience and Remote Sensing* 43(7), 1507–1516, 2005. With permission.)

Figure 7.7 and Figure 7.8 present the performance of indirect QR signal estimation followed by an energy detector. The data are measured for both explosive and nonexplosive samples in the presence of RFI. The two different explosive compounds are referred to as Type A explosive and Type B explosive. Data II and Data III-1 are Type A explosive and Data III-2 is Type B explosive. Indirect QR signal estimation provides almost perfect

TABLE 7.1
Training–Testing Strategies for Data II and Data III

Training	Testing
First 50%	Last 50%
Last 50%	First 50%
Odd 50%	Even 50%
Even 50%	Odd 50%

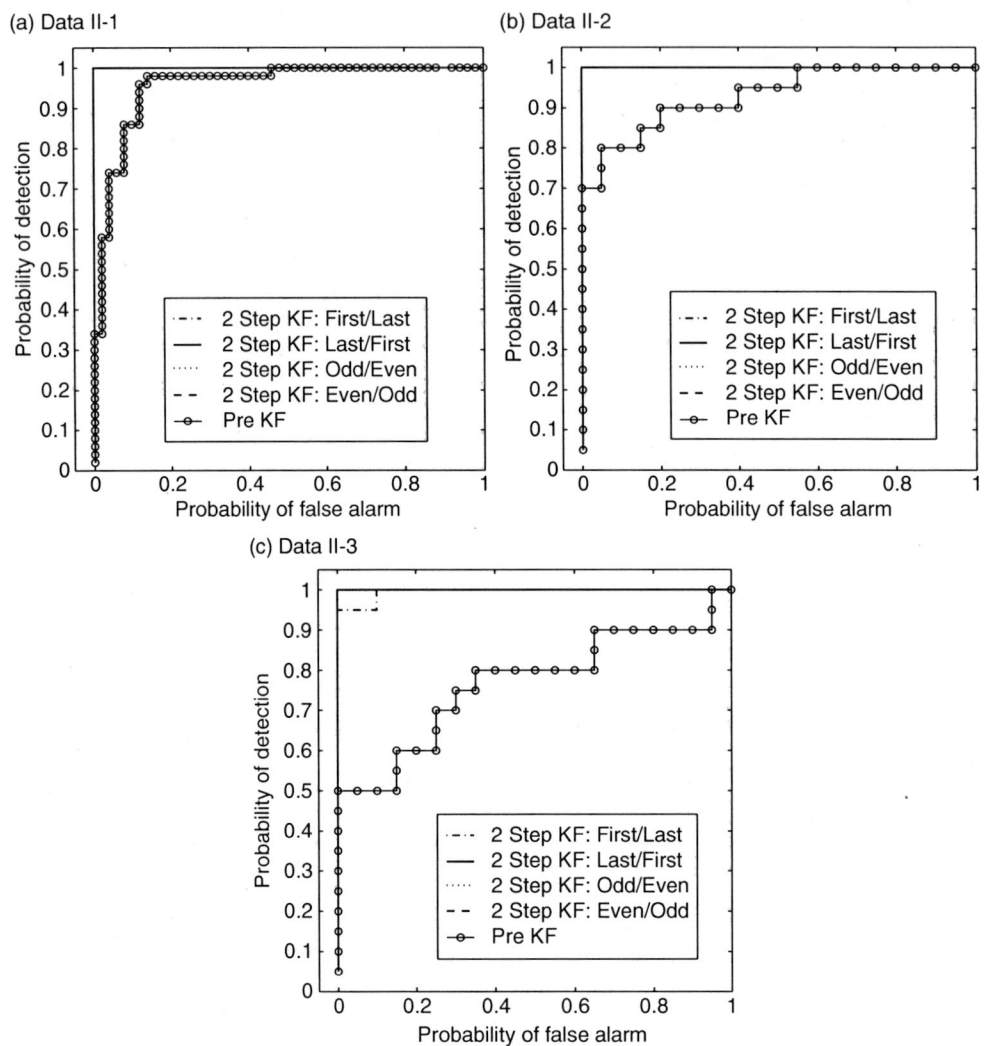

FIGURE 7.7
Indirect estimation of QR signal tested on Data II (true data). Four different training–testing strategies are plotted together. (From Tan et al., *IEEE Transactions on Geoscience and Remote Sensing* 43(7), 1507–1516, 2005. With permission.)

detection for the Type A explosive. Although detection is not near-perfect for the Type B explosive, the detection performance following the application of the Kalman filter is better than the performance prior to applying the Kalman filter.

7.7 Summary

The detectability of weak signals in remote sensing applications can be hindered by the presence of interference signals. In situations where it is not possible to record the measurement without the interference, adaptive filtering is an appropriate method to mitigate the interference in the measured signal. Adaptive filtering, however, may not remove all the interference from the measured signal if the reference signals are not

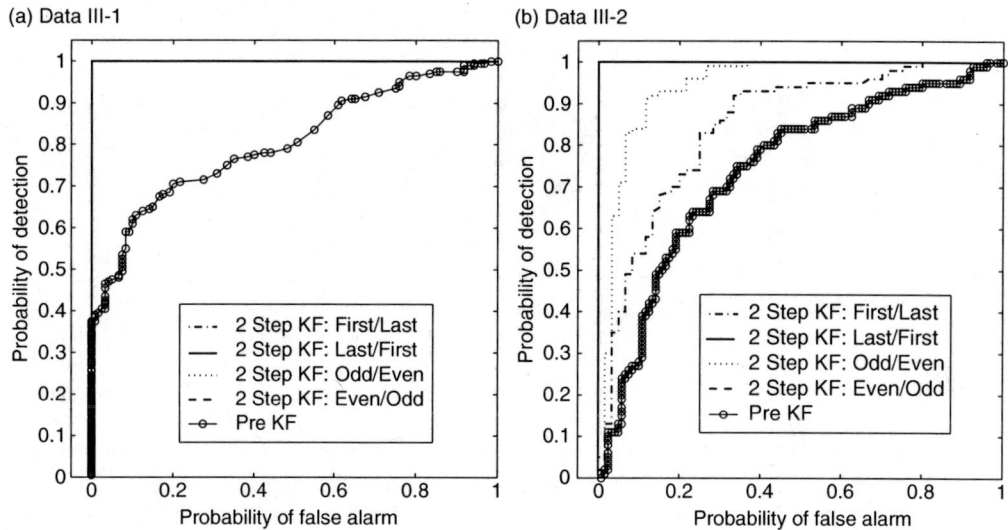

FIGURE 7.8
Indirect estimation of QR signal tested on Data III (true data). Four different training–testing strategies are plotted together. (From Tan et al., *IEEE Transactions on Geoscience and Remote Sensing* 43(7), 1507–1516, 2005. With permission.)

completely correlated with the primary measured signal. One approach for subsequent processing to detect the signal of interest when residual interference remains after the adaptive noise cancellation is Kalman filtering.

An accurate signal model is necessary for Kalman filters to perform well. It is so critical that even small deviations may cause very poor performance. The harmonic signal of interest may be sensitive to the external environment, which may then restrict the signal model accuracy. To overcome this limitation, an adaptive two-step algorithm, employing Kalman filters, is proposed to estimate the signal of interest indirectly.

The utility of this approach is illustrated by applying it to QR signal estimation for landmine detection. QR technology provides promising explosive detection efficiency because it can detect the "fingerprint" of explosives. In applications such as humanitarian demining, QR has proven to be highly effective if the QR sensor is not exposed to RFI. Although adaptive RFI mitigation removes most of RFI, additional signal processing algorithms applied to the postmitigation signal are still necessary to improve landmine detection. Indirect signal estimation is compared to direct signal estimation using Kalman filters and is shown to be more effective. The results of this study indicate that indirect QR signal estimation provides robust detection performance.

References

1. B.D.O. Anderson and J.B. Moore, *Optimal Filtering*, Prentice-Hall, Englewood Cliffs, NJ, 1979.
2. S.J. Norton and I.J. Won, Identification of buried unexploded ordnance from broadband electromagnetic induction data, *IEEE Transactions on Geoscience and Remote Sensing*, 39(10), 2253–2261, 2001.
3. K. O'Neill, Discrimination of UXO in soil using broadband polarimetric GPR backscatter, *IEEE Transactions on Geoscience and Remote Sensing*, 39(2), 356–367, 2001.

4. N. Garroway, M.L. Buess, J.B. Miller, B.H. Suits, A.D. Hibbs, G.A. Barrall, R. Matthews, and L.J. Burnett, Remote sensing by nuclear quadrupole resonance, *IEEE Transactions on Geoscience and Remote Sensing*, 39(6), 1108–1118, 2001.
5. J.A.S. Smith, Nuclear quadrupole resonance spectroscopy, general principles, *Journal of Chemical Education*, 48, 39–49, 1971.
6. S. Haykin, *Adaptive Filter Theory*, Prentice-Hall, Englewood Cliffs, NJ, 1991.
7. C.F. Cowan and P.M. Grant, *Adaptive Filters*, Prentice-Hall, Englewood Cliffs, NJ, 1985.
8. G.E.P. Box and G.M. Jenkins, *Time Series Analysis: Forecasting and Control*, Holden-Day, San Francisco, CA, 1970.
9. B. Picinbono and P. Bondon, Second-order statistics of complex signals, *IEEE Transactions on Signal Processing*, SP-45(2), 411–420, 1979.
10. P.M.T. Broersen, The ABC of autoregressive order selection criteria, *Proceedings of SYSID SICE*, 231–236, 1997.
11. S. Haykin, B.W. Currie, and S.B. Kesler, Maximum-entropy spectral analysis of radar clutter, *Proceedings of the IEEE*, 70(9), 953–962, 1982.
12. A.A. Beex and M.D.A. Raham, On averaging Burg spectral estimators for segments, *IEEE Transactions on Acoustics, Speech, and Signal Processing*, ASSP-34, 1473–1484, 1986.
13. D. Tjostheim and J. Paulsen, Bias of some commonly-used time series estimates, *Biometrika*, 70(2), 389–399, 1983.
14. S. de Waele and P.M.T. Broersen, The Burg algorithm for segments, *IEEE Transactions on Signal Processing*, SP-48, 2876–2880, 2000.
15. H. Akaike, A new look at the statistical model identification, *IEEE Transactions on Automatic Control*, AC-19(6), 716–723, 1974.
16. M. Wax and T. Kailath, Detection of signals by information theoretic criteria, *IEEE Transactions on Acoustics, Speech, and Signal Processing*, ASSP-33(2), 387–392, 1985.
17. Y. Tan, S.L. Tantum, and L.M. Collins, Kalman filtering for enhanced landmine detection using quadrupole resonance, *IEEE Transactions on Geoscience and Remote Sensing*, 43(7), 1507–1516, 2005.
18. J.D. Gibson, B. Koo, and S.D. Gray, Filtering of colored noise for speech enhancement and coding, *IEEE Transactions on Signal Processing*, 39(8), 1732–1742, 1991.
19. A.P. Sage and J.L. Melsa, *Estimation Theory with Applications to Communications and Control*, McGraw-Hill, New York, 1971.
20. P.S. Maybeck, *Stochastic Models, Estimation, and Control*, Academic Press, New York, 1979.
21. X.R. Li, C. Han, and J. Wang, Discrete-time linear filtering in arbitrary noise, *Proceedings of the 39th IEEE Conference on Decision and Control, 2000*, 1212–1217, 2000.
22. R.K. Mehra, Approaches to adaptive filtering, *IEEE Transactions on Automatic Control*, AC-17, 693–398, 1972.
23. Y. Tan, S.L. Tantum, and L.M. Collins, Landmine detection with nuclear quadrupole resonance, In *IGARSS 2002 Proceedings*, 1575–1578, July 2002.

8

Relating Time-Series of Meteorological and Remote Sensing Indices to Monitor Vegetation Moisture Dynamics

J. Verbesselt, P. Jönsson, S. Lhermitte, I. Jonckheere, J. van Aardt, and P. Coppin

CONTENTS

8.1 Introduction 151
8.2 Data 153
    8.2.1 Study Area 153
    8.2.2 Climate Data 154
    8.2.3 Remote Sensing Data 155
8.3 Serial Correlation and Time-Series Analysis 156
    8.3.1 Recognizing Serial Correlation 156
    8.3.2 Cross-Correlation Analysis 158
    8.3.3 Time-Series Analysis: Relating Time-Series and Autoregression ... 160
8.4 Methodology 161
    8.4.1 Data Smoothing 161
    8.4.2 Extracting Seasonal Metrics from Time-Series and Statistical Analysis 163
8.5 Results and Discussion 164
    8.5.1 Temporal Analysis of the Seasonal Metrics 164
    8.5.2 Regression Analysis Based on Values of Extracted Seasonal Metrics 165
    8.5.3 Time-Series Analysis Techniques 166
8.6 Conclusions 167
Acknowledgments 168
References 168

8.1 Introduction

The repeated occurrence of severe wildfires, which affect various fire-prone ecosystems of the world, has highlighted the need to develop effective tools for monitoring fire-related parameters. Vegetation water content (VWC), which influences the biomass burning processes, is an example of one such parameter [1–3]. The physical definitions of VWC vary from water volume per leaf or ground area (equivalent water thickness) to water mass per mass of vegetation [4]. Therefore, VWC could also be used to infer vegetation water stress and to assess drought conditions that linked with fire risk [5]. Decreases in VWC due to the seasonal decrease in available soil moisture can induce severe fires in most ecosystems. VWC is particularly important for determining the behavior of fires in savanna ecosystems because the herbaceous layer becomes especially flammable during the dry season when the VWC is low [6,7].

Typically, VWC in savanna ecosystems is measured using labor-intensive vegetation sampling. Several studies, however, indicated that VWC can be characterized temporally and spatially using meteorological or remote sensing data, which could contribute to the monitoring of fire risk [1,4]. The meteorological Keetch–Byram drought index (KBDI) was selected for this study. This index was developed to incorporate soil water content in the root zone of vegetation and is able to assess the seasonal trend of VWC [3,8]. The KBDI is a cumulative algorithm for the estimation of fire potential from meteorological information, including daily maximum temperature, daily total precipitation, and mean annual precipitation [9,10]. The KBDI also has been used for the assessment of VWC for vegetation types with shallow rooting systems, for example, the herbaceous layer of the savanna ecosystem [8,11].

The application of drought indices, however, presents specific operational challenges. These challenges are due to the lack of meteorological data for certain areas, as well as spatial interpolation techniques that are not always suitable for use in areas with complex terrain features. Satellite data provide sound alternatives to meteorological indices in this context. Remotely sensed data have significant potential for monitoring vegetation dynamics at regional to global scale, given the synoptic coverage and repeated temporal sampling of satellite observations (e.g., SPOT VEGETATION or NOAA AVHRR) [12,13]. These data have the advantage of providing information on remote areas where ground measurements are impossible to obtain on a regular basis.

Most research in the scientific community using optical sensors (e.g., SPOT VEGETATION) to study biomass burning has focused on two areas [4]: (1) the direct estimation of VWC and (2) the estimation of chlorophyll content or degree of drying as an alternative to the estimation of VWC. Chlorophyll-related indices are related to VWC based on the hypothesis that the chlorophyll content of leaves decreases proportionally to the VWC [4]. This assumption has been confirmed for selected species with shallow rooting systems (e.g., grasslands and understory forest vegetation) [14–16], but cannot be generalized to all ecosystems [4]. Therefore, chlorophyll-related indices, such as the normalized difference vegetation index (NDVI), only can be used in regions where the relationship among chlorophyll content, degree of curing, and water content has been established.

Accordingly, a remote sensing index that is directly coupled to the VWC is used to investigate the potential of hyper-temporal satellite imagery to monitor the seasonal vegetation moisture dynamics. Several studies [4,16–18] have demonstrated that VWC can be estimated directly through the normalized difference of the near infrared reflectance (NIR, 0.78–0.89 µm) ρ_{NIR}, influenced by the internal structure and the dry matter, and the shortwave infrared reflectance (SWIR, 1.58–1.75 µm) ρ_{SWIR}, influenced by plant tissue water content:

$$\text{NDWI} = \frac{\rho_{NIR} - \rho_{SWIR}}{\rho_{NIR} + \rho_{SWIR}} \tag{8.1}$$

The NDWI or normalized difference infrared index (NDII) [19] is similar to the global vegetation moisture index (GVMI) [20].

The relationship between NDWI and KBDI time-series, both related to VWC dynamics, is explored. Although the value of time-series data for monitoring vegetation moisture dynamics has been firmly established [21], only a few studies have taken serial correlation into account when correlating time-series [6,22–25]. Serial correlation occurs when data collected through time contain values at time t, which are correlated with observations at time $t-1$. This type of correlation in time-series, when related to VWC dynamics, is mainly caused by the seasonal variation (dry–wet cycle) of vegetation [26]. Serial correlation can be used to forecast future values of the time-series by modeling the dependence between observations but affects correlations between variables measured in time and violates the basic regression assumption of independence [22]. Correlation coefficients of serially correlated data cannot be used as indicators of goodness-of-fit of a model as the correlation coefficients are artificially inflated [22,27].

The study of the relationship between NDWI and KBDI is a nontrivial task due to the effect of serial correlation. Remedies for serial correlation include sampling or aggregating the data over longer time intervals, as well as further modeling, which can include techniques such as weighted regression [25,28]. However, it is difficult to account for serial correlation in time-series related to VWC dynamics using extended regression techniques. The time-series related to VWC dynamics often exhibit high non-Gaussian serial correlation and are more significantly affected by outliers and measurement errors [28]. A sampling technique therefore is proposed, which accounts for serial correlation in seasonal time-series, to study the relationship between different time-series. The serial correlation effect in time-series is assumed to be minimal when extracting one metric per season (e.g., start of the dry season). The extracted seasonal metrics are then utilized to study the relationship between time-series at a specific moment in time (e.g., start of the dry season).

The aim of this chapter is to address the effect of serial correlation when studying the relationship between remote sensing and meteorological time-series related to VWC by comparing nonserially correlated seasonal metrics from time-series. This chapter therefore has three defined objectives. Firstly, an overview of time-series analysis techniques and concepts (e.g., stationarity, autocorrelation, ARIMA, etc.) is presented and the relationship between time-series is studied using cross-correlation and ordinary least square (OLS) regression analysis. Secondly, an algorithm for the extraction of seasonal metrics is optimized for satellite and meteorological time-series. Finally, the temporal occurrence and values of the extracted nonserially correlated seasonal metrics are analyzed statistically to define the quantitative relationship between NDWI and KBDI time-series. The influence of serial correlation is illustrated by comparing results from cross-correlation and OLS analysis with the results from the investigation of correlation between extracted metrics.

8.2 Data

8.2.1 Study Area

The Kruger National Park (KNP), located between latitudes 23°S and 26°S and longitudes 30°E and 32°E in the low-lying savanna of the northeastern part of South Africa, was selected for this study (Figure 8.1). Elevations range from 260 to 839 m above sea level, and mean annual rainfall varies between 350 mm in the north and 750 mm in the south. The rainy season within the annual climatic season can be confined to the summer months (i.e., November to April), and over a longer period can be defined by alternating wet and dry seasons [7]. The KNP is characterized by an arid savanna dominated by thorny, fine-leafed trees of the families Mimosaceae and Burseraceae. An exception is the northern part of the KNP where the Mopane, a broad-leafed tree belonging to the Ceasalpinaceae, almost completely dominates the tree layer.

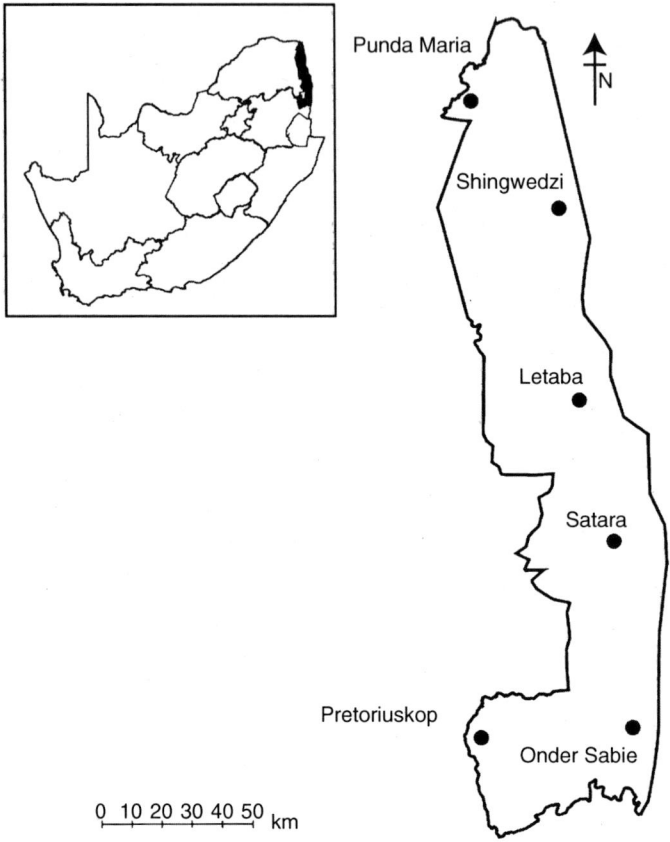

FIGURE 8.1
The Kruger National Park (KNP) study area with the weather stations used in the analysis (right). South Africa is shown with the borders of the provinces and the study area (top left).

8.2.2 Climate Data

Climate data from six weather stations in the KNP with similar vegetation types were used to estimate the daily KBDI (Figure 8.1). KBDI was derived from daily precipitation and maximum temperature data to estimate the net effect on the soil water balance [3]. Assumptions in the derivation of KBDI include a soil water capacity of approximately 20 cm and an exponential moisture loss from the soil reservoir. KBDI was initialized during periods of rainfall events (e.g., rainy season) that result in soils with maximized field capacity and KBDI values of zero [8]. The preprocessing of KBDI was done using the method developed by Janis et al. [10]. Missing daily maximum temperatures were replaced with interpolated values of daily maximum temperatures, based on a linear interpolation function [30]. Missing daily precipitation, on the other hand, was assumed to be zero. A series of error logs were automatically generated to indicate missing precipitation values and associated estimated daily KBDI values. This was done because zeroing missing precipitation may lead to an increased fire potential bias in KBDI. The total percentage of missing data gaps in rainfall and temperature series was maximally 5% during the study period for each of the six weather stations. The daily KBDI time-series were transformed into 10-daily KBDI series, similar to the SPOT VEGETATION S10 dekads (i.e., 10-day periods), by taking the maximum of

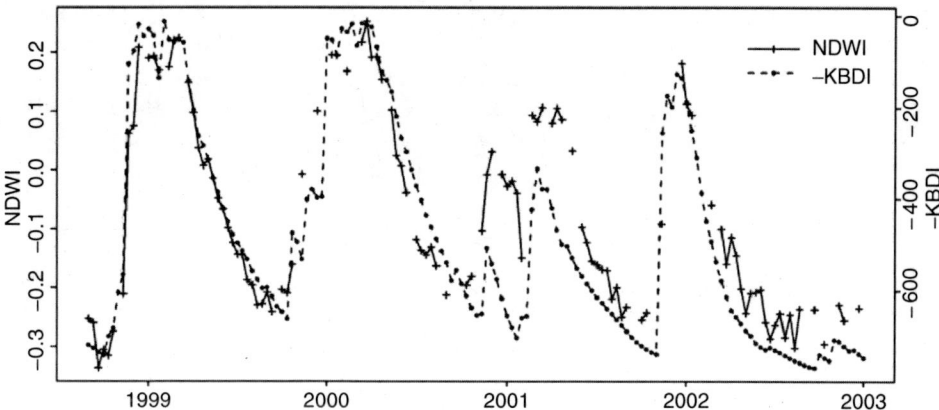

FIGURE 8.2
The temporal relationship between NDWI and −KBDI time-series for the "Satara" weather station (Figure 8.1).

each dekad. The negative of the KBDI time-series (i.e., −KBDI) was analyzed in this chapter such that the temporal dynamics of KBDI and NDWI were related (Figure 8.2). The −KBDI and NDWI are used throughout this chapter. The Satara weather station, centrally positioned in the study area, was selected to represent the temporal vegetation dynamics. The other weather stations in the study area demonstrate similar temporal vegetation dynamics.

8.2.3 Remote Sensing Data

The data set used is composed of 10-daily SPOT VEGETATION (SPOT VGT) composites (S10 NDVI maximum value syntheses) acquired over the study area for the period April 1998 to December 2002. SPOT VGT can provide local to global coverage on a regular basis (e.g., daily for SPOT VGT). The syntheses result in surface reflectance in the blue (0.43–0.47 μm), red (0.61–0.68 μm), NIR (0.78–0.89 μm), and SWIR (1.58–1.75 μm) spectral regions. Images were atmospherically corrected using the simplified method for atmospheric correction (SMAC) [30]. The geometrically and radiometrically corrected S10 images have a spatial resolution of 1 km.

The S10 SPOT VGT time-series were preprocessed to detect data that erroneously influence the subsequent fitting of functions to time-series, necessary to define and extract metrics [6]. The image preprocessing procedures performed were:

- Data points with a satellite viewing zenith angle (VZA) above 50° were masked out as pixels located at the very edge of the image (VZA > 50.5°) swath are affected by re-sampling methods that yield erroneous spectral values.
- The aberrant SWIR detectors of the SPOT VGT sensor, flagged by the status mask of the SPOT VGT S10 synthesis, also were masked out.
- A data point was classified as cloud-free if the blue reflectance was less than 0.07 [31]. The developed threshold approach was applied to identify cloud-free pixels for the study area.

NDWI time-series were derived by selecting savanna pixels, based on the land cover map of South Africa [32], for a 3 × 3 pixel window centered at each of the meteorological stations to reduce the effect of potential spatial misregistration (Figure 8.1). Median values of the 9-pixel windows were then retained instead of single pixel values [33]. The median was preferred to average values as it is less affected by extreme values and therefore is less sensitive to potentially undetected data errors.

8.3 Serial Correlation and Time-Series Analysis

Serial correlation affects correlations between variables measured in time, and violates the basic regression assumption of independence. Techniques that are used to recognize serial correlation therefore are discussed by applying them to the NDWI and –KBDI time-series. Cross-correlation analysis is illustrated and used to study the relationship between time-series of –KBDI and NDWI. Fundamental time-series analysis concepts (e.g., stationarity and seasonality) are introduced and a brief overview is presented of the most frequently used method for time-series analysis to account for serial correlation, namely autoregression(AR).

8.3.1 Recognizing Serial Correlation

This chapter focuses on discrete time-series, which contain observations made at discrete time intervals (e.g., 10 daily time steps of –KBDI and NDWI time-series). Time-series are defined as a set of observations, x_t, recorded at a specific time, t [26]. Time-series of –KBDI and NDWI contain a seasonal variation which is illustrated in Figure 8.2 by a smooth increase or decrease of the series related to vegetation moisture dynamics. The gradual increase or decrease of the graph of a time-series is generally indicative of the existence of a form of dependence or serial correlation among observations.

The presence of serial correlation systematically biases regression analysis when studying the relationship between two or more time-series [25]. Consider the OLS regression line with a slope and an intercept:

$$Y(t) = a_0 + a_1 X(t) + e(t) \tag{8.2}$$

where t is time, a_0 and a_1 are the respective OLS regression intercept and slope parameter, $Y(t)$ the dependent variable, $X(t)$ the independent variable, and $e(t)$ the random error term. The standard error(SE) of each parameter is required for any regression model to define the confidence interval(CI) and derive the significance of parameters in the regression equation. The parameters a_0, a_1, and the CIs, estimated by minimizing the sum of the squared "residuals" are valid only if certain assumptions related to the regression and $e(t)$ are met [25]. These assumptions are detailed in statistical textbooks [34] but are not always met or explicitly considered in real-world applications. Figure 8.3 illustrates the biased CIs of the OLS regression model at a 95% confidence level. The SE term of the regression model is underestimated due to serially correlated residuals and explains the biased confidence interval, where CI = mean $\pm$ 1.96 $\times$ SE.

The Gauss–Markov theorem states that the OLS parameter estimate is the best linear unbiased estimate (BLUE); that is, all other linear unbiased estimates will have a larger variance, if the error term, $e(t)$, is stationary and exhibits no serial correlation. The Gauss–Markov theorem consequently points to the error term and not to the time-series themselves as the critical consideration [35]. The error term is defined as stationary when

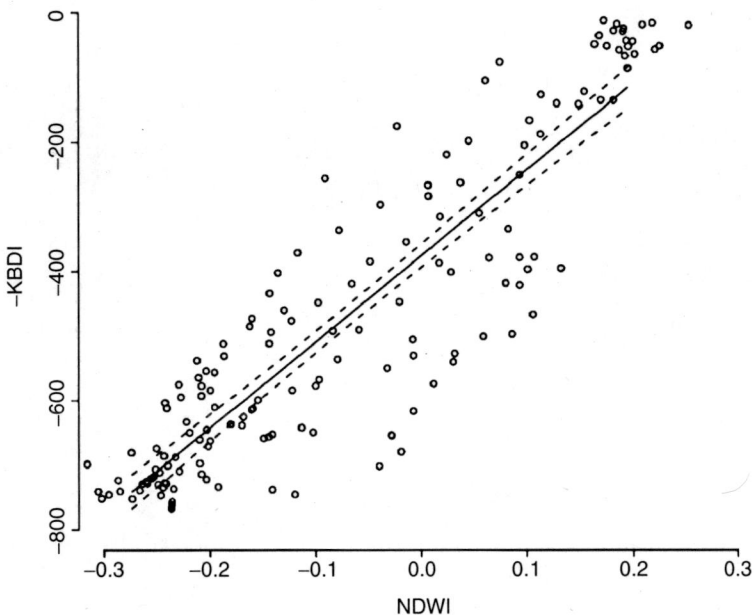

FIGURE 8.3
Result of the OLS regression fit between −KBDI and NDWI as dependent and independent variables, respectively, for the Satara weather station ($n = 157$). Confidence intervals (- - -) at a 95% confidence level are shown, but are "narrowed" due to serial correlation in the residuals.

it does not present a trend and the variance remains constant over time [27]. It is possible that the residuals are serially correlated if one of the dependent or independent variables also is serially correlated, because the residuals constitute a linear combination of both types of variables. Both dependent and independent variables of the regression model are serially correlated (KBDI and NDWI), which explains the serial correlation observed in the residuals.

A sound practice used to verify serial correlation in time-series is to perform multiple checks by both graphical and diagnostic techniques. The autocorrelation function (ACF) can be viewed as a graphical measure of serial correlation between variables or residuals. The sample ACF is defined when $x_1, \ldots, x_n$ are observations of a time-series. The sample mean of $x_1, \ldots, x_n$ is [26]:

$$\bar{x} = \frac{1}{n} \sum_{t=1}^{n} x_t \tag{8.3}$$

The sample autocovariance function with lag h and time t is

$$\hat{\gamma}(h) = n^{-1} \sum_{t=1}^{n-|h|} (x_{t+|h|} - \bar{x})(x_t - \bar{x}), -n < h < n \tag{8.4}$$

The sample ACF is

$$\hat{\rho} = \frac{\hat{\gamma}(h)}{\hat{\gamma}(0)}, -n < h < n \tag{8.5}$$

Figure 8.4 illustrates the ACF for time-series of −KBDI and NDWI presented from the Kruger park data. The ACF clearly indicates a significant autocorrelation in the

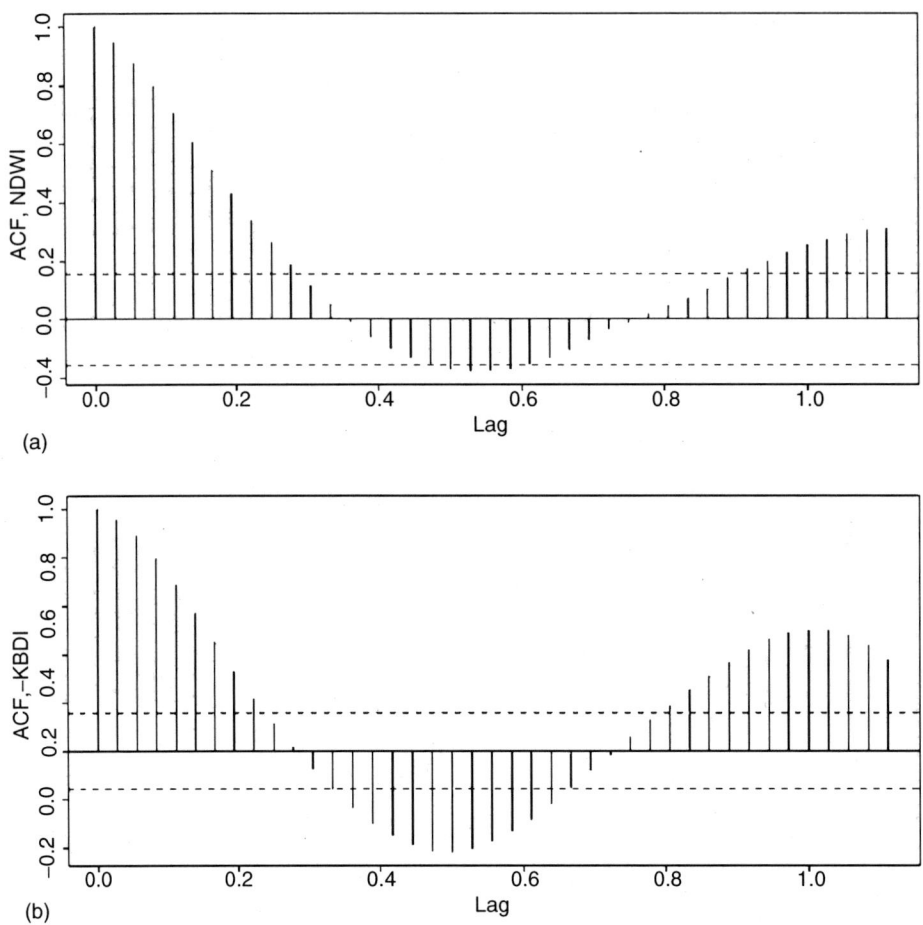

FIGURE 8.4
The autocorrelation function (ACF) for (a) −KBDI and (b) NDWI time-series for the Satara weather station. The horizontal lines on the graph are the bounds = $\pm 1.96/\sqrt{n}$ ($n = 157$).

time-series, as more than 5% of the sample autocorrelations fall outside the significance bounds = $\pm 1.96/\sqrt{n}$ [26]. There are also formal tests available to detect autocorrelation such as the Ljung–Box test statistic and the Durbin–Watson statistic [25,26].

8.3.2 Cross-Correlation Analysis

The cross-correlation function (CCF) can be derived between two time-series utilizing a technique similar to the ACF applied for one time-series [27]. Cross-correlation is a measure of the degree of linear relationship existing between two data sets and can be used to study the connection between time-series. The CCF, however, can only be used if the time-series is stationary [27]. For example, when all variables are increasing in value over time, cross-correlation results will be spurious and subsequently cannot be used to study the relationship between time-series.

Nonstationary time-series can be transformed to stationary time-series by implementing one of the following techniques:

- Differencing the time-series by a period d can yield a series that satisfies the assumption of stationarity (e.g., $x_t - x_{t-1}$ for $d=1$). The differenced series will

contain one point less than the original series. Although a time-series can be differenced more than once, one difference is usually sufficient.
- Lower order polynomials can be fitted to the series when the data contains a trend or seasonality that needs to be subtracted from the original series. Seasonal time-series can be represented as the sum of a specified trend, and seasonal and random terms. For example, for statistical interpretation results, it is important to recognize the presence of seasonal components and remove them to avoid confusion with long-term trends. Figure 8.5 illustrates the seasonal trend decompositioning method using locally weighted regression for the NDWI time-series [36].
- The logarithm or square root of the series may stabilize the variance in the case of a nonconstant variance.

Figure 8.6 illustrates the cross-correlation plot for stationary series of –KBDI and NDWI. –KBDI and NDWI time-series became stationary after differencing with $d = 1$. The stationarity was confirmed using the "augmented Dickey–Fuller" test for stationarity [26,29] at a confidence level of 95% ($p < 0.01$; with stationarity as the alternative hypothesis). Note that approximately 95% confidence limits are shown for the autocorrelation plots of an independent series. These limits must be regarded with caution, since there exists an *a priori* expectation of serial correlation for time-series [37].

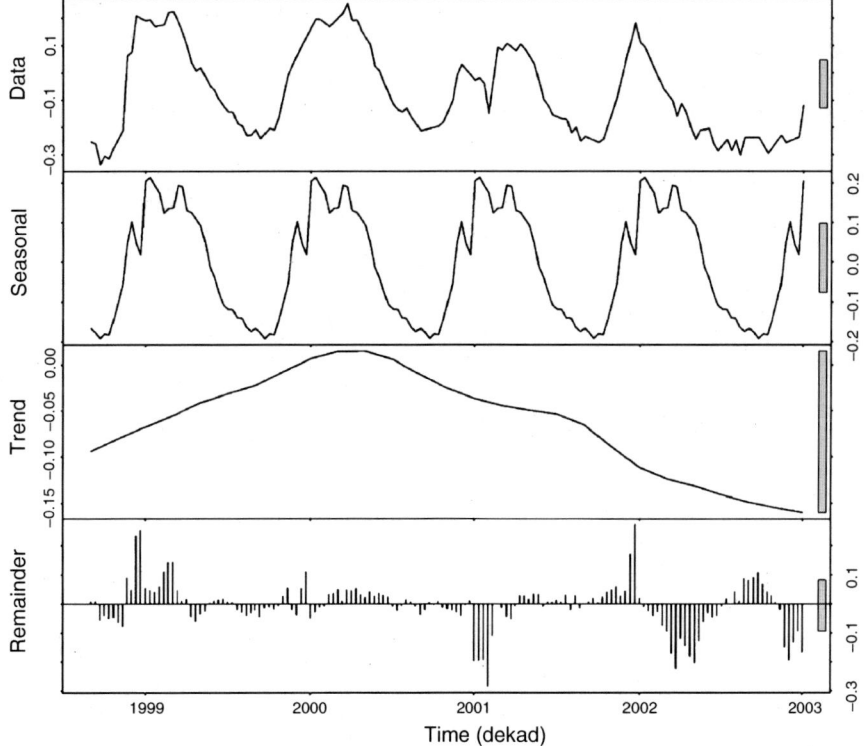

FIGURE 8.5
The results of the seasonal trend decomposition (STL) technique for NDWI time-series of Satara weather station. The original series can be reconstructed by summing the seasonal, trend, and remainder. In the *y*-axes the NDWI values are indicated. The gray bars at the right-hand side of the plots illustrate the relative data range of the time-series.

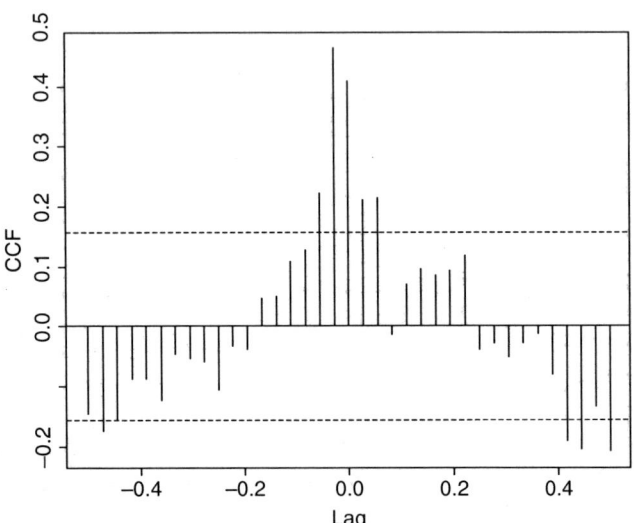

FIGURE 8.6
The cross-correlation plot between stationary −KBDI and NDWI time-series of the Satara weather station, where CCF indicates results of the cross-correlation function. The horizontal lines on the graph are the bounds ($= \pm 1.96/\sqrt{n}$) of the approximate 95% confidence interval.

Table 8.1 illustrates the coefficients of determination (i.e., multiple R^2) of the OLS regression analysis with serially correlated residuals −KBDI as dependent and NDWI as independent variable for all six weather stations in the study area. The Durbin–Watson statistic indicated that the residuals were serially correlated at a 95% confidence level ($p < 0.01$). These results will be compared with the method presented in Section 8.5. Table 8.1 also indicates the time lags at which correlation between time-series was maximal, as derived from the cross-correlation plot. A negative lag indicates that −KBDI reacts prior to NDWI, for example, in the cases of Punda Maria and Shingwedzi weather stations, and subsequently can be used to predict NDWI. This is logical since weather conditions, for example, rainfall and temperature, change before vegetation reacts. NDWI, which is related to the amount of water in the vegetation, consequently lags behind the −KBDI. The major vegetation type in savanna vegetation is the herbaceous layer, which has a shallow rooting system. This explains why the vegetation in the study area quickly follows climatic changes and NDWI did not lag behind −KBDI for the other four weather stations.

8.3.3 Time-Series Analysis: Relating Time-Series and Autoregression

A remedy for serial correlation, apart from applying variations in sampling strategy, is modeling of the time dependence in the error structure by AR. AR most often is used for

TABLE 8.1

Coefficients of Determination of the OLS Regression Model between −KBDI and NDWI ($n = 157$).

Station	R^2	Time Lag
Punda Maria	0.74	−1
Letaba	0.88	0
Onder Sabie	0.72	0
Pretoriuskop	0.31	0
Shingwedzi	0.72	−1
Satara	0.81	0

Note: The time expressed in dekads of maximum correlation of the cross-correlation between −KBDI and NDWI is also indicated.

purposes of forecasting and modeling of a time-series [25]. The simplest AR model for Equation 8.2, where ρ is the result of the sample ACF at lag 1, is

$$e_t = \rho e_{t-1} + \varepsilon_t \qquad (8.6)$$

where ε_t is a series of serially independent numbers with mean zero and constant variance. The Gauss–Markov theorem cannot be applied and therefore OLS is not an efficient estimator of the model parameters if ρ is not zero [35].

Many different AR models are available in statistical software systems that incorporate time-series modules. One of the most frequently used models to account for serial correlation is the autoregressive-integrated-moving-average model (ARIMA) [26,37]. Briefly stated, ARIMA models can have an AR term of order p, a differencing (integrating) term (I) of order d, and a moving average (MA) term of order q. The notation for specific models takes the form of (p,d,q) [27]. The order of each term in the model is determined by examining the raw data and plots of the ACF of the data. For example, a second-order AR ($p = 2$) term in the model would be appropriate if a series has significant autocorrelation coefficients between x_t, and x_{t-1}, and x_{t-2}. ARIMA models that are fitted to time-series data using AR and MA parameters, p and q, have coefficients Φ and θ to describe the serial correlation. An underlying assumption of ARIMA models is that the series being modeled is stationary [26–27].

8.4 Methodology

The TIMESAT program is used to extract nonserially correlated metrics from remote sensing and meteorological time-series [38,39]. These metrics are utilized to study the relationship between time-series at specific moments in time. The relationship between time-series, in turn, is evaluated using statistical analysis of extracted nonserially correlated seasonal metrics from time-series (–KBDI and NDWI).

8.4.1 Data Smoothing

It often is necessary to generate smooth time-series from noisy satellite sensors or meteorological data to extract information on seasonality. The smoothing can be achieved by applying filters or by function fitting. Methods based on Fourier series [40–42] or least-square fits to sinusoidal functions [43–45] are known to work well in most instances. These methods, however, are not capable of capturing a sudden, steep rise or decrease of remote sensing or meteorological data values that often occur in arid and semiarid environments. Alternative smoothing and fitting methods have been developed to overcome these problems [38]. An adaptive Savitzky–Golay filtering method, implemented in the TIMESAT processing package developed by Jönsson and Eklundh [39], is used in this chapter. The filter is based on local polynomial fits. Suppose we have a time-series (t_i, y_i), $i = 1, 2, \ldots, N$. For each point i, a quadratic polynomial

$$f(t) = c_1 + c_2 t + c_3 t^2 \qquad (8.7)$$

is fit to all 2k+1 points for a window from $n = i - k$ to $m = i + k$ by solving the system of normal equations

$$\mathbf{A}^T \mathbf{A} c = \mathbf{A}^T b \tag{8.8}$$

where

$$\mathbf{A} = \begin{pmatrix} w_n & w_n t_n & w_n t_n^2 \\ w_{n+1} & w_{n+1} t_{n+1} & w_{n+1} t_{n+1}^2 \\ & \vdots & \\ w_m & w_m t_m & w_m t_m^2 \end{pmatrix} \text{ and } b = \begin{pmatrix} w_n y_n \\ w_{n+1} y_{n+1} \\ \vdots \\ w_m y_m \end{pmatrix} \tag{8.9}$$

The filtered value is set to the value of the polynomial at point i. Weights are designated as w in the above expression, with weights assigned to all of the data values in the window. Data values that were flagged in the preprocessing are assigned weight "zero" in this application and thus do not influence the result. The clean data values all have weights "one." Residual negatively biased noise (e.g., clouds) may occur for the remote sensing data and accordingly the fitting was performed in two steps [6]. The first fit was conducted using weights obtained from the preprocessing. Data points above the resulting smoothed function from the first fit are regarded more important, and in the second step the normal equations are solved using the weight of these data values, but increased by a factor 2. This multistep procedure leads to a smoothed function that is adapted to the upper envelope of the data (Figure 8.7). Similarly, the ancillary metadata of the meteorological data from the preprocessing also were used in the iterative fitting to the upper envelope of the –KBDI time-series [6].

The width of the fitting window determines the degree of smoothing, but it also affects the ability to follow a rapid change. It is sometimes necessary to locally tighten the window even when the global setting of the window performs well. A typical situation occurs in savanna ecosystems where vegetation, associated remote sensing, and meteorological indices respond rapidly to vegetation moisture dynamics. A small fitting window can be used to capture the corresponding sudden rise in data values.

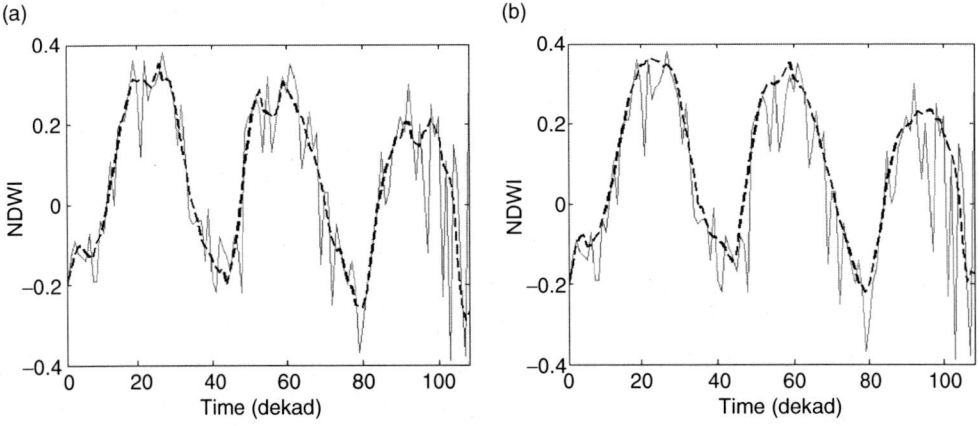

FIGURE 8.7
The Savitzky–Golay filtering of NDWI (——) is performed in two steps. Firstly, the local polynomials are fitted using the weights from the preprocessing (a). Data points above the resulting smoothed function (– – –) from the first fit are attributed a greater importance. Secondly, the normal equations are solved with the weights of these data values increased by a factor 2 (b).

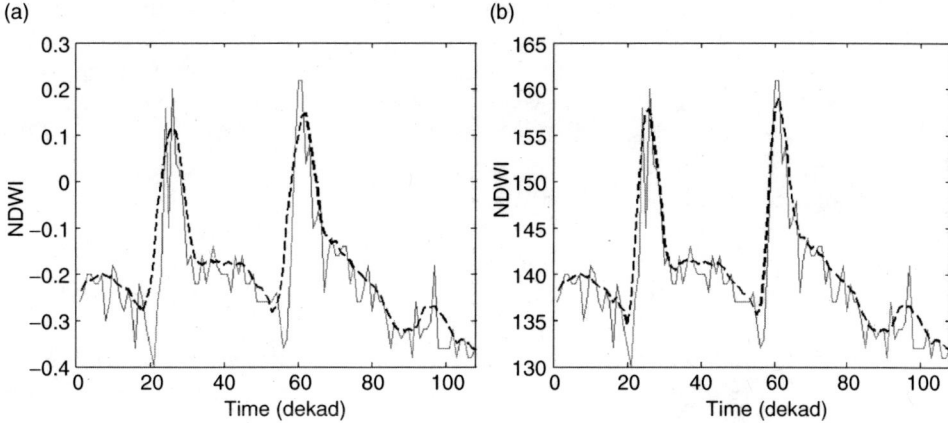

FIGURE 8.8
The filtering of NDWI (——) in (a) is done with a window that is too large to allow the filtered data (- - -) to follow sudden increases and decreases of underlying data values. The data in the window are scanned and if there is a large increase or decrease, an automatic decrease in the window size will result. The filtering is then repeated using the new locally adapted size (b). Note the improved fit at rising edges and narrow peaks.

The data in the window are scanned and if a large increase or decrease is observed, the adaptive Savitzky–Golay method applied an automatic decrease in the window size. The filtering is then repeated using the new locally adapted size. Savitzky–Golay filtering with and without the adaptive procedure is illustrated in Figure 8.8. In the figure it is shown that the adaptation of the window improves the fit at the rising edges and at narrow seasonal peaks.

8.4.2 Extracting Seasonal Metrics from Time-Series and Statistical Analysis

Four seasonal metrics were extracted for each of the rainy seasons. Figure 8.9 illustrates the different metrics per season for NDWI and KBDI time-series. The beginning of a season, that is, 20% left of the rainy season, is defined from the final function fit as the point in time for which the index value has increased by 20% of the distance between the left minimum level and the maximum. The end of the season is defined in a similar way

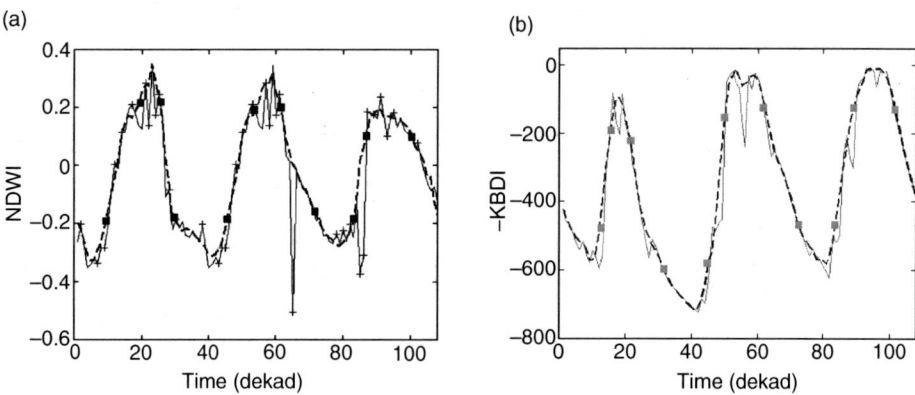

FIGURE 8.9
The final fit of the Savitzky–Golay function (- - -) to the NDWI (a) and –KBDI (b) series (——), with the four defined metrics, that is, 20% left and right, and 80% left and right (■), overlaid on the graph. Points with flagged data errors (+) were assigned weights of zero and did not influence the fit. A dekad is defined as a 10-day period.

as the point 20% right of the rain season. The 80% left and right points are defined as the points for which the function fit has increased to 80% of the distance between, respectively, the left and right minimum levels and the maximum. The current technique used to define metrics also is used by Verbesselt et al. [6] to define the beginning of the fire season.

The temporal occurrence and the value of each metric were extracted for further exploratory statistical analysis to study the relationship between time-series. The SPOT VGT S10 time-series consisted of four seasons (1998–2002) from which four occurrences and values per metric type were extracted. Twenty-four occurrence–value combinations per metric type ultimately were available for further analysis since six weather stations were used.

Serial correlation that occurs in remote sensing and climate-based time-series invalidates inferences made by standard parametric tests, such as the Student's t-test or the Pearson correlation. All extracted occurrence–value combinations per metric type were tested for autocorrelation using the Ljung–Box autocorrelation test [26]. Robust non-parametric techniques, such as the Wilcoxon's signed rank test were used in case of non-normally distributed data. The normality of the data was verified using the Shapiro–Wilkinson normality test [29].

Firstly, the distribution of the temporal occurrence of each metric was visualized and evaluated based on whether or not there was a significant difference between the temporal occurrence of the four metric types extracted from –KBDI and NDWI time-series. Next, the strength and significance of the relationship between –KBDI and NDWI values of the four metric types were assessed with an OLS regression analysis.

8.5 Results and Discussion

Figure 8.9 illustrates the optimized function fit and the defined metrics for the –KBDI and NDWI. Notice that the Savitzky–Golay function could properly define the behavior of the different time-series. The function was fitted to the upper envelope of the data by using the uncertainty information derived during the preprocessing step. The results of the statistical analysis based on the extracted metrics for –KBDI and NDWI are presented. The Ljung–Box statistic indicated that the extracted occurrences and values were not significantly autocorrelated at a 95% confidence level. All p-values were greater than 0.1, failing to reject the null hypothesis of independence.

8.5.1 Temporal Analysis of the Seasonal Metrics

Figure 8.10 illustrates the temporal distribution of temporal occurrence of extracted metrics from time-series of –KBDI and NDWI. The occurrences of extracted metrics were significantly non-normally distributed at a 95% confidence level ($p > 0.1$), indicating that the Wilcoxon's signed rank can be used. The Wilcoxon's signed rank test showed that –KBDI and NDWI occurrences of the 80% left and right, and 20% right were not significantly different from each other at a 95% confidence level ($p > 0.1$). This confirmed that –KBDI and NDWI were temporally related. It also corroborated the results of Burgan [11] and Ceccato et al. [4] who found that both –KBDI and NDWI were related to the seasonal vegetation moisture dynamics, as measured by VWC.

Figure 8.10, however, illustrates that the start of the rainy season (i.e., 20% left occurrence), derived from the –KBDI and NDWI time-series, was different. The Wilcoxon's

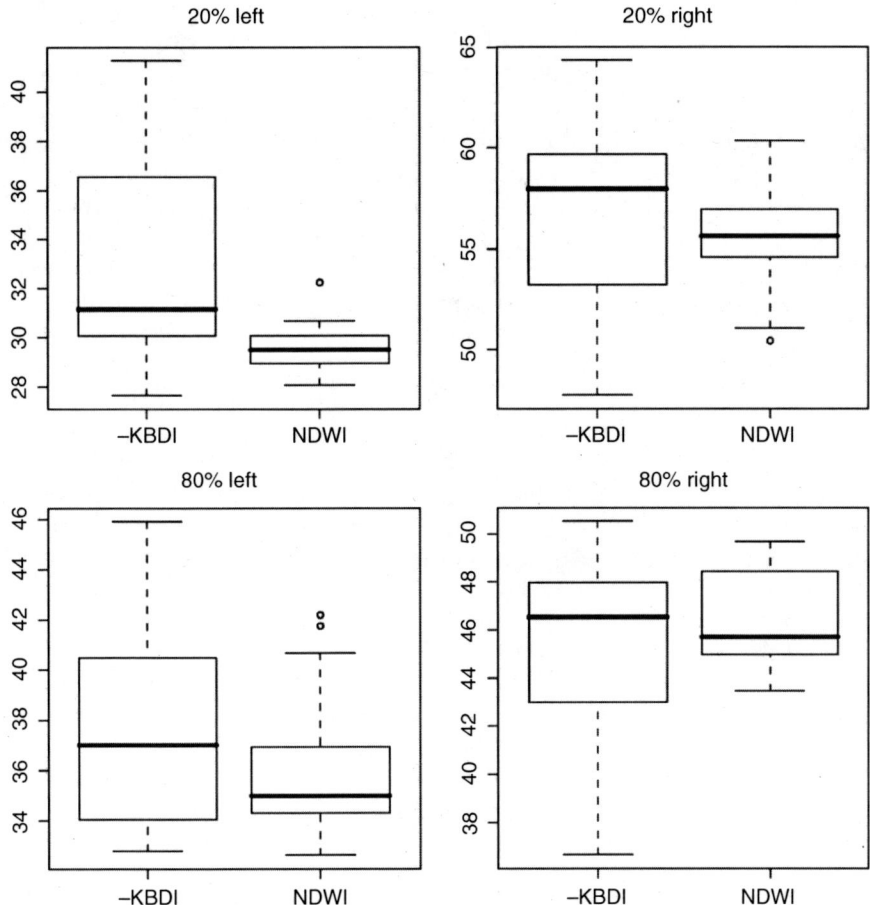

FIGURE 8.10
Box plots of the temporal occurrence of the four defined metrics, that is, 20% left and right, and 80% left and right, extracted from time-series of −KBDI and NDWI. The dekads (10-day period) are shown on the y-axis and are indicative of the temporal occurrence of the metric. The upper and lower boundaries of the boxes indicate upper and lower quartiles. The median is indicated by the solid line (—) within each box. The whiskers connect the extremes of the data, which were defined as 1.5 times the inter-quartile range. Outliers are represented by (o).

signed rank test confirmed that the −KBDI and NDWI differed significantly from each other at a 95% confidence level ($p < 0.01$). This phenomenon can be explained by the fact that vegetation in the study area starts growing before the rainy season starts, due to an early change in air temperature (N. Govender, Scientific Service Kruger National Park, South Africa, personal communication). This explained why the NDWI reacted before the change in climatic conditions as measured by the −KBDI, given that the NDWI is directly related to vegetation moisture dynamics [4].

8.5.2 Regression Analysis Based on Values of Extracted Seasonal Metrics

The assumptions of the OLS regression models between values of metrics extracted from −KBDI and NDWI time-series were verified. The Wald test statistic showed nonlinearity to be not significant at a 95% confidence level ($p > 0.15$). The Shapiro–Wilkinson normality test confirmed that the residuals were normally distributed at a

95% confidence level ($p < 0.01$) [6,29]. Table 8.2 illustrates the results of the OLS regression analysis between values of metrics extracted from –KBDI and NDWI time-series. The values extracted at the "20% right" position of the –KBDI and NDWI time-series showed a significant relationship at a 95% confidence level ($p < 0.01$). The other extracted metrics did not exhibit significant relationships at a 95% confidence level ($p > 0.1$). A significant relationship between the –KBDI and NDWI time-series was observed only at the moment when savanna vegetation was completely cured (i.e., 20% right-hand side). The savanna vegetation therefore reacted differently to changes in climate parameters such as rainfall and temperature, as measured by KBDI, depending on the phenological growing cycle. This phenomenon could be explained because a living plant uses defense mechanisms to protect itself from drying out, while a cured plant responds to climatic conditions [46]. These results consequently indicated that the relationship between extracted values of –KBDI and NDWI was influenced by seasonality. This is in corroboration with the results of Ji and Peters [23], who indicated that seasonality had a significant effect on the relationship between vegetation as measured by a remote sensing index and drought index. These results further illustrated that the seasonal effect needs to be taken into account when regression techniques are used to quantify the relationship between time-series related to vegetation moisture dynamics. The seasonal effect also can be accounted for by utilizing autoregression models with seasonal dummy variables, which take the effect of serial correlation and seasonality into account [23,26]. However, the proposed method to account for serial correlation by sampling at specific moments in time had an additional advantage; the influence of seasonality could be studied by extracting metrics at the specified moments, besides the fact that serial correlation was taken into account.

Furthermore, it was shown that serial correlation caused an overestimation of the correlation coefficient is when results from Table 8.1 and Table 8.2 were compared. All the coefficients of determination (R^2) of Table 8.1 were significant with an average value of 0.7, while in Table 8.2 only the correlation coefficient at the end of the rainy season (20% right-hand side) was significant ($R^2 = 0.49$). This confirmed the importance of accounting for serial correlation and seasonality in the residuals of a regression model, when studying the relationship between two time-series.

8.5.3 Time-Series Analysis Techniques

Time-series analysis models most often are used for purposes of describing current conditions and forecasting [25]. The models use the serial correlation in time-series as a

TABLE 8.2

Coefficients of Determination of the OLS Regression Models (NDWI $\sim$ –KBDI) for the Four Extracted Seasonal Metric Values between –KBDI and NDWI Time-Series ($n = 24$ per Metric)

NDWI $\sim$ –KBDI	R^2	p-Values
20% left	0.01	0.66
20% right	0.49	<0.01
80% left	0.00	0.97
80% right	0.01	0.61

tool to relate temporal observations. Future observations can be predicted by modeling the serial correlation structure of a time-series [26]. Time-series analysis techniques (e.g., ARIMA) can be used to model a time-series by using other, independent time-series [27]. ARIMA subsequently can be used to study the relationship between –KBDI and NDWI. However, there are constraints that have to be considered before ARIMA models can be applied to a time-series (e.g., drought index) by using other predictor time-series (e.g., satellite index).

Firstly, the goodness-of-fit of an ARIMA model will not be significant when changes in the satellite index precede or coincide with those in the drought index. The CCF can be used in this context to verify how time-series are related to each other. Time lag results in Table 8.1 indicate that the –KBDI (drought index) precedes or coincides with the NDWI time-series. This illustrates that ARIMA models cannot directly be used to predict the KBDI, with NDWI as the predictor variable. Consequently, other more advanced time-series analysis techniques are needed to model vegetation dynamics because they will precede or coincide with the dynamics monitored by remote sensing indices in most of the cases. Such more advanced time-series analysis techniques, however, are not discussed since they are outside the scope of this chapter.

Secondly, availability of data is limited for time-series analysis, namely, from 1998 to 2002. This is an important constraint because two separate data sets are needed to parameterize and evaluate an ARIMA model. One set is needed for parameterization, while the other is used to forecast and validate the ARIMA model through comparison of the observed and expected values. Accordingly, it is necessary to interpolate missing satellite data that were masked out during preprocessing to ensure adequate data are available for parameterization.

Thirdly, the proposed sampling strategy made investigation of the time lag and correlation at a defined instant in time possible, as opposed to ARIMA or cross-correlation analysis, through which only the overall relationship between time-series can be studied [25]. The applied sampling strategy is thus ideally suited to study the relationship between time-series of climate and remote sensing data, characterized by seasonality and serial correlation. The sampling of seasonal metrics minimized the influence of serial correlation, thereby making the study of seasonality possible.

8.6 Conclusions

Serial correlation problems are not unknown in the field of statistical or general meteorology. However, the presence of serial correlation, found during analysis of a variable sampled sequentially at regular time intervals, seems to be disregarded by many agricultural meteorologists and remote sensing scientists. This is true despite abundant documentation available in the traditional meteorological and statistical literature. Therefore, an overview of the most important time-series analysis techniques and concepts was presented, namely, stationarity, autocorrelation, differencing, decomposition, autoregression, and ARIMA.

A method was proposed to study the relationship between a meteorological drought index (KBDI) and remote sensing index (NDWI), both related to vegetation moisture dynamics, by accounting for the serial correlation effect. The relationship between –KBDI and NDWI was studied by extracting nonserially correlated seasonal metrics, for example, 20% and 80% left- and right-hand side metrics of the rainy season, based on a Savitzky–Golay fit to the upper envelope of the time-series. Serial correlation between the extracted metrics was shown to be minimal and seasonality was an important factor influencing the relationship between NDWI and −KBDI time-series. Statistical analysis using the temporal occurrence of the extracted metrics revealed that NDWI and −KBDI time-series are temporally connected, except at the beginning of the rainy season. The fact that the savanna vegetation starts re-greening before the start of the rainy season explains this inability to detect the beginning of the rainy season. The values of the extracted seasonal metrics of NDWI and −KBDI were significantly related only at the end of the rainy season, namely, at the 20% right-hand side value of the fitted curve. The savanna vegetation at the end of the rainy season was cured and responded strongly to changes in climatic conditions monitored by the −KBDI, such as rain and temperature. The relationship between −KBDI and NDWI consequently changes during the season, which indicates that seasonality is an important factor that needs to be taken into account. Moreover, it was shown that correlation coefficients estimated by OLS regression analysis were overestimated due to the influence of serial correlation in the residuals. This confirmed the importance of taking serial correlation of the residuals into account by sampling nonserially correlated seasonal metrics when studying the relationship between time-series.

The serial correlation effect consequently was taken into account by the extraction of seasonal metrics from time-series. The seasonal metrics in turn could be used to study the relationship between remote sensing and ground-based time-series, such as meteorological or field measurements. A better understanding of the relationship between remote sensing and *in situ* observations at regular time intervals will contribute to the use of remotely sensed data for the development of an index that represents seasonal vegetation moisture dynamics.

Acknowledgments

The SPOT VGT S10 data sets were generated by the Flemish Institute for Technological Development. The climate data were provided by the Weather Services of South Africa, whereas the National Land Cover Map (1995) was supplied by the Agricultural Research Centre of South Africa. We acknowledge the support of L. Eklundh, as well as the Crafoord and Bergvall foundations. The authors wish to thank N. Govender from the Scientific Services of Kruger National Park for scientific input.

References

1. Camia, A. et al., Meteorological fire danger indices and remote sensing, in *Remote Sensing of Large Wildfires in the European Mediterranean Basin*, E. Chuvieco, Ed., Springer-Verlag, New York, 1999, p. 39.
2. Ceccato, P. et al., Estimation of live fuel moisture content, in *Wildland Fire Danger Estimation and Mapping: The Role of Remote Sensing Data*, E. Chuvieco, Ed., World Scientific Publishing, New York, 2003, p. 63.
3. Dennison, P.E. et al., Modeling seasonal changes in live fuel moisture and equivalent water thickness using a cumulative water balance index, *Remote Sens. Environ.*, 88, 442, 2003.

4. Ceccato, P. et al., Detecting vegetation leaf water content using reflectance in the optical domain, *Remote Sens. Environ.*, 77, 22, 2001.
5. Jackson, T.J. et al., Vegetation water content mapping using landsat data derived normalized difference water index for corn and soybeans, *Remote Sens. Environ.*, 92, 475, 2004.
6. Verbesselt, J. et al., Evaluating satellite and climate data derived indices as fire risk indicators in savanna ecosystems, *IEEE Trans. Geosci. Remote Sens.*, 44; 1622–1632; 2006.
7. Van Wilgen, B.W. et al., Response of Savanna fire regimes to changing fire-management policies in a large African national park, *Conserv. Biol.*, 18, 1533, 2004.
8. Dimitrakopoulos, A.P. and Bemmerzouk, A.M., Predicting live herbaceous moisture content from a seasonal drought index, *Int. J. of Biometeorol.*, 47, 73, 2003.
9. Keetch, J.J. and Byram, G.M., A drought index for forest fire control, *U.S.D.A. Forest Service, Asheville NC*, SE-38, 1988.
10. Janis, M.J., Johnson, M.B., and Forthun, G., Near-real time mapping of Keetch–Byram drought index in the south-eastern United States, *Int. J. Wildland Fire*, 11, 281, 2002.
11. Burgan, E.R., Correlation of plant moisture in Hawaii with the Keetch–Byram drought index, *USDA. Forest Service Research Note*, PSW-307, 1976.
12. Chuvieco, E. et al., Combining NDVI and surface temperature for the estimation of live fuel moisture content in forest fire danger rating, *Remote Sens. Environ.*, 92, 322, 2004.
13. Maki, M., Ishiahra, M., and Tamura, M., Estimation of leaf water status to monitor the risk of forest fires by using remotely sensed data, *Remote Sens. Environ.*, 90, 441, 2004.
14. Paltridge, G.W. and Barber, J., Monitoring grassland dryness and fire potential in Australia with NOAA AVHRR data, *Remote Sens. Environ.*, 25, 381, 1988.
15. Chladil, M.A. and Nunez, M., Assessing grassland moisture and biomass in Tasmania—the application of remote-sensing and empirical-models for a cloudy environment, *Int. J. Wildland Fire*, 5, 165, 1995.
16. Hardy, C.C. and Burgan, R.E., Evaluation of NDVI for monitoring live moisture in three vegetation types of the western US, *Photogramm. Eng. Remote Sens.*, 65, 603, 1999.
17. Chuvieco, E. et al., Estimation of fuel moisture content from multitemporal analysis of LANDSAT thematic mapper reflectance data: applications in fire danger assessment, *Int. J. Remote Sens.*, 23, 2145, 2002.
18. Fensholt, R. and Sandholt, I., Derivation of a shortwave infrared water stress index from MODIS near- and shortwave infrared data in a semiarid environment, *Remote Sens. Environ.*, 87, 111, 2003.
19. Hunt, E.R., Rock, B.N., and Nobel, P.S., Measurement of leaf relative water content by infrared reflectance, *Remote Sens. Environ.*, 22, 429, 1987.
20. Ceccato, P., Flasse, S., and Gregoire, J.M., Designing a spectral index to estimate vegetation water content from remote sensing data—Part 2: Validation and applications, *Remote Sens. Environ.*, 82, 198, 2002.
21. Myneni, R.B. et al., Increased plant growth in the northern high latitudes from 1981 to 1991, *Nature*, 386, 698, 1997.
22. Eklundh, L., Estimating relations between AVHRR NDVI and rainfall in East Africa at 10-day and monthly time scales, *Int. J. Remote Sens.*, 19, 563, 1998.
23. Ji, L. and Peters, A.J., Assessing vegetation response to drought in the northern great plains using vegetation and drought indices, *Remote Sens. Environ.*, 87, 85, 2003.
24. De Beurs, K.M. and Henebry, G.M., Land surface phenology, climatic variation, and institutional change: analyzing agricultural land cover change in Kazakhstan, *Remote Sens. Environ.*, 89, 497, 2004.
25. Meek, D.W. et al., A note on recognizing autocorrelation and using autoregression, *Agricult. Forest Meterol.*, 96, 9, 1999.
26. Brockwell, P.J. and Davis, R.A., *Introduction to Time Series and Forecasting*, 2nd ed., Springer-Verlag, New York, 2002, p. 31.
27. Ford, C.R. et al., Modeling canopy transpiration using time series analysis: a case study illustrating the effect of soil moisture deficit on Pinus Taeda, *Agricult. Forest Meterol.*, 130, 163, 2005.
28. Montanari, A., Deseasonalisation of hydrological time series through the normal quantile transform, *J. Hydrol.*, 313, 274, 2005.

29. R Development Core Team, R: a language and environment for statistical computing, R Foundation for Statistical Computing, Vienna, Austria, ISBN 3-900051-07-0, URL http://www.R-project.org, 2005.
30. Rahman, H. and Dedieu, G., SMAC—a simplified method for the atmospheric correction of satellite measurements in the solar spectrum, *Int. J. Remote Sens.*, 15, 123, 1994.
31. Stroppiana, D. et al., An algorithm for mapping burnt areas in Australia using SPOT-VEGETATION data, *IEEE Trans. Geosci. Remote Sens.*, 41, 907, 2003.
32. Thompson, M.A., Standard land-cover classification scheme for remote-sensing applications in South Africa, *South Afr. J. Sci.*, 92, 34, 1996.
33. Aguado, I. et al., Assessment of forest fire danger conditions in southern Spain from NOAA images and meteorological indices, *Int. J. Remote Sens.*, 24, 1653, 2003.
34. von Storch, H. and Zwiers, F.W., *Statistical Analysis in Climate Research*, Cambridge University Press, Cambridge, 1999, p. 483.
35. Thejll, P. and Schmith, T., Limitations on regression analysis due to serially correlated residuals: application to climate reconstruction from proxies, *J. Geophys. Res. Atmos.*, 110, 2005.
36. Cleveland, R.B. et al., STL: a seasonal-trend decomposition procedure based on Loess., *J. Off. Stat.*, 6, 3, 1990.
37. Venables, W.N. and Ripley, B.D., *Modern Applied Statistics with S*, 4th ed., Springer-Verlag, New York, 2003, p. 493.
38. Jönsson, P. and Eklundh, L., Seasonality extraction by function fitting to time-series of satellite sensor data, *IEEE Trans. Geosci. Remote Sens.*, 40, 1824, 2002.
39. Jönsson, P. and Eklundh, L., TIMESAT—a program for analyzing time-series of satellite sensor data, *Comput. Geosci.*, 30, 833, 2004.
40. Menenti, M. et al., Mapping agroecological zones and time-lag in vegetation growth by means of Fourier-analysis of time-series of NDVI images, *Adv. Space Res.*, 13, 233, 1993.
41. Azzali, S. and Menenti, M., Mapping vegetation–soil–climate complexes in Southern Africa using temporal Fourier analysis of NOAA-AVHRR NDVI data, *Int. J. Remote Sens.*, 21, 973, 2000.
42. Olsson, L. and Eklundh, L., Fourier-series for analysis of temporal sequences of satellite sensor imagery, *Int. J. Remote Sens.*, 15, 3735, 1994.
43. Cihlar, J., Identification of contaminated pixels in AVHRR composite images for studies of land biosphere, *Remote Sens. Environ.*, 56, 149, 1996.
44. Sellers, P.J. et al., A global 1-degrees-by-1-degrees NDVI data set for climate studies. The generation of global fields of terrestrial biophysical parameters from the NDVI, *Int. J. Remote Sens.*, 15, 3519, 1994.
45. Roerink, G.J., Menenti, M., and Verhoef, W., Reconstructing cloudfree NDVI composites using Fourier analysis of time series, *Int. J. Remote Sens.*, 21, 1911, 2000.
46. Pyne, S.J., Andrews, P.L., and Laven, R.D., *Introduction to Wildland Fire*, 2nd ed., John Wiley & Sons, New York, 1996, 117.

9

Use of a Prediction-Error Filter in Merging High- and Low-Resolution Images

Sang-Ho Yun and Howard Zebker

CONTENTS

9.1 Image Descriptions 172
    9.1.1 TOPSAR DEM 172
    9.1.2 SRTM DEM 173
9.2 Image Registration 174
9.3 Artifact Elimination 175
9.4 Prediction-Error (PE) Filter 176
    9.4.1 Designing the Filter 176
    9.4.2 1D Example 177
    9.4.3 The Effect of the Filter 177
9.5 Interpolation 178
    9.5.1 PE Filter Constraint 178
    9.5.2 SRTM DEM Constraint 179
    9.5.3 Inversion with Two Constraints 179
    9.5.4 Optimal Weighting 179
    9.5.5 Simulation of the Interpolation 181
9.6 Interpolation Results 181
9.7 Effect on InSAR 183
9.8 Conclusion 185
References 186

A prediction-error (PE) filter is an array of numbers designed to interpolate missing parts of data such that the interpolated parts have the same spectral content as the existing parts. The data can be a one-dimensional time series, two-dimensional image, or a three-dimensional quantity such as subsurface material property. In this chapter, we discuss the application of a PE filter to recover missing parts of an image when a low-resolution image of the missing parts is available.

One of the research issues on PE filter is improving the quality of image interpolation for nonstationary images, in which the spectral content varies with position. Digital elevation models (DEMs) are in general nonstationary. Thus, PE filter alone cannot guarantee the success of image recovery. However, the quality of the image recovery of a high-resolution image can be improved with independent data set such as a low-resolution image that has valid pixels for the missing regions of the high-resolution image. Using a DEM as an example image, we introduce a systematic method to use a

PE filter incorporating the low-resolution image as an additional constraint, and show the improved quality of the image interpolation.

High-resolution DEMs are often limited in spatial coverage; they also may possess systematic artifacts when compared to comprehensive low-resolution maps. We correct artifacts and interpolate regions of missing data in topographic synthetic aperture radar (TOPSAR) DEMs using a low-resolution shuttle radar topography mission (SRTM) DEM. Then PE filters are to interpolate and fill missing data so that the interpolated regions have the same spectral content as the valid regions of the TOPSAR DEM. The SRTM DEM is used as an additional constraint in the interpolation. Using cross-validation methods one can obtain the optimal weighting for the PE filter and the SRTM DEM constraints.

9.1 Image Descriptions

InSAR is a powerful tool for generating DEMs [1]. The TOPSAR and SRTM sensors are primary sources for the academic community for DEMs derived from single-pass interferometric data. Differences in system parameters such as altitude and swath width (Table 9.1) result in very different properties for derived DEMs. Specifically, TOPSAR DEMs have better resolution, while SRTM DEMs have better accuracy over larger areas. TOPSAR coverage is often not spatially complete.

9.1.1 TOPSAR DEM

TOPSAR DEMs are produced from cross-track interferometric data acquired with NASA's AIRSAR system mounted on a DC-8 aircraft. Although the TOPSAR DEMs have a higher resolution than other existing data, they sometimes suffer from artifacts and missing data due to roll of the aircraft, layover, and flight planning limitations. The DEMs derived from the SRTM have lower resolution, but fewer artifacts and missing data than TOPSAR DEMs. Thus, the former often provides information in the missing regions of the latter.

We illustrate joint use of these data sets using DEMs acquired over the Galápagos Islands. Figure 9.1 shows the TOPSAR DEM used in this study. The DEM covers Sierra Negra volcano on the island of Isabela. Recent InSAR observations reveal that the volcano has been deforming relatively rapidly [2,3]. InSAR analysis can require use of a DEM to produce a simulated interferogram required to isolate ground deformation. The effect of artifact elimination and interpolation for deformation studies is discussed later in this chapter.

TABLE 9.1

TOPSAR Mission versus SRTM Mission

Mission	TOPSAR	SRTM
Platform	DC-8 aircraft	Space shuttle
Nominal	Altitude 9 km	233 km
Swath width	10 km	225 km
Baseline	2.583 m	60 m
DEM resolution	10 m	90 m
DEM coord. system	None	Lat/Long

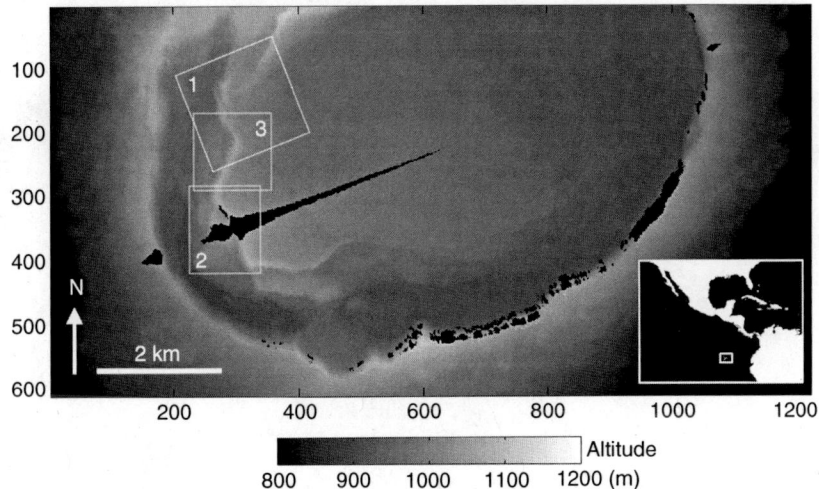

FIGURE 9.1
The original TOPSAR DEM of Sierra Negra volcano in Galápagos Islands (inset for location). The pixel spacing of the image is 10 m. The boxed areas are used for illustration later in this paper. Note that there are a number of regions of missing data with various shapes and sizes. Artifacts are not identifiable due to the variation in topography. (From Yun, S.-H., Ji, J., Zebker, H., and Segall, P., *IEEE Trans. Geosci. Rem. Sens.*, 43(7), 1682, 2005. With permission.)

The TOPSAR DEMs have a pixel spacing of about 10 m, sufficient for most geodetic applications. However, regions of missing data are often encountered (Figure 9.1), and significant residual artifacts are found (Figure 9.2). The regions of missing data are caused by layover of the steep volcanoes and flight planning limitations. Artifacts are large-scale and systematic and most likely due to uncompensated roll of the DC-8 aircraft [4]. Attempts to compensate this motion include models of piecewise linear imaging geometry [5] and estimating imaging parameters that minimize the difference between the TOPSAR DEM and an independent reference DEM [6]. We use a nonparameterized direct approach by subtracting the difference between the TOPSAR and SRTM DEMs.

9.1.2 SRTM DEM

The recent SRTM mission produced nearly worldwide topographic data at 90 m posting. SRTM topographic data are in fact produced at 30 m posting (1 arcsec); however, high-resolution data sets for areas outside of the United States are not available to the public at this time. Only DEMs at 90 m posting (3 arcsec) are available to download.

For many analyses, finer scale elevation data are required. For example, a typical pixel spacing in a spaceborne SAR image is 20 m. If the SRTM DEMs are used for topography removal in spaceborne interferometry, the pixel spacing of the final interferograms would be limited by the topography data to at best 90 m. Despite the lower resolution, the SRTM DEM is useful because it has fewer motion-induced artifacts than the TOPSAR DEM. It also has fewer data holes.

The merits and demerits of the two DEMs are in many ways complementary to each other. Thus, a proper data fusion method can overcome the shortcomings of each and produce a new DEM that combines the strengths of the two data sets: a DEM that has a

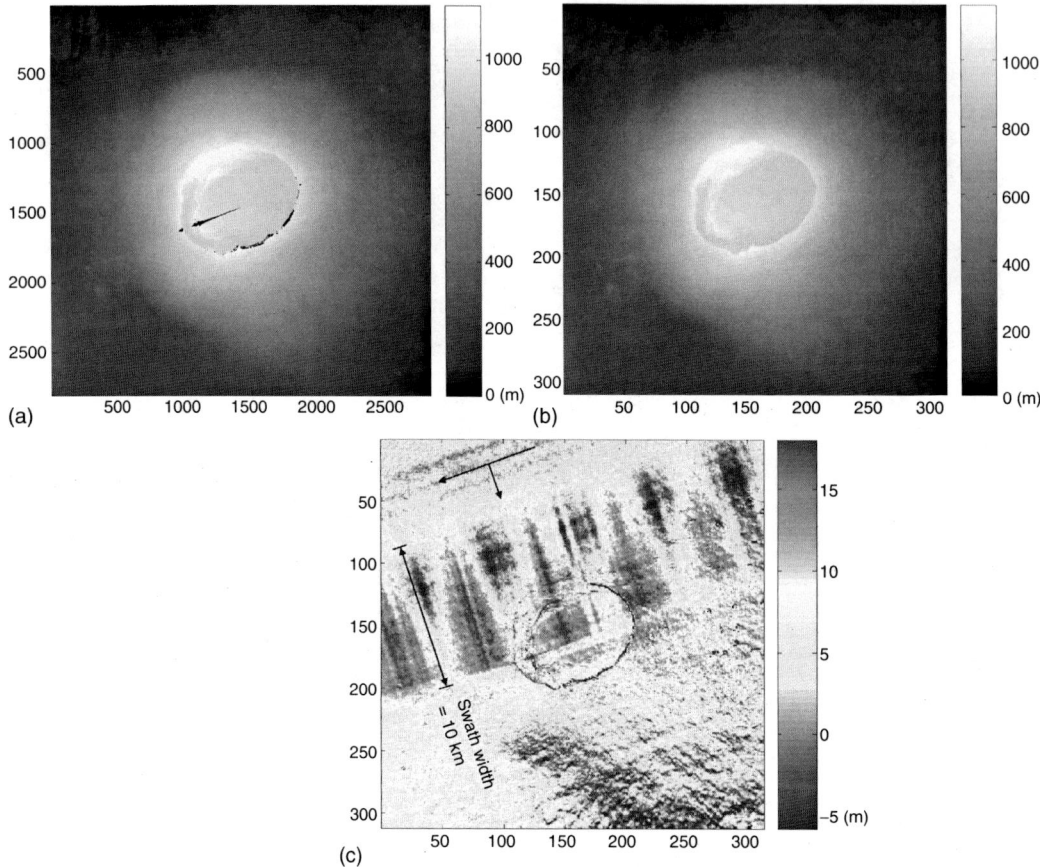

FIGURE 9.2 (See color insert following page 178.)
(a) TOPSAR DEM and (b) SRTM DEM. The tick labels are pixel numbers. Note the difference in pixel spacing between the two DEMs. (c) Artifacts obtained by subtracting the SRTM DEM from the TOPSAR DEM. The flight direction and the radar look direction of the aircraft associated with the swath with the artifact are indicated with long and short arrows, respectively. Note that the artifacts appear in one entire TOPSAR swath, while they are not as serious in other swaths.

resolution of the TOPSAR DEM and large-scale reliability of the SRTM DEM. In this chapter, we present an interpolation method that uses both TOPSAR and SRTM DEMs as constraints.

9.2 Image Registration

The original TOPSAR DEM, while in ground-range coordinates, is not georeferenced. Thus, we register the TOPSAR DEM to the SRTM DEM, which is already registered in a latitude–longitude coordinate system. The image registration is carried out between the DEM data sets using an affine transformation. Although the TOPSAR DEM is not georeferenced, it is already on the ground coordinate system. Thus, scaling and rotation are the two most important components. We have seen that skewing component was negligible. Any higher order transformation between the two DEMs would also be negligible. The affine transformation is as follows:

Use of a Prediction-Error Filter in Merging High- and Low-Resolution Images

$$\begin{bmatrix} x_S \\ y_S \end{bmatrix} = \begin{bmatrix} a & b \\ c & d \end{bmatrix} \begin{bmatrix} x_T \\ y_T \end{bmatrix} + \begin{bmatrix} e \\ f \end{bmatrix} \tag{9.1}$$

where $\begin{bmatrix} x_S \\ y_S \end{bmatrix}$ and $\begin{bmatrix} x_T \\ y_T \end{bmatrix}$ are tie points in the SRTM and TOPSAR DEM coordinate systems, respectively. Since [a b e] and [c d f] are estimated separately, at least three tie points are required to uniquely determine them. We picked 10 tie points from each DEM based on topographic features and solved for the six unknowns in a least-square sense.

Given the six unknowns, we choose new georeferenced sample locations that are uniformly spaced; every ninth sample location corresponds to the sample location of SRTM DEM. Those sample locations from $\begin{bmatrix} x_S \\ y_S \end{bmatrix}$ and $\begin{bmatrix} x_T \\ y_T \end{bmatrix}$ are calculated. Then, the nearest TOPSAR DEM value is selected and put into the corresponding new georeferenced sample location. The intermediate values are filled in from the TOPSAR map to produce the georeferenced 10-m data set.

It should be noted that it is not easy to determine the tie points in DEM data sets. Enhancing the contrast of the DEMs facilitated the process. In general, fine registration is important for correctly merging different data sets. The two DEMs in this study have different pixel spacings. It is difficult to pick tie points with higher precision than the pixel spacing of the coarser image. In our method, however, the SRTM DEM, the coarser image, is treated as an averaged image of the TOPSAR DEM, the finer image. In our inversion, only the 9-by-9 averaged values of the TOPSAR DEM are compared with the pixel values of the SRTM DEM. Thus, the fine registration is less critical in this approach than in the case where a one-to-one match is required.

9.3 Artifact Elimination

Examination of the georeferenced TOPSAR DEM (Figure 9.2a) shows motion artifacts when compared to the SRTM DEM (Figure 9.2b). The artifacts are not clearly discernible in Figure 9.2a because their magnitude is small in comparison to the overall data values. The artifacts are identified by downsampling the registered TOPSAR DEM and subtracting the SRTM DEM. Large-scale anomalies that periodically fluctuate over an entire swath are visible in Figure 9.2c. The periodic pattern is most likely due to uncompensated roll of the DC-8 aircraft. The spaceborne data are less likely to exhibit similar artifacts, because the spacecraft is not greatly affected by the atmosphere. Note that the width of the anomalies corresponds to the width of a TOPSAR swath. Because the SRTM swath is much larger than that of the TOPSAR system (Table 9.1), a larger area is covered under consistent conditions, reducing the number of parallel tracks required to form an SRTM DEM.

The maximum amplitude of the motion artifacts in our study area is about 20 m. This would result in substantial errors in many analyses if not properly corrected. For example, if this TOPSAR DEM is used for topography reduction in repeat-pass InSAR using ERS-2 data with a perpendicular baseline of about 400 m, the resulting deformation interferogram would contain one fringe (= 2.8 cm) of spurious signal.

To remove these artifacts from the TOPSAR DEM, we up-sample the difference image with bilinear interpolation by a factor of 9 so that its pixel spacing matches the TOPSAR DEM. The difference image is subtracted from the TOPSAR DEM. This process is described with a flow diagram in Figure 9.3. Note that the lower branch

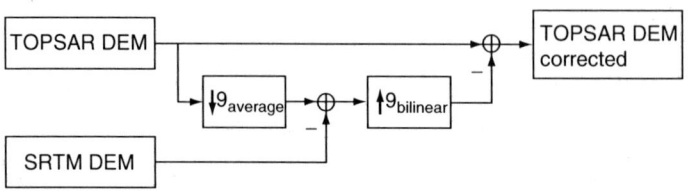

FIGURE 9.3
The flow diagram of the artifact elimination. (From Yun, S.-H., Ji, J., Zebker, H., and Segall, P., *IEEE Trans. Geosci. Rem. Sens.*, 43(7), 1682, 2005. With permission.)

undergoes two low-pass filter operations when averaging and bilinear interpolation are implemented, while the upper branch preserves the high frequency contents of the TOPSAR DEM. In this way we can eliminate the large-scale artifacts while retaining details in the TOPSAR DEM.

9.4 Prediction-Error (PE) Filter

The next step in the DEM process is to fill in missing data. We use a PE filter operating on the TOPSAR DEM to fill these gaps. The basic idea of the PE filter constraint [7,8] is that missing data can be estimated so that the restored data yield minimum energy when the PE filter is applied. The PE filter is derived from training data, which are normally valid data surrounding the missing regions. The PE filter is selected so that the missing data and the valid data share the same spectral content. Hence, we assume that the spectral content of the missing data in the TOPSAR DEM is similar to that of the regions with valid data surrounding the missing regions.

9.4.1 Designing the Filter

We generate a PE filter such that it rejects data with statistics found in the valid regions of the TOPSAR DEM. Given this PE filter, we solve for data in the missing regions such that the interpolated data are also nullified by the PE filter. This concept is illustrated in Figure 9.4.

The PE filter, f_{PE}, is found by minimizing the following objective function,

$$\|f_{PE} * x_e\|^2 \tag{9.2}$$

where x_e is the existing data from the TOPSAR DEM, and $*$ represents convolution. This expression can be rewritten in a linear algebraic form using the following matrix operation:

$$\|F_{PE}\, x_e\|^2 \tag{9.3}$$

or equivalently

$$\|X_e\, f_{PE}\|^2 \tag{9.4}$$

where F_{PE} and X_e are the matrix representations of f_{PE} and x_e for convolution operation. These matrix and vector expressions are used to indicate their linear relationship.

Use of a Prediction-Error Filter in Merging High- and Low-Resolution Images

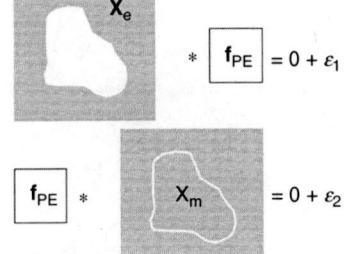

FIGURE 9.4
Concept of PE filter. The PE filter is estimated by solving an inverse problem constrained with the remaining part, and the missing part is estimated by solving another inverse problem constrained with the filter. The ε_1 and ε_2 are white noise with small amplitude.

9.4.2 1D Example

The procedure of acquiring the PE filter can be explained with a 1D example. Suppose that a data set, $\mathbf{x} = [x_1, \ldots, x_n]$ (where $n \gg 3$) is given, and we want to compute a PE filter of length 3, $\mathbf{f}_{PE} = [1 \; f_1 \; f_2]$. Then we form a system of linear equations as follows:

$$\begin{bmatrix} x_3 & x_2 & x_1 \\ x_4 & x_3 & x_2 \\ \vdots & \vdots & \vdots \\ x_n & x_{n-1} & x_{n-2} \end{bmatrix} \begin{bmatrix} 1 \\ f_1 \\ f_2 \end{bmatrix} \approx 0 \quad (9.5)$$

The first element of the PE filter should be equal to one to avoid the trivial solution, $\mathbf{f}_{PE} = 0$. Note that Equation 9.5 is the convolution of the data and the PE filter. After simple algebra and with

$$\mathbf{d} \equiv \begin{bmatrix} x_3 \\ \vdots \\ x_n \end{bmatrix} \quad \text{and} \quad \mathbf{D} \equiv \begin{bmatrix} x_2 & x_1 \\ \vdots & \vdots \\ x_{n-1} & x_{n-2} \end{bmatrix}$$

we get

$$\mathbf{D} \begin{bmatrix} f_1 \\ f_2 \end{bmatrix} \approx -\mathbf{d} \quad (9.6)$$

and its normal equation becomes

$$\begin{bmatrix} f_1 \\ f_2 \end{bmatrix} = (\mathbf{D}^T \mathbf{D})^{-1} \mathbf{D}^T (-\mathbf{d}) \quad (9.7)$$

Note that Equation 9.7 minimizes Equation 9.2 in a least-square sense. This procedure can be extended to 2D problems, and more details are described in Refs. [7] and [8].

9.4.3 The Effect of the Filter

Figure 9.5 shows the characteristics of the PE filter in the spatial and Fourier domains. Figure 9.5a is the sample DEM chosen from Figure 9.1 (numbered box 1) for demonstration. It contains various topographic features and has a wide range of spectral content (Figure 9.5d). Figure 9.5b is the 5-by-5 PE filter derived from Figure 9.5a by solving the inverse problem in Equation 9.3. Note that the first three elements in the first column of the filter coefficients are 0 0 1. This is the PE filter's unique constraint that ensures the filtered output to be white noise [7]. In the filtered output (Figure 9.5c) all the variations in the DEM were effectively suppressed. The size (order) of the PE filter is based on the

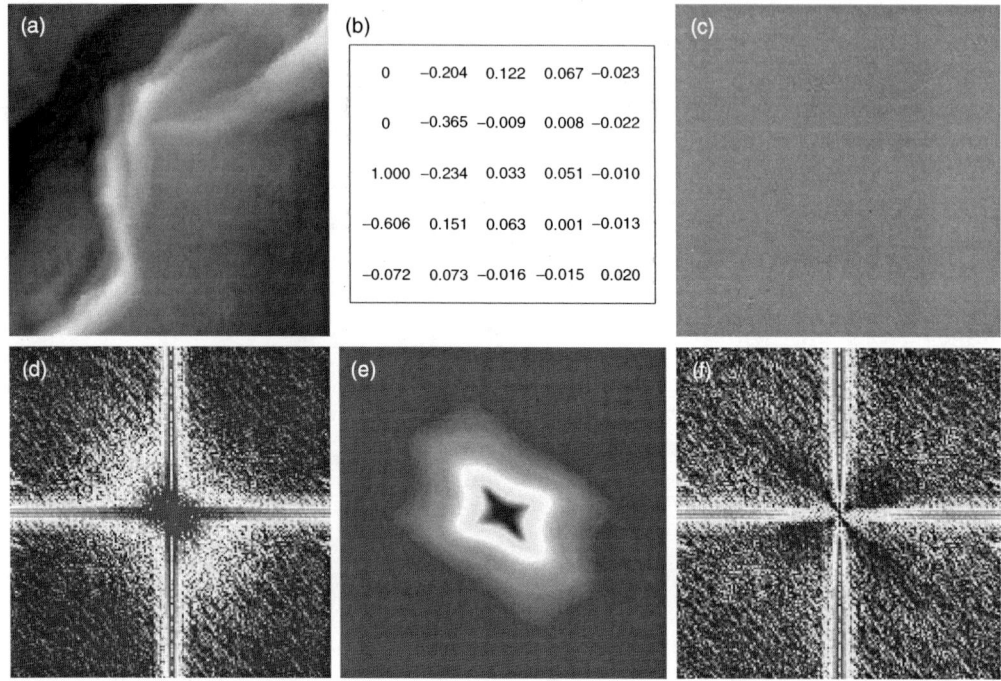

FIGURE 9.5
The effect of a PE filter. (a) original DEM; (b) a 2D PE filter found from the DEM; (c) DEM filtered with the PE filter; and (d), (e), and (f) the spectra of (a), (b), and (c), respectively, plotted in dB. (a) and (c) are drawn with the same color scale. Note that in (c) the variation of image (a) was effectively suppressed by the filter. The standard deviations of (a) and (c) are 27.6 m and 2.5 m, respectively. (From Yun, S.-H., Ji, J., Zebker, H., and Segall, P., *IEEE Trans. Geosci. Rem. Sens.*, 43(7), 1682, 2005. With permission.)

complexity of the spectrum of the DEM. In general, as the spectrum becomes more complex, a larger size filter is required. After testing various sizes of the filter, we found a 5-by-5 size appropriate for the DEM used in our study. Figure 9.5d and Figure 9.5e show the spectra of the DEM and the PE filter, respectively. These illustrate the inverse relationship of the PE filter to the corresponding DEM in the Fourier domain, such that their product is minimized (Figure 9.5f). This PE filter constrains the interpolated data in the DEM to similar spectral content to the existing data.

All inverse problems in this study were derived using the conjugate gradient method, where forward and adjoint functional operators are used instead of the explicit inverse operators [7], saving computer memory space.

9.5 Interpolation

9.5.1 PE Filter Constraint

Once the PE filter is determined, we next estimate the missing parts of the image. As depicted in Figure 9.4, interpolation using the PE filter requires that the norm of the filtered output be minimized. This procedure can be formulated as an inverse computation minimizing the following objective function:

$$\|\mathbf{F}_{\text{PE}}\,\mathbf{x}\|^2 \tag{9.8}$$

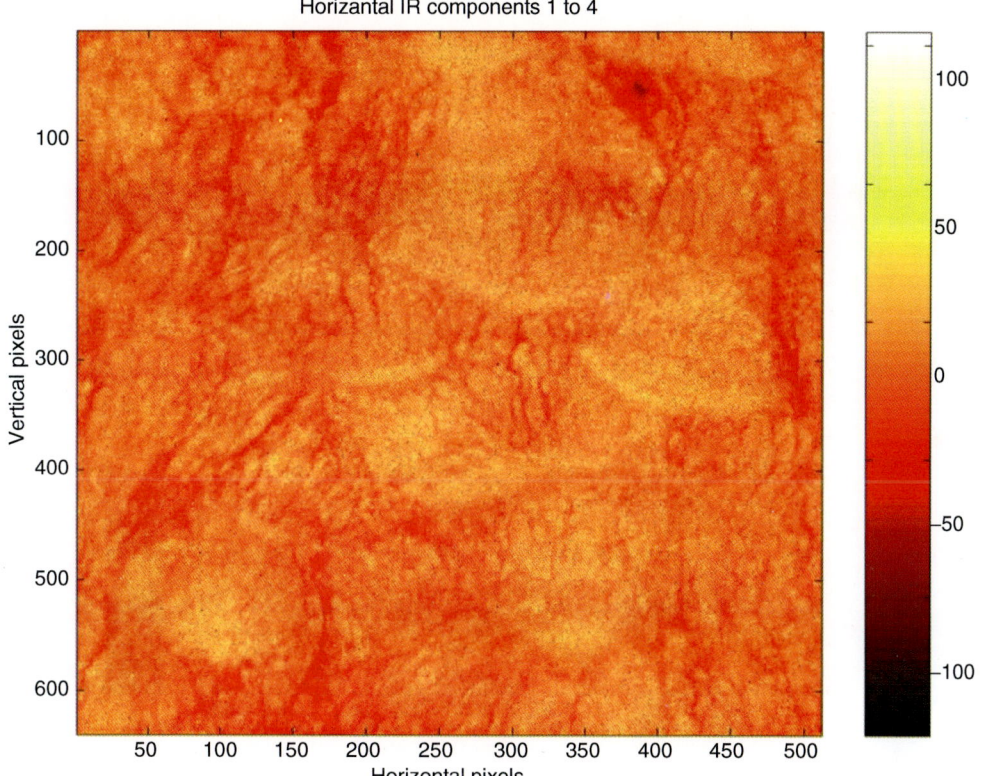

FIGURE 1.13
Component images 1 to 4 from the horizontal rows used to produce a composite image representing the shortest scales.

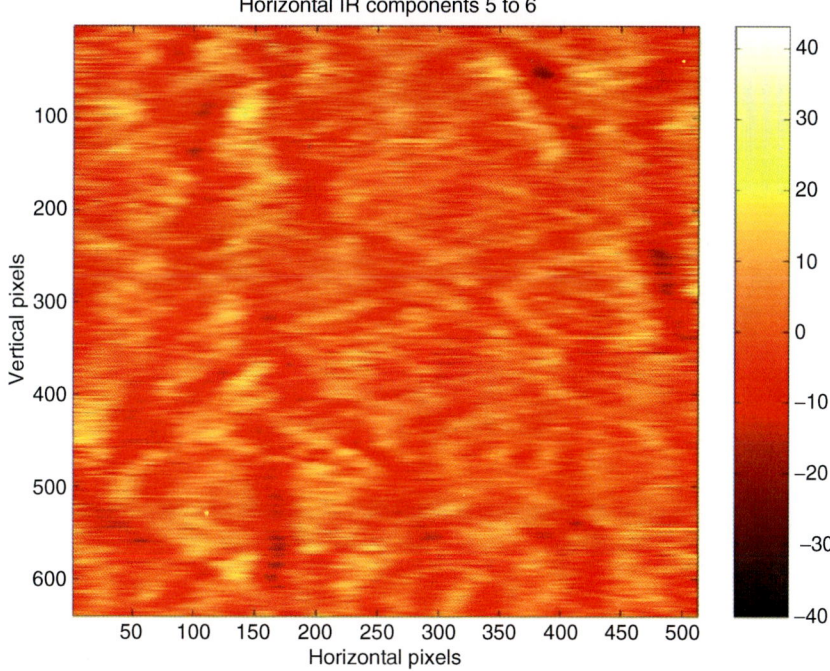

FIGURE 1.14
Component images 5 to 6 from the horizontal rows used to produce a composite image representing the longer scales.

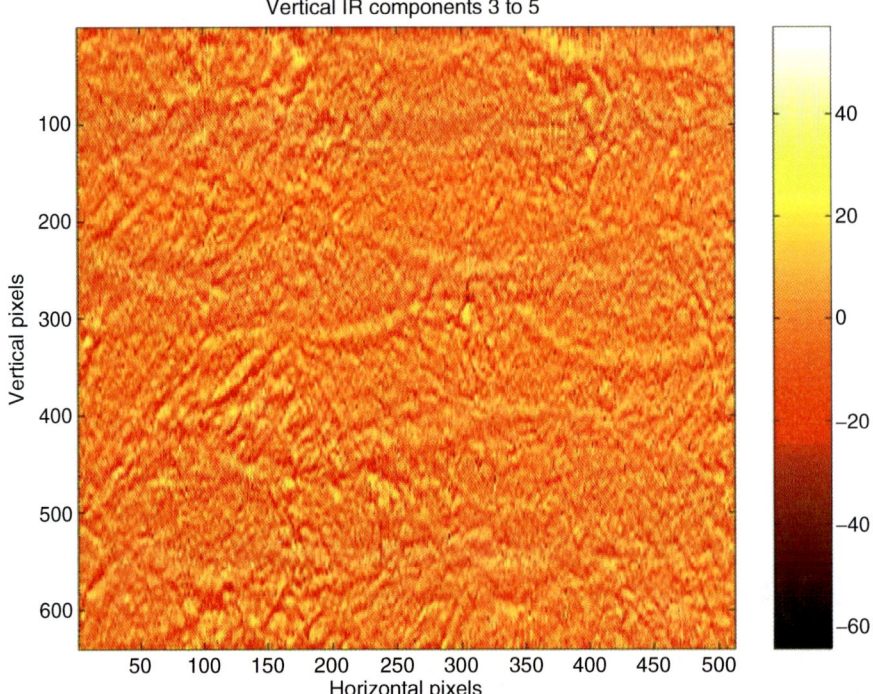

FIGURE 1.15
Component images 3 to 5 from the vertical rows here combined to produce a composite image representing the midrange scales.

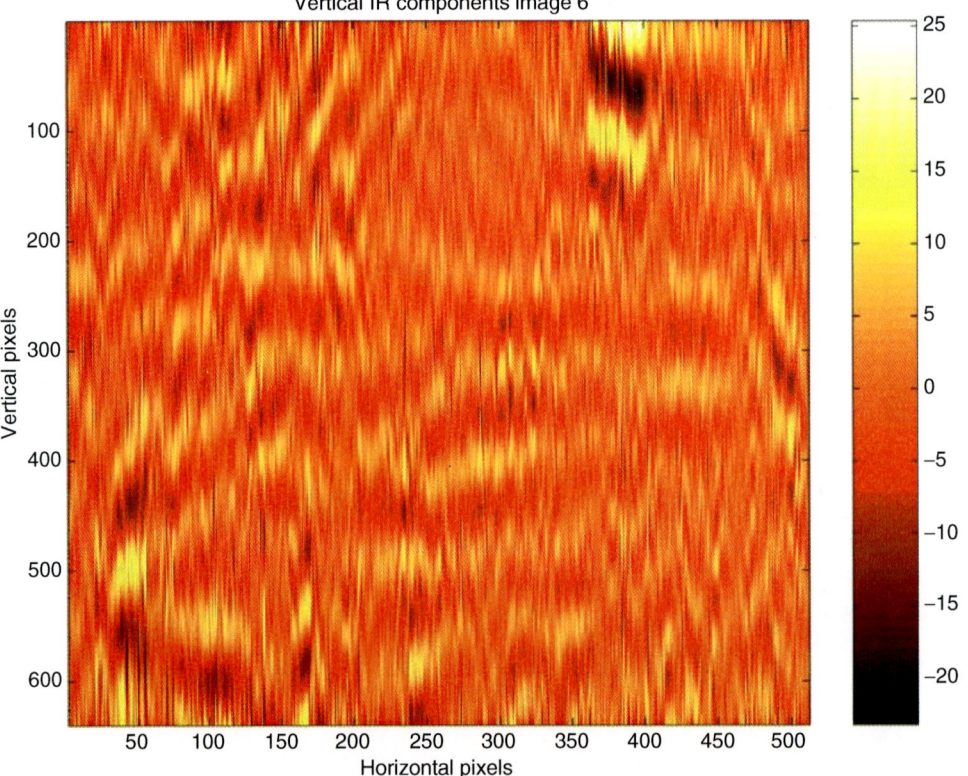

FIGURE 1.16
Component image 6 from the vertical row used to produce a composite image representing the longer scale.

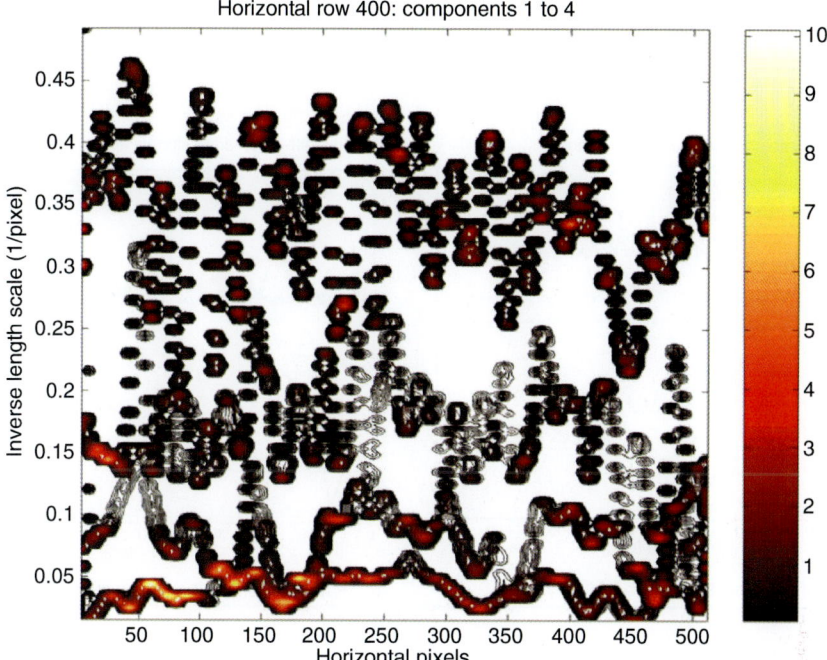

FIGURE 1.17
The results from the NEMD/NHT computation on horizontal row 400 for components 1 to 4, which resulted from Figure 1.13. Note the apparent influence of surface waves on the IR information. The most intense IR radiation can be seen at the smaller values of inverse length scale, denoting the longer scales in components 3 and 4. A wavelike influence can be seen at all scales.

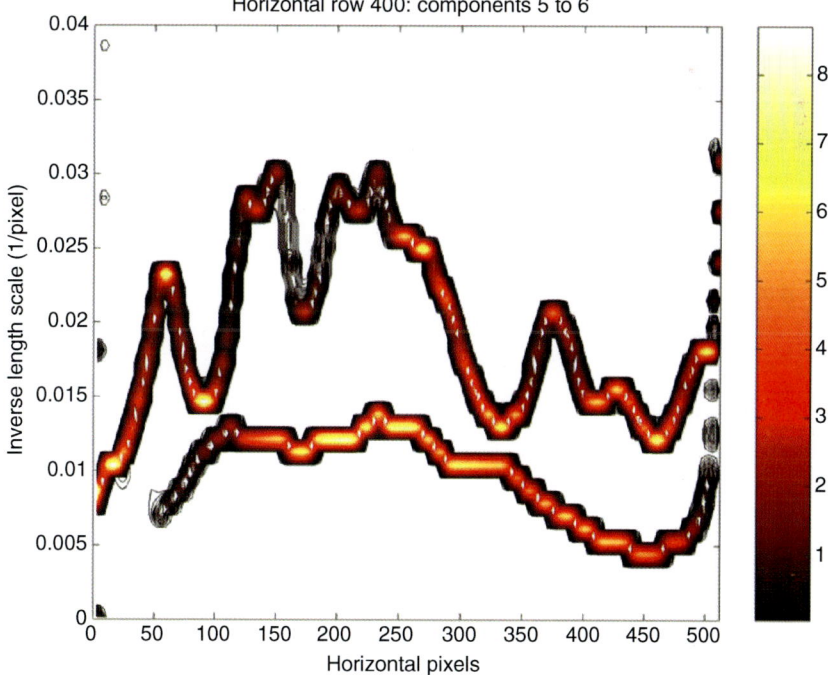

FIGURE 1.18
The results from the NEMD/NHT computation on horizontal row 400 for components 5 to 6, which resulted from Figure 1.14. Even at the longer scales, an apparent influence of surface waves on the IR information can still be seen.

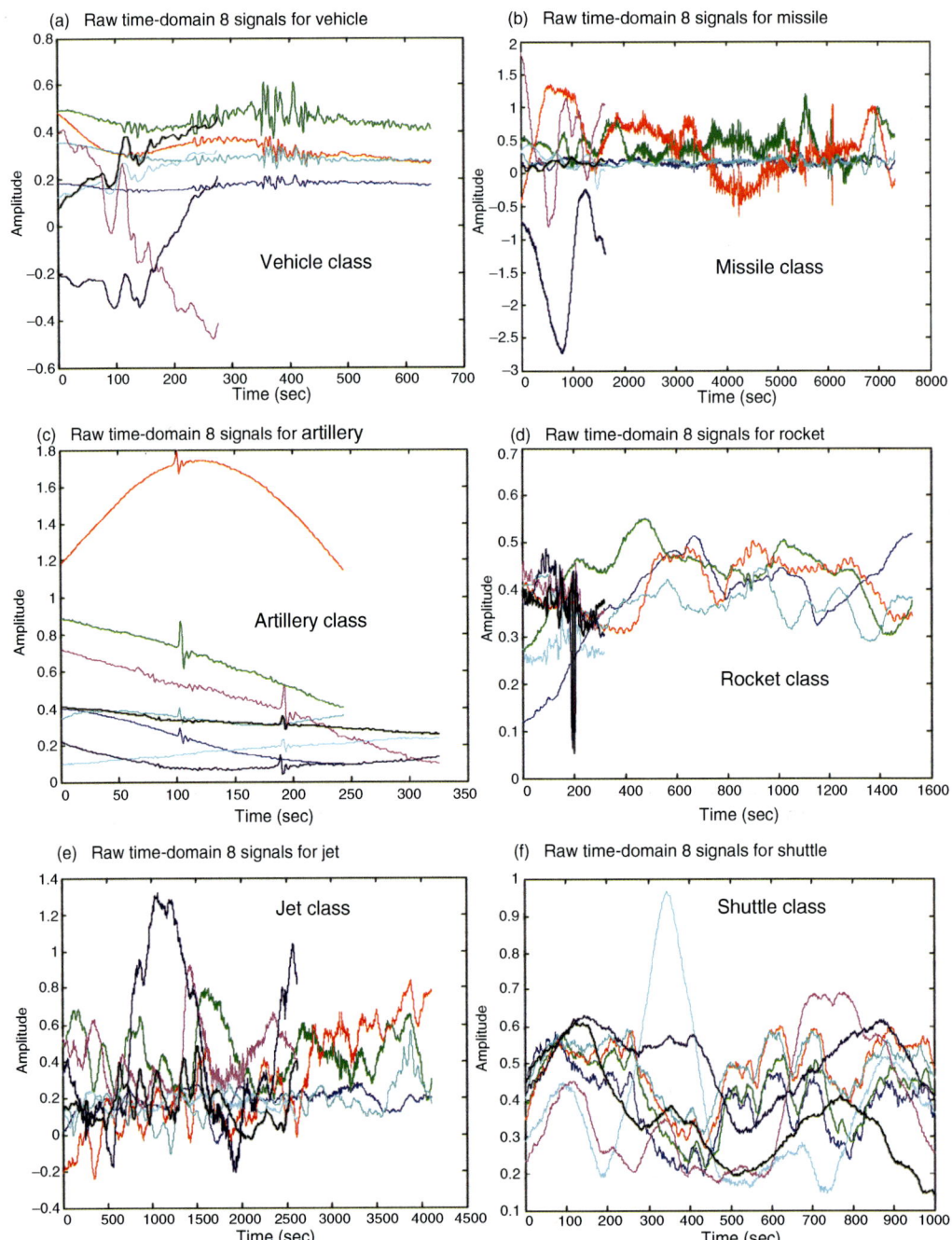

FIGURE 3.7
Infrasound signals for six classes.

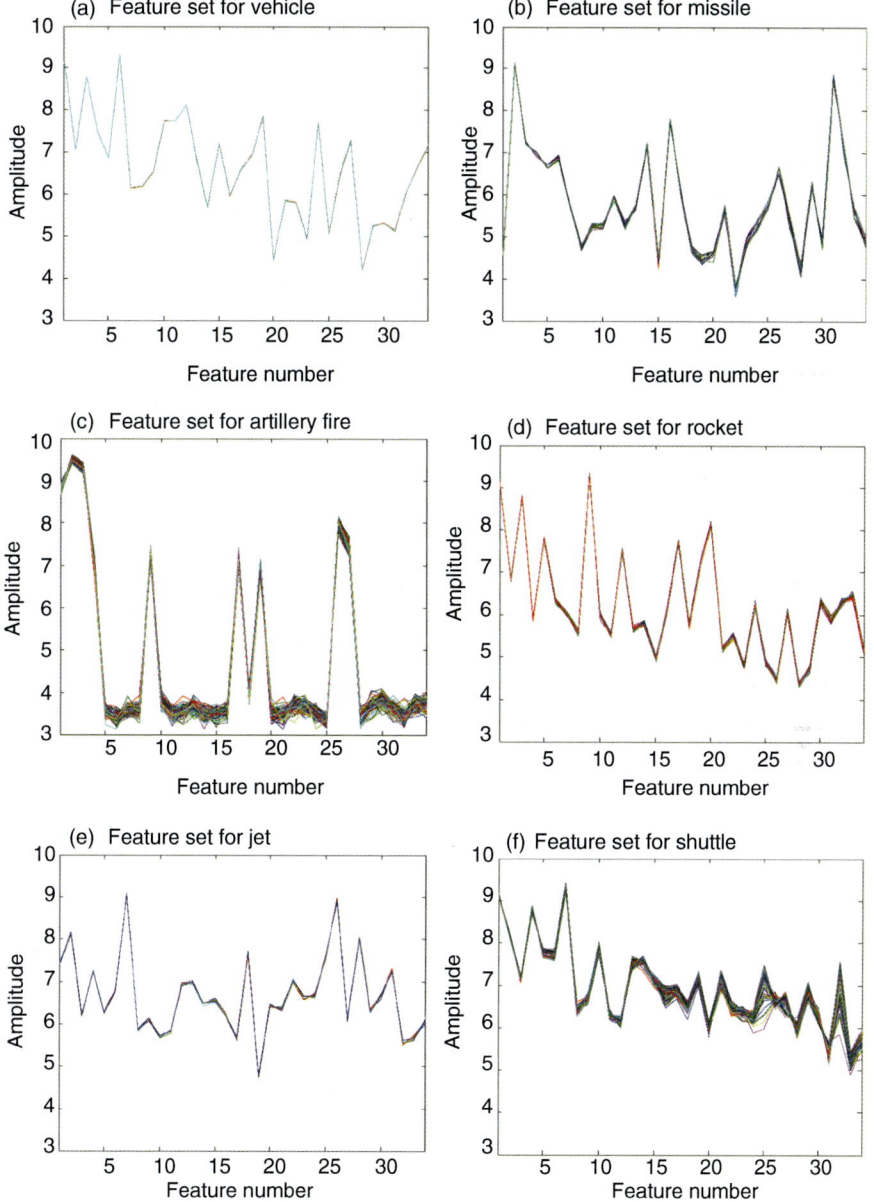

FIGURE 3.8
Infrasound signals for the six class different classes.

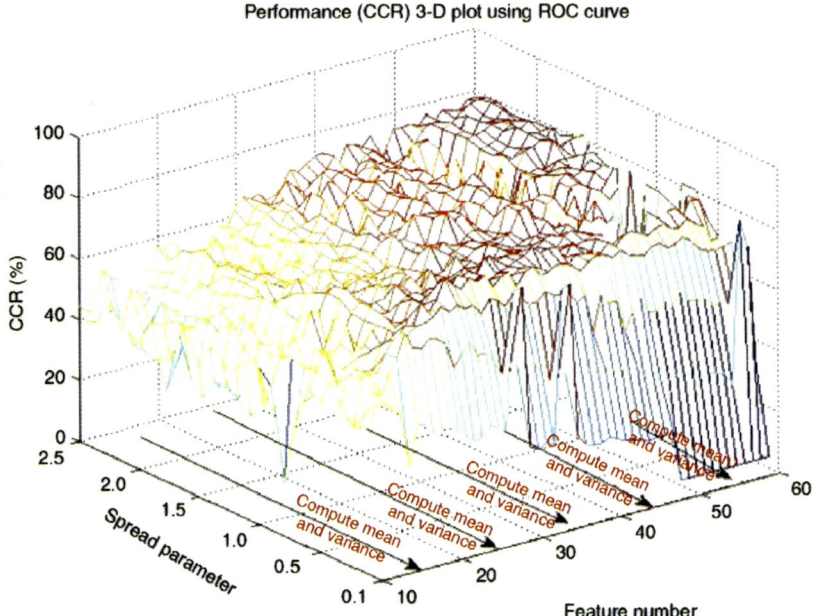

FIGURE 3.9
Performance plot used to determine the optimal number components in the feature vector. Ill conditioning occurs for the feature number less than 10, and for the feature number greater than 60, the CCR dramatically declines.

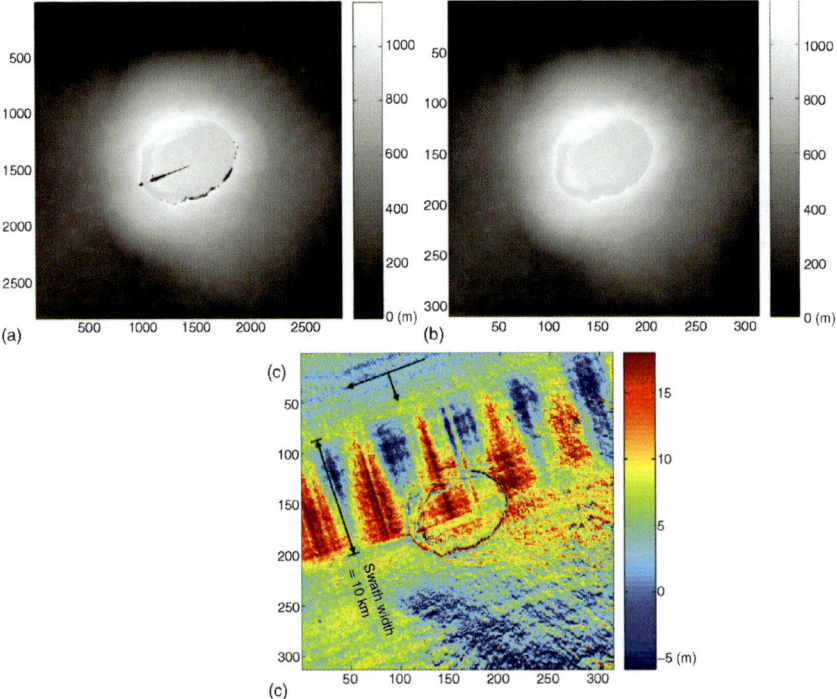

FIGURE 9.2
(a) TOPSAR DEM and (b) SRTM DEM. The tick labels are pixel numbers. Note the difference in pixel spacing between the two DEMs. (c) Artifacts obtained by subtracting the SRTM DEM from the TOPSAR DEM. The flight direction and the radar look direction of the aircraft associated with the swath with the artifact are indicated with long and short arrows, respectively. Note that the artifacts appear in one entire TOPSAR swath, while they are not as serious in other swaths.

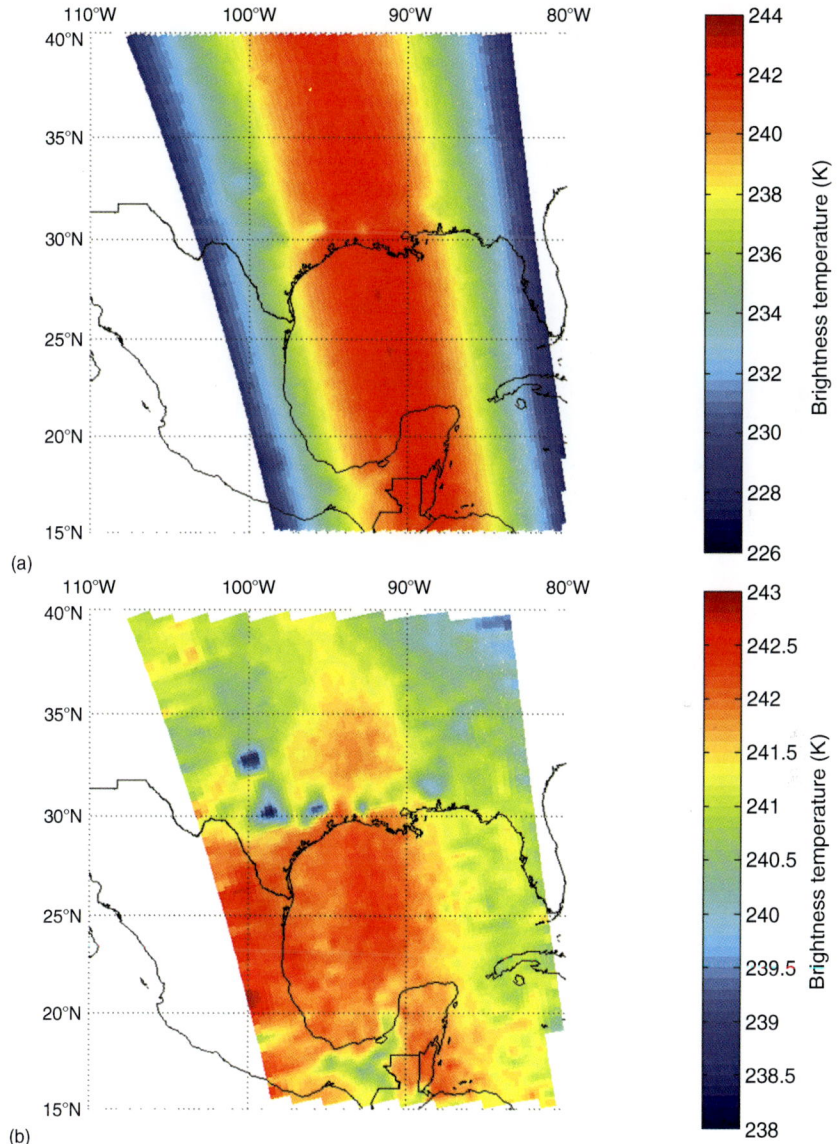

FIGURE 12.13
NOAA-15 AMSU-A 54.4-GHz brightness temperatures for a northbound track on September 13, 2000. (a) Uncorrected and (b) limb and surface corrected.

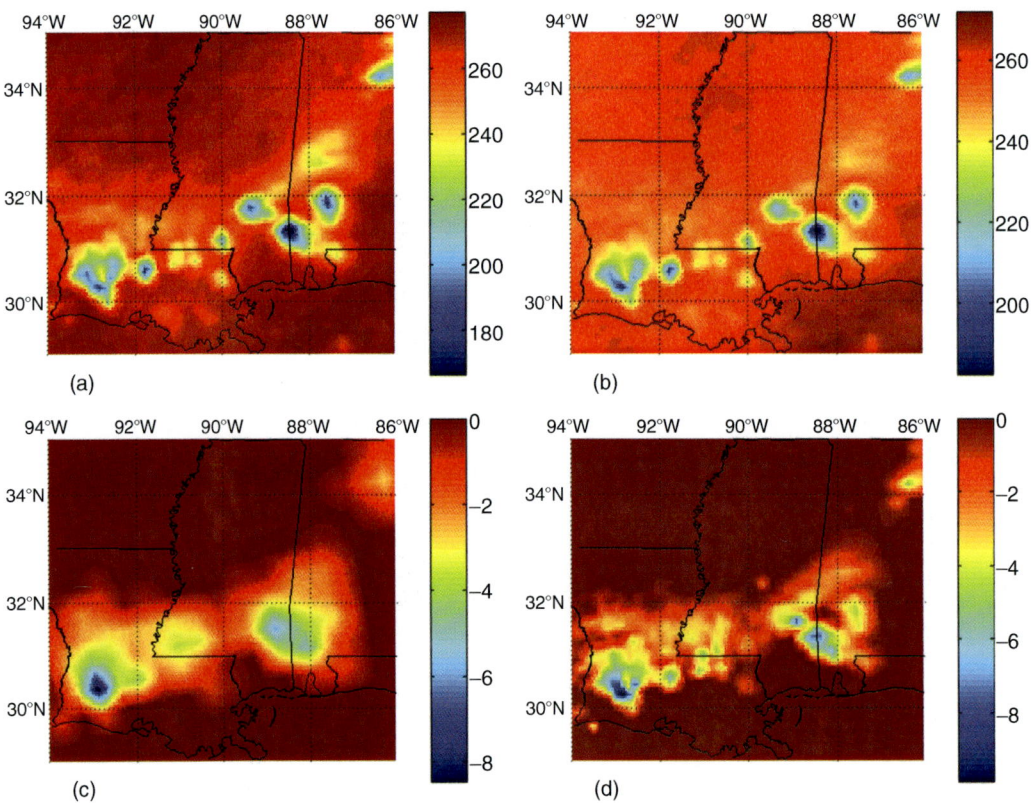

FIGURE 12.14
Frontal system on September 13, 2000, 0130 UTC. (a) Brightness temperatures (K) near 183 ± 7 GHz. (b) Brightness temperatures (K) near 183 ± 3 GHz. (c) Brightness temperature perturbations (K) near 52.8 GHz. (d) Inferred 15-km-resolution brightness temperature perturbations (K) near 52.8 GHz.

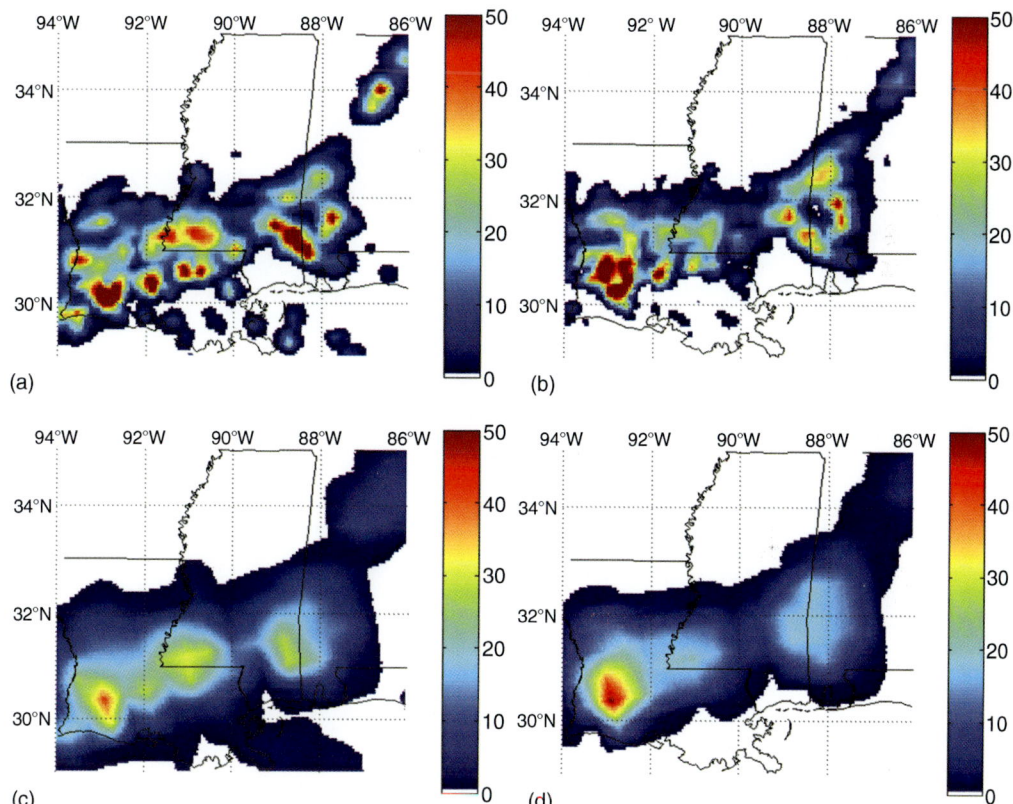

FIGURE 12.15
Precipitation rates (mm/h) above 0.5 mm/h observed on September 13, 2000, 0130 UTC. (a) 15-km-resolution NEXRAD retrievals, (b) 15-km-resolution AMSU retrievals, (c) 50-km-resolution NEXRAD retrievals, and (d) 50-km-resolution AMSU retrievals.

where $\mathbf{F}_{PE}$ is the matrix representation of the PE filter convolution, and $\mathbf{x}$ represents the entire data set including the known and the missing regions. In the inversion process we only update the missing region, without changing the known region. This guarantees seamless interpolation across the boundaries between the known and missing regions.

9.5.2 SRTM DEM Constraint

As previously stated, 90-m posting SRTM DEMs were generated from 30-m posting data. This downsampling was done by calculating three "looks" in both the easting and northing directions. To use the SRTM DEM as a constraint to interpolate the TOPSAR DEM, we posit the following relationship between the two DEMs: each pixel value in a 90-m posting SRTM DEM can be considered equivalent to the averaged value of a 9-by-9 pixel window in a 10-m posting TOPSAR DEM centered at the corresponding pixel in the SRTM DEM.

The solution using the constraint of the SRTM DEM to find the missing data points in the TOPSAR DEM can be expressed as minimizing the following objective function:

$$\|\mathbf{y} - \mathbf{A}\mathbf{x}_m\|^2 \tag{9.9}$$

where $\mathbf{y}$ is an SRTM DEM expressed as a vector that covers the missing regions of the TOPSAR DEM, and $\mathbf{A}$ is an averaging operator generating nine looks, and $\mathbf{x}_m$ represents the missing regions of the TOPSAR DEM.

9.5.3 Inversion with Two Constraints

By combining two constraints, one derived from the statistics of the PE filter and one from the SRTM DEM, we can interpolate the missing data optimally with respect to both criteria. The PE filter guarantees that the interpolated data will have the same spectral properties as the known data. At the same time the SRTM constraint forces the interpolated data to have average height near the corresponding SRTM DEM. We formulate the inverse problem as a minimization of the following objective function:

$$\lambda^2 \|\mathbf{F}_{PE}\ \mathbf{x}_m\|^2 + \|\mathbf{y} - \mathbf{A}\mathbf{x}_m\|^2 \tag{9.10}$$

where λ set the relative effect of each criterion. Here $\mathbf{x}_m$ has the dimensions of the TOPSAR DEM, while $\mathbf{y}$ has the dimensions of the SRTM DEM. If regions of missing data are localized in an image, the entire image does not have to be used for generating a PE filter. We implement interpolation in subimages to save time and computer memory space. An example of such a subimage is shown in Figure 9.6. The image is a part of Figure 9.1 (numbered box 2). Figure 9.6a and Figure 9.6b are examples of $\mathbf{x}_e$ in Equation 9.3 and $\mathbf{y}$, respectively.

The multiplier λ determines the relative weight of the two terms in the objective function. As $\lambda \to \infty$, the solution satisfies the first constraint only, and if $\lambda = 0$, the solution satisfies the second constraint only.

9.5.4 Optimal Weighting

We used cross-validation sum of squares (CVSS) [9] to determine the optimal weights for the two terms in Equation 9.10. Consider a model $\mathbf{x}_m$ that minimizes the following quantity:

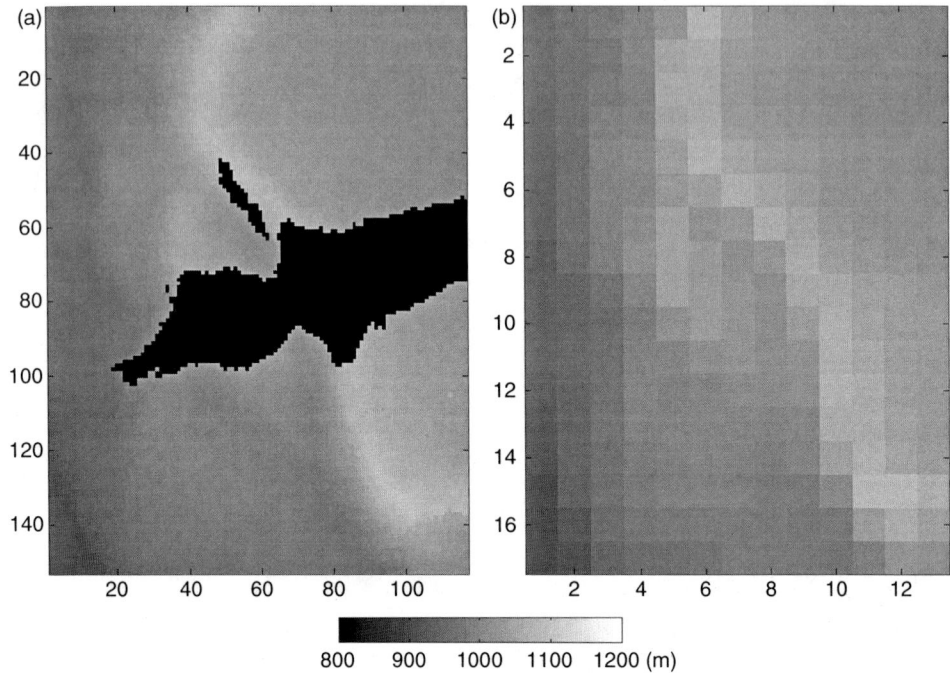

FIGURE 9.6
Example subimages of (a) TOPSAR DEM showing regions of missing data (black), and (b) SRTM DEM of the same area. These subimages are engaged in one implementation of the interpolation. The grayscale is altitude in meters. (From Yun, S.-H., Ji, J., Zebker, H., and Segall, P., *IEEE Trans. Geosci. Rem. Sens.*, 43(7), 1682, 2005. With permission.)

$$\lambda^2 \|\mathbf{F}_{PE}\ \mathbf{x}_m\|^2 + \|\mathbf{y}^{(k)} - \mathbf{A}^{(k)}\ \mathbf{x}_m\|^2 \quad (k = 1, \ldots, N) \tag{9.11}$$

where $\mathbf{y}^{(k)}$ and $\mathbf{A}^{(k)}$ are the $\mathbf{y}$ and the $\mathbf{A}$ in Equation 9.10 with the k-th element and the k-th row omitted, respectively, and N is the number of elements in $\mathbf{y}$ that fall into the missing region. Denote this model $\mathbf{x}_m^{(k)}(\lambda)$. Then we compute the CVSS defined as follows:

$$\text{CVSS}(\lambda) = \frac{1}{N} \sum_{k=1}^{N} (y_k - A_k \mathbf{x}_m^{(k)}(\lambda))^2 \tag{9.12}$$

where y_k is the omitted element from the vector $\mathbf{y}$ and A_k is the omitted row vector from the matrix $\mathbf{A}$ when the $\mathbf{x}_m^{(k)}(\lambda)$ was estimated. Thus, $A_k \mathbf{x}_m^{(k)}(\lambda)$ is the prediction based on the other $N - 1$ observations. Finally, we minimize CVSS(λ) with respect to λ to obtain the optimal weight (Figure 9.7).

In the case of the example shown in Figure 9.6, the minimum CVSS was obtained for $\lambda = 0.16$ (Figure 9.7). The effect of varying λ is shown in Figure 9.8. It is apparent (see Figure 9.8) that the optimal weight is a more "plausible" result than either of the end members, preserving aspects of both constraints.

In Figure 9.8a the interpolation uses only the PE filter constraint. This interpolation does not recover the continuity of the ridge running across the DEM in north–south direction, which is observed in the SRTM DEM (Figure 9.6b). This follows from a PE filter obtained such that it eliminates the overall variations in the image. The variations include not only the ridge but also the accurate topography in the DEM.

The other end member, Figure 9.8c, shows the result for applying zero weight to the PE filter constraint. Since the averaging operator $\mathbf{A}$ in Equation 9.10 is applied independently

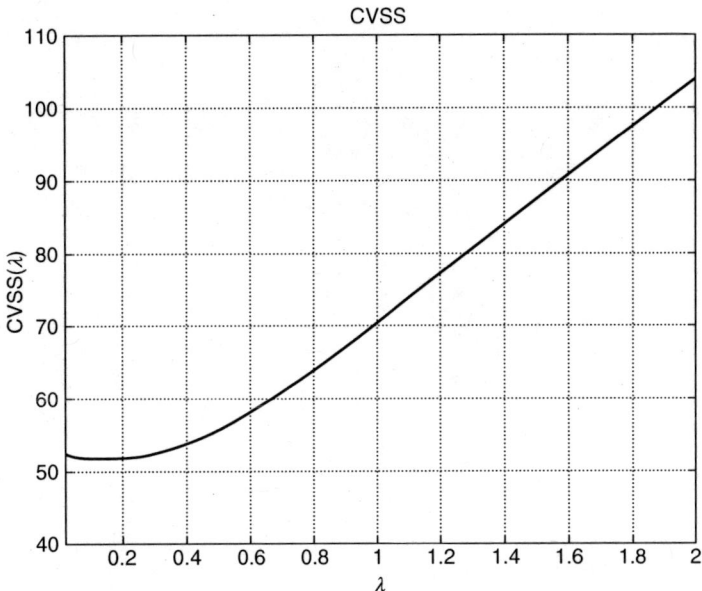

FIGURE 9.7
Cross-validation sum of squares. The minimum occurs when $\lambda = 0.16$. (From Yun, S.-H., Ji, J., Zebker, H., and Segall, P., *IEEE Trans. Geosci. Rem. Sens.*, 43(7), 1682, 2005. With permission.)

for each 9-by-9 pixel group, it is equivalent to simply filling the regions of missing data with 9-by-9 identical values that are the same as the corresponding SRTM DEM (Figure 9.6b).

9.5.5 Simulation of the Interpolation

The quality of cross-validation in this study is itself validated by simulating the interpolation process with known subimages that do not contain missing data. For example, if a known subimage is selected from Figure 9.1 (numbered box 3), we can remove some data and apply our recovery algorithm. The subimage is similar in topographic features to the area shown in Figure 9.6. The process is illustrated in Figure 9.9. We introduce a hole as shown in Figure 9.9b and calculate the CVSS (Figure 9.9d) for each λ ranging from 0 to 2. Then we use the estimated λ, which minimizes the CVSS, for the interpolation process to obtain the image in Figure 9.9c. For each value of λ we also calculate the RMS error between the known and the interpolated images. The RMS error is plotted against λ in Figure 9.9e. The CVSS is minimized for $\lambda = 0.062$, while the RMS error has a minimum at $\lambda = 0.065$. This agreement suggests that minimizing the CVSS is a useful method to balance the constraints. Note that the minimum RMS error in Figure 9.9e is about 5 m. This value is smaller than the relative vertical height accuracy of the SRTM DEM, which is about 10 m.

9.6 Interpolation Results

The method presented in the previous section was applied to the entire image of Figure 9.1. The registered TOPSAR DEM contains missing data in regions of various sizes. Small subimages were extracted from the DEM. Each subimage is interpolated, and the

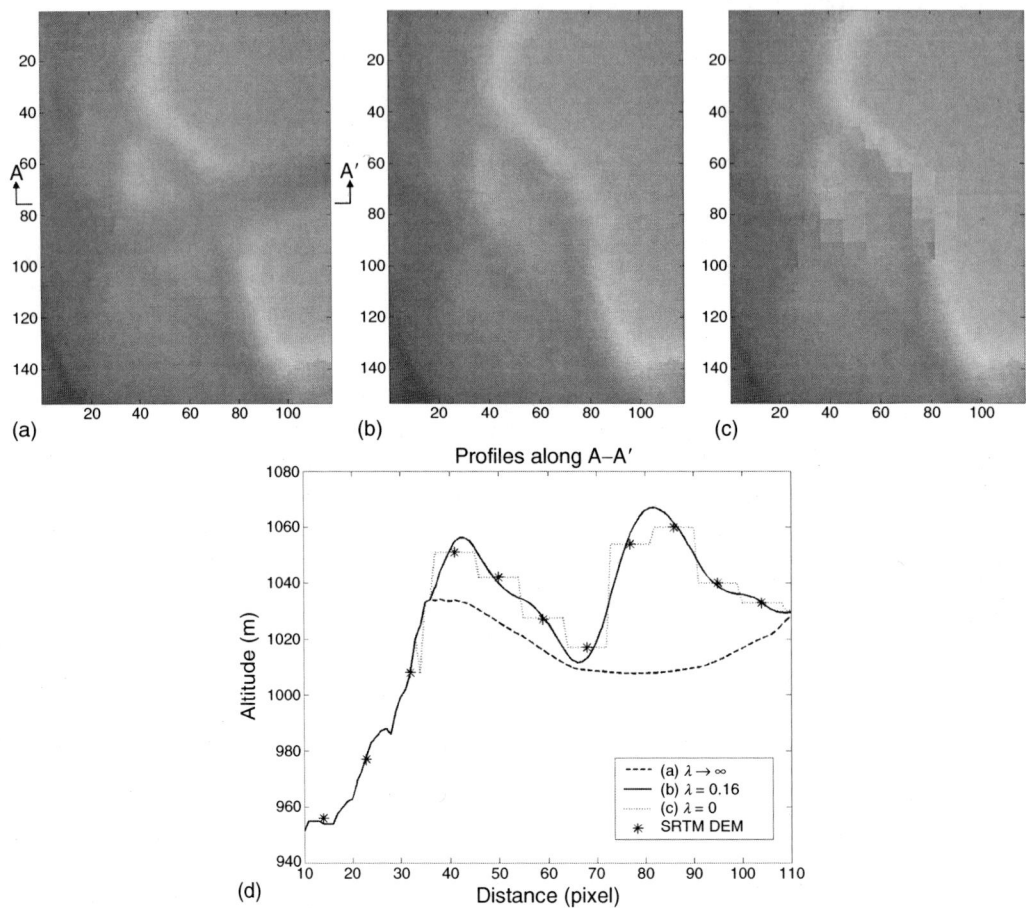

FIGURE 9.8
The results of interpolation applied to DEMs in Figure 9.6, with various weights. (a) $\lambda \to \infty$, (b) $\lambda = 0.16$, and (c) $\lambda = 0$. Profiles along A–A' are shown in the plot (d). (From Yun, S.-H., Ji, J., Zebker, H., and Segall, P., *IEEE Trans. Geosci. Rem. Sens.*, 43(7), 1682, 2005. With permission.)

results are reinserted into the large DEM. The locations and sizes of the subimages are indicated with white boxes in Figure 9.10a. Note the largest region of missing data in the middle of the caldera. This region is not only a simple large gap but also a gap between two swaths. The interpolation is an iterative process and fills up regions of missing data starting from the boundary. If valid data along the boundary (boundaries of a swath for example) contain edge effects, error tends to propagate through the interpolation process. In this case, expanding the region of missing data by a few pixels before interpolation produces better results. If there is a large region of missing data, the spectral content information of valid data can fade out as the interpolation proceeds toward the center of the gap. In this case, sequentially applying the interpolation to parts of the gap is one solution. Due to edge effects along the boundary of the large gap, the interpolation result does not produce topography that matches the surrounding terrain well. Hence, we expand the gap by three pixels to eliminate edge effects. We divided the gap into multiple subimages, and each subimage was interpolated individually.

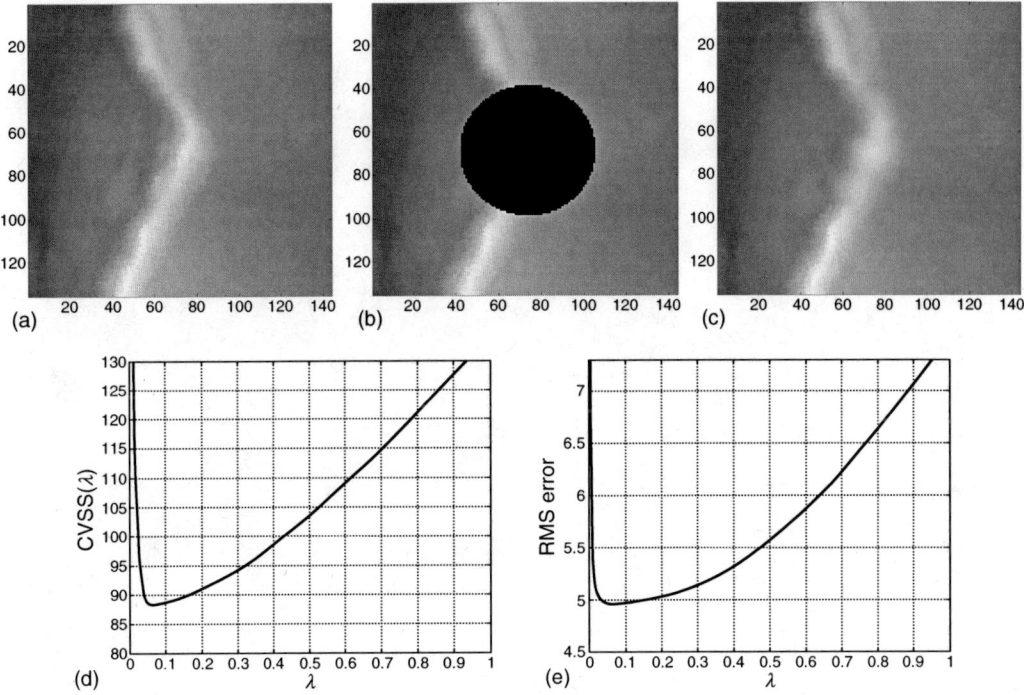

FIGURE 9.9
The quality of the CVSS, (a) a sample image that does not have a hole, (b) a hole was made, (c) interpolated image with an optimal weight, (d) CVSS as a function of λ. The CVSS has a minimum when $\lambda = 0.062$, and (e) RMS error between true image (a) and the interpolated image (c). The minimum occurs when $\lambda = 0.065$. (From Yun, S.-H., Ji, J., Zebker, H., and Segall, P., *IEEE Trans. Geosci. Rem. Sens.*, 43(7), 1682, 2005. With permission.)

9.7 Effect on InSAR

Finally, we can investigate the effect of the artifact elimination and the interpolation on simulated interferograms. It is often easier to see differences in elevation in simulated interferograms than in conventional contour plots. In addition, simulated interferograms provide a measure of how sensitive the interferogram is to the topography. Figure 9.11 shows georeferenced simulated interferograms from three DEMs: the registered TOPSAR DEM, the TOPSAR DEM after the artifact elimination, and the TOPSAR DEM after the interpolation. In all interferograms, a C-band wavelength is used, and we assume a 452 m perpendicular baseline between two satellite positions. This perpendicular baseline is realistic [2]. The fringe lines in the interferograms are approximately height contour lines. The interval of the fringe lines is inversely proportional to the perpendicular baseline [10], and in this case one color cycle of the fringes represents about 20 m. Note in Figure 9.11a that the fringe lines are discontinuous across the long region of missing data inside the caldera. This is due to artifacts in the original TOPSAR DEM. After eliminating these artifacts the discontinuity disappears (Figure 9.11b). Finally, the missing data regions are interpolated in a seamless manner (Figure 9.11c).

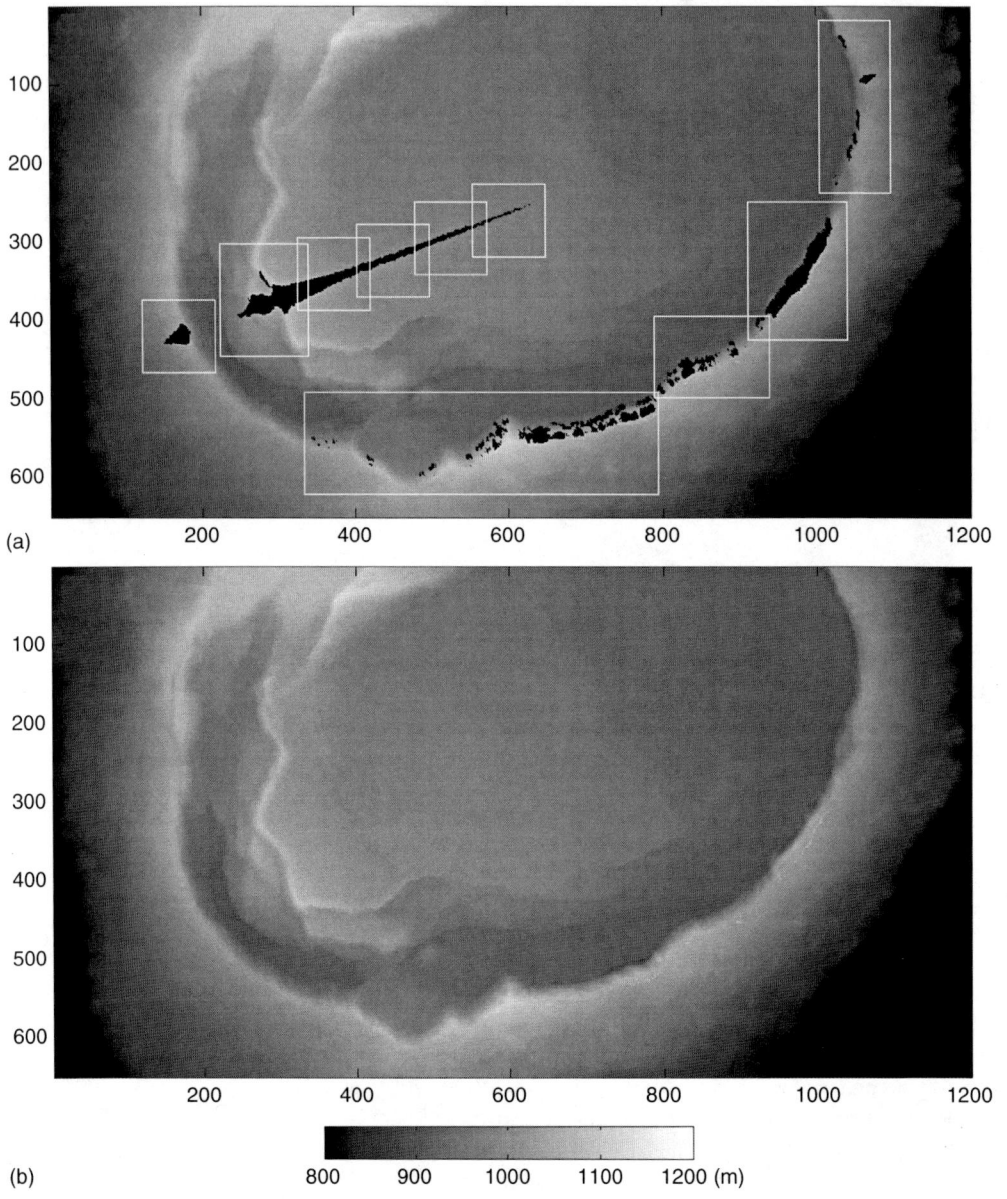

FIGURE 9.10
The original TOPSAR DEM (a) and the reconstructed DEM (b) after interpolation with PE filter and SRTM DEM constraints. The grayscale is altitude in meters, and the spatial extent is about 12 km across the image. (From Yun, S.-H., Ji, J., Zebker, H., and Segall, P., *IEEE Trans. Geosci. Rem. Sens.*, 43(7), 1682, 2005. With permission.)

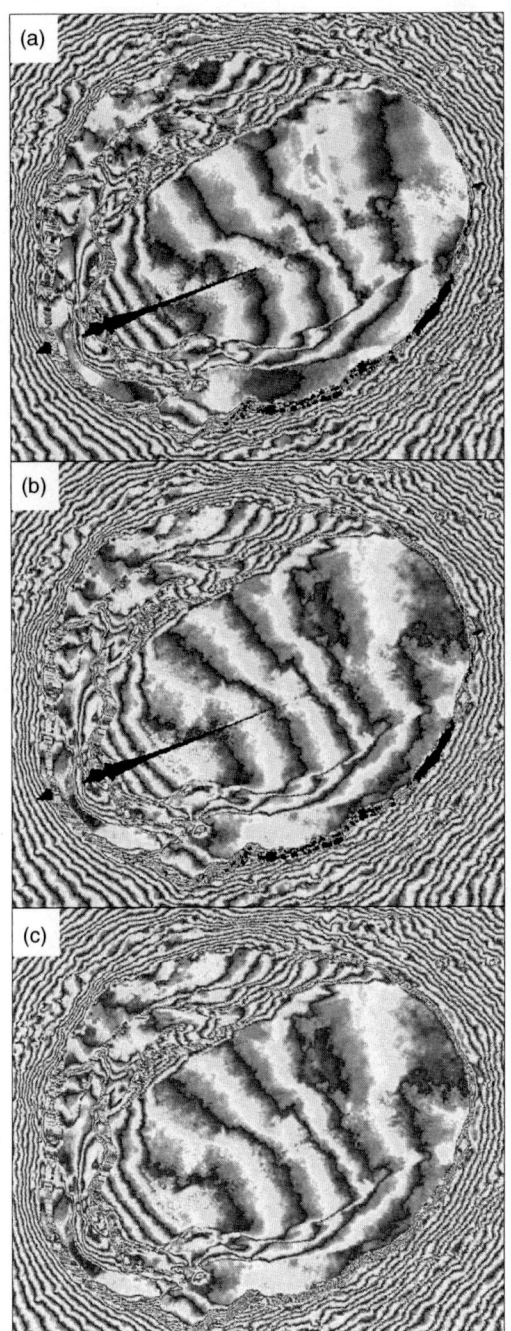

FIGURE 9.11
Simulated interferograms from (a) the original registered TOPSAR DEM, (b) the DEM after the artifact was removed, and (c) the DEM interpolated with PE filter and the SRTM DEM. All the interferograms were simulated with the C-band wavelength (5.6 cm) and a perpendicular baseline of 452 m. Thus, one color cycle represents 20 m height difference. (From Yun, S.-H., Ji, J., Zebker, H., and Segall, P., *IEEE Trans. Geosci. Rem. Sens.*, 43(7), 1682, 2005. With permission.)

9.8 Conclusion

The aircraft roll artifacts in the TOPSAR DEM were eliminated by subtracting the difference between the TOPSAR and SRTM DEMs. A 2D PE filter derived from the existing data and the SRTM DEM for the same region are then used as interpolation constraints.

Solving the inverse problem constrained with both the PE filter and the SRTM DEM produces a high-quality interpolated map of elevation. Cross-validation works well to select optimal constraint weighting in the inversion. This objective criterion results in less biased interpolation and guarantees the best fit to the SRTM DEM. The quality of many other TOPSAR DEMs can be improved similarly.

References

1. H.A. Zebker and R.M. Goldstein, Topographic mapping from interferometric synthetic aperture radar observations, *Journal of Geophysical Research*, 91(B5), 4993–4999, 1986.
2. F. Amelung, S. Jónsson, H. Zebker, and P. Segall, Widespread uplift and 'trapdoor' faulting on Galápagos volcanoes observed with radar interferometry, *Nature*, 407(6807), 993–996, 2000.
3. S. Yun, P. Segall, and H. Zebker, Constraints on magma chamber geometry at Sierra Negra volcano, Galápagos Islands, based on InSAR observations, *Journal of Volcanology and Geothermal Research*, 150, 232–243, 2006.
4. H.A. Zebker, S.N. Madsen, J. Martin, K.B. Wheeler, T. Miller, Y.L. Lou, G. Alberti, S. Vetrella, and A. Cucci, The TOPSAR interferometric radar topographic mapping instrument, *IEEE Transactions on Geoscience and Remote Sensing*, 30(5), 933–940, 1992.
5. S.N. Madsen, H.A. Zebker, and J. Martin, Topographic mapping using radar interferometry: processing techniques, *IEEE Transactions on Geoscience and Remote Sensing*, 31(1), 246–256, 1993.
6. Y. Kobayashi, K. Sarabandi, L. Pierce, and M.C. Dobson, An evaluation of the JPL TOPSAR for extracting tree heights, *IEEE Transactions on Geoscience and Remote Sensing*, 38(6), 2446–2454, 2000.
7. J.F. Claerbout, *Earth Sounding Analysis: Processing versus Inversion*, Blackwell, 1992 [Online], http://sepwww.stanford.edu/sep/prof/index.html.
8. J.F. Claerbout and S. Fomel, *Image Estimation by Example: Geophysical Soundings Image Construction (Class Notes)*, 2002 [Online], http://sepwww.stanford.edu/sep/prof/index.html.
9. G. Wahba, *Spline Models for Observational Data*, ser. No. 59, Applied Mathematics, Philadelphia, PA, SIAM, 1990.
10. H.A. Zebker, P.A. Rosen, and S. Hensley, Atmospheric effects in interferometric synthetic aperture radar surface deformation and topographic maps, *Journal of Geophysical Research*, 102(B4), 7547–7563, 1997.

10

Blind Separation of Convolutive Mixtures for Canceling Active Sonar Reverberation

Fengyu Cong, Chi Hau Chen, Shaoling Ji, Peng Jia, and Xizhi Shi

CONTENTS

10.1 Introduction 187
10.2 Problem Description 188
10.3 BSCM Algorithm 190
10.4 Experiments and Analysis 193
      10.4.1 Backward Matrix of Active Sonar Data for Approximating the BSCM Model 193
      10.4.2 Examples of Canceling Real Sea Reverberation 194
            10.4.2.1 Example 1: Simulation 1 194
            10.4.2.2 Example 2: Simulation 2 194
            10.4.2.3 Example 3: Simulation 3 195
            10.4.2.4 Example 4: Real Target Experiment 198
            10.4.2.5 Example 5: Simulation 4 198
10.5 Conclusions 199
References 200

10.1 Introduction

Under heavy oceanic reverberation, it can be difficult to detect a target echo accurately with an active sonar. To resolve the problem, researchers apply two methods that use reverberation models and specialized processing techniques [1]. For the first method, receivers with differing range resolutions may encounter different statistics for a given waveform. Furthermore, a given receiver may encounter different statistics at different ranges [2]. Researchers have used Weibull, log-normal, Rician, multi-modal Rayleigh, and non-Rayleigh distributions to describe sonar reverberations [3,4]. For the second method, it is usually assumed that reverberation is a sum of returns issued from the transmitted signal. Under this assumption, a data matrix is first generated from the data received by the active sonar data [5,6]. The principal component inverse (PCI) [7–13] is primarily used to separate reverberation and target echoes from the data matrix. However, important prior knowledge such as the target power should be provided [11]. In Refs. [11–13], PCI and other methods have performed very well in Doppler cases, and the authors have also shown that PCI still performs well when the Doppler effect is not introduced. Provided that prior knowledge is hard to obtain and the Doppler effect does not exist, it becomes very desirable to cancel reverberation with easily obtainable but minimal prior knowledge even in more complicated undersea situations. The chapter focuses on this case.

The essence of PCI is separation by rank reduction. Nevertheless, the problem of separation is an old one in electrical engineering and many algorithms exist depending on the nature of signals. Blind signal separation (BSS) [14] is a significant statistical signal processing method that has been developed in the past 15 years. The advantage of BSS is that it does not need much prior knowledge and makes full use of the simple and apparent statistical properties, such as non-Gaussianity, nonstationarity, colored character, uncorrelatedness, independence, and so on. We studied BSS on canceling reverberation in Ref. [15]. From the perspective of BSS, the data received by active sonar is the convolutive mixture [16], while the instantaneous mixture model was only discussed in Ref. [15]. Consequently, we perform blind separation of convolutive mixtures (BSCM) to nullify oceanic reverberation in this contribution. In Ref. [15], the data waveform is only described in time-domain, and here, we will provide more illustrations on the matched filter outputs under different signal-to-reverberation ratios (SRR). We provide more examples for better explanation.

The rest of this chapter is organized as follows: Section 10.2 presents the problem description; Section 10.3 introduces the BSCM algorithm, which is based on the reverberation characters; Section 10.4 provides examples of canceling real sea reverberation as the main content; and finally, Section 10.5 summarizes the above contents.

10.2 Problem Description

In this chapter, we assume that reverberation is a sum of returns generated from the transmitted signal. In Ref. [16], the active sonar data series $d(t)$ is expressed as

$$d(t) = \underbrace{h_1(t-\tau_1) \times e(t)}_{\text{target}} + \sum_{k=2}^{K} \underbrace{h_k(t-\tau_k) \times e(t)}_{\text{reverberation}} + \underbrace{n(t)}_{\text{noise}} \qquad (10.1)$$

where τ_k is the propagation delay, $h_k(t)$ is the path impulse response, and $e(t)$ is the transmitted signal. In the reverberation dominant circumstance, we can decompose the active sonar data into the target echo component $\tilde{d}(t)$ and the reverberation component $r(t)$, as defined in Equation 10.2. Extracting the target and nullifying the reverberation are done simultaneously

$$\tilde{d}(t) = \underbrace{h_1(t-\tau_1) \times e(t)}_{\text{target}} + \underbrace{n(t)}_{\text{noise}}$$

$$r(t) = \sum_{k=2}^{K} \underbrace{h_k(t-\tau_k) \times e(t)}_{\text{reverberation}} + \underbrace{n(t)}_{\text{noise}} \qquad (10.2)$$

It is apparent that $\tilde{d}(t)$ and $r(t)$ are of different non-Gaussianity. Next, we show an example of real sea reverberation.

For simplicity, we do not introduce the experimental detail in this section. The transmitted signal is sinusoid with a frequency of 1750 Hz, a duration of 200 msec, and a sampling frequency of 6000 Hz. Figure 10.1 contains only the target echo and the background noise,

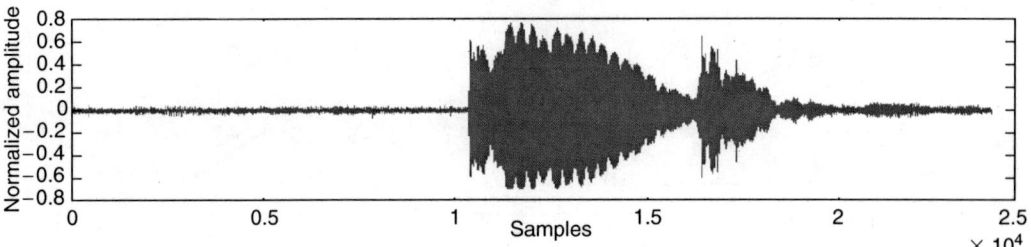

FIGURE 10.1
Target echo waveform.

and most reverberation generated by a sine wave is included in Figure 10.2 where no target echo exists and the background noise is embedded. The two waveforms in Figure 10.1 and Figure 10.2 are the time-domain descriptions of $\tilde{d}(t)$ and $r(t)$ with corresponding standard kurtosis [17] 4.1 and 1.6, respectively. Here, the time-domain non-Gaussianity is enough to discriminate $\tilde{d}(t)$ and $r(t)$. It may appear that detection for $\tilde{d}(t)$ must be evident. However, in the real world problem, it is not simple to cancel the reverberation to the degree as in Figure 10.1. To explore different effects of reverberation on detection in different SRRs, we give several examples of real sea reverberation as follows.

In Figure 10.3, the first upper plot is reverberation, and in the remaining three plots, different targets are simulated and added to the reverberation with the SRR = 0 dB, −7 dB, and −14 dB, respectively, from the upper to lower plots. The target echo is located between the 9,000th and 11,000th samples. The target does not move, hence no Doppler effect is produced. The matched filter outputs follow in the next figure.

In Figure 10.4, the middle two plots show that the matched filter results are satisfactory even in the case that the SRR is quite small, and the lowest plot gives too many false alarms. The task for us is to cancel the reverberation to the degree that the detection is possible, similar to the middle two plots. In Equation 10.2, $\tilde{d}(t)$ does not contain any reverberation and just covers only the target echo and the background noise. In real world case, that is impossible. However, it does not harm the detection. Figure 10.4 also reminds us that even if

$$\tilde{d}(t) = \underbrace{h_1(t - \tau_1)}_{\text{target}} \times e(t) + \sum_{k=2}^{\tilde{K}} \underbrace{h_k(t - \tau_k)}_{\text{reverberation}} \times e(t) + \underbrace{n(t)}_{\text{noise}} \qquad (10.3)$$

where $\tilde{K} < K$, and as long as the reverberation is reduced to some satisfactory level, the detection can also be reliable. After comparing the first and the third plots in Figure 10.3 and Figure 10.4, respectively, we find that Figure 10.3 shows that the reverberation waveform with small SRR target added is nearly identical to the waveform of

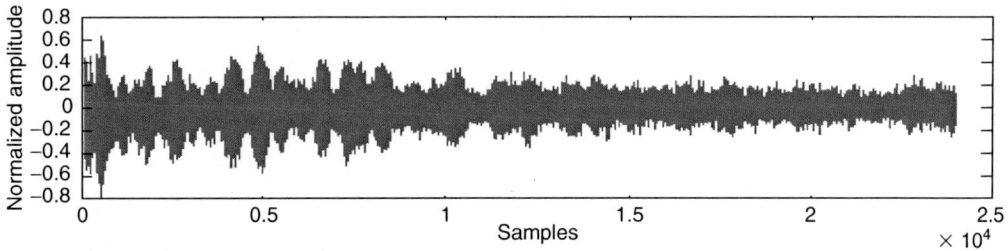

FIGURE 10.2
Reverberation waveform.

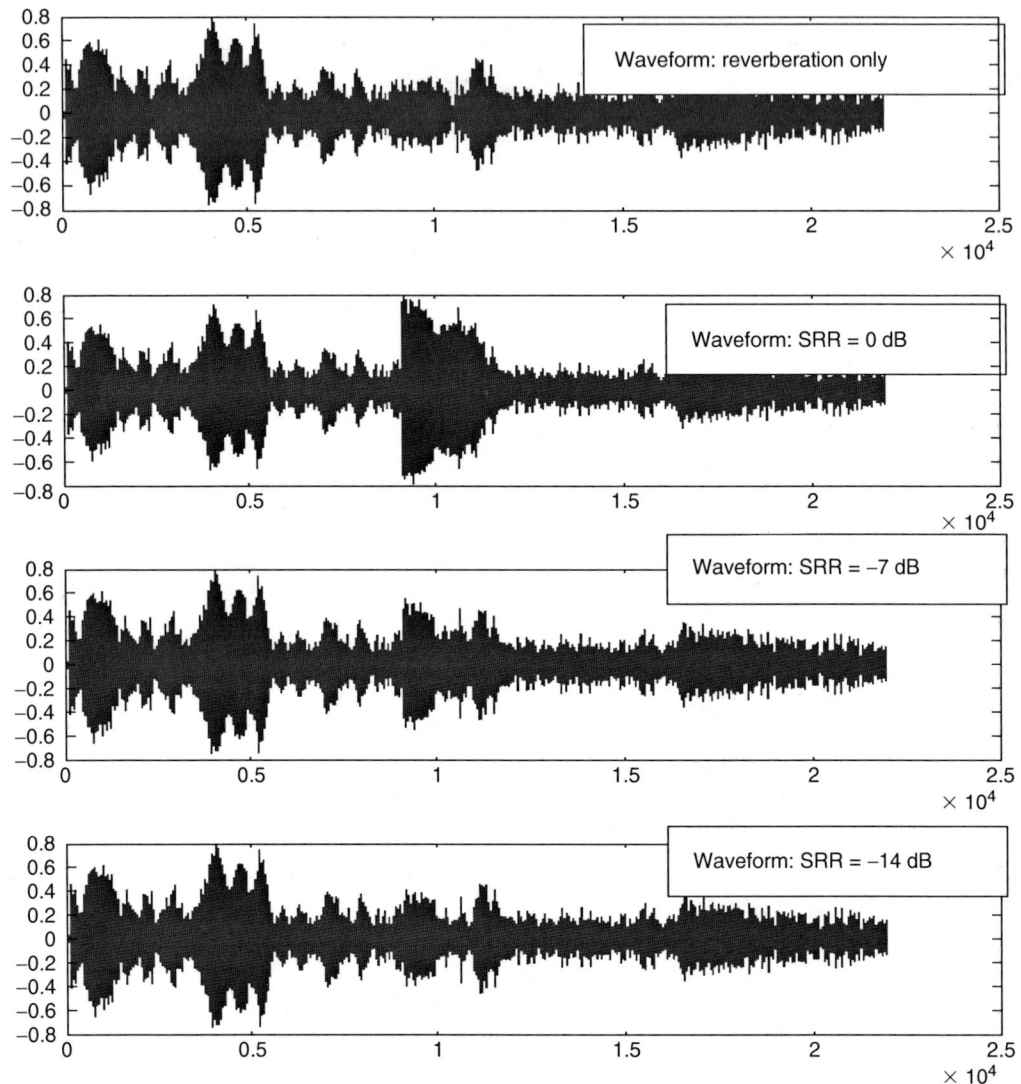

FIGURE 10.3
Real reverberation with different simulated targets.

reverberation, but Figure 10.4 implies that the corresponding matched filter outputs are definitely of different non-Gaussianity. This information is useful to our method for distinguishing the target echoes from the reverberation echoes.

10.3 BSCM Algorithm

In the past several years, different researchers developed several BSCM algorithms according to signal properties. Generally speaking, for a given source–receiver pair, the waveform arrives at different travel times, owing to varying path lengths [18]. All of these make reverberations nonstationary. The transmitted signal is sinusoid here, so the

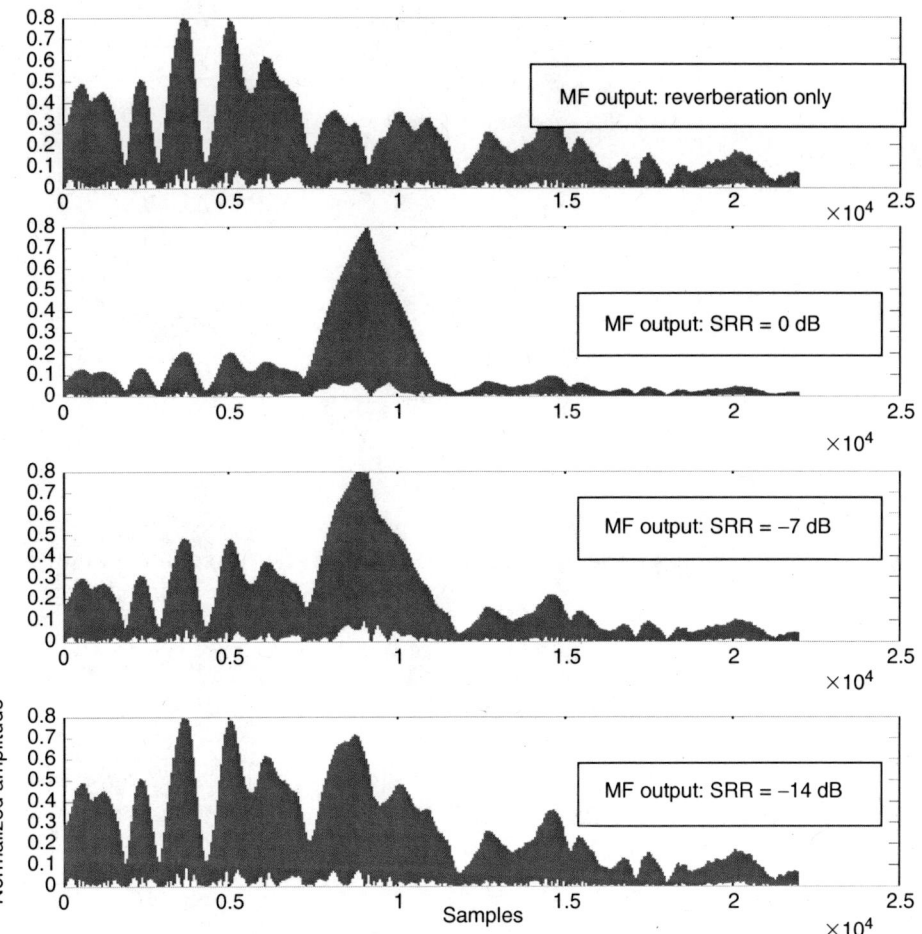

FIGURE 10.4
Outputs of matched filter on the data in Figure 10.3.

reverberations must retain the colored feature. Therefore, we say reverberations are nonstationary and colored. In this section, we will give only a brief introduction on the BSCM algorithm.

Consider a speech scenario, where J microphones receive multiple filtered copies of statistically uncorrelated or independent signals. Mathematically, the received signals can be expressed as a convolution, that is,

$$x_j(t) = \sum_{i=1}^{I} \sum_{p=0}^{P-1} h_{ji}(p) s_i(t-p), \quad j = 1, \ldots, J0; \quad t = 1, \ldots, L \quad (10.4)$$

where $h_{ji}(p)$ models the P-point impulse response from source to microphone and L is the length of the received signal. In a more compact matrix–vector notation, Equation 10.4 can be stated as

$$\mathbf{x}(t) = \sum_{p=0}^{P-1} \mathbf{H}(p) \mathbf{s}(t-p)$$

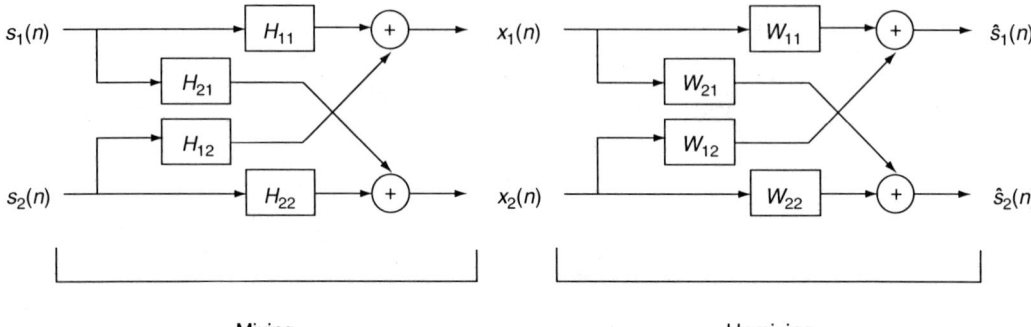

FIGURE 10.5
Specific (2 × 2) blind separation of convolutive mixtures problem.

$$= \mathbf{H}(p) * \mathbf{s}(t) \tag{10.5}$$

where $\mathbf{x}(t) = [x_1(t), \ldots, x_J(t)]^T$ is the received signal vector, $\mathbf{s}(t) = [s_1(t), \ldots, s_I(t)]^T$ is the source vector, $\mathbf{H}(p)$ is the mixing filter matrix, $*$ denotes the convolution, and $[\bullet]^T$ means the transpose operation. Without loss of generality, we assume $J = I$ in this chapter. With this problem setup, the objective of the BSCM techniques is to find an unmixing filter $\mathbf{W}(q)$ of length $\mathbf{Q}$ for deconvolution, that is,

$$\hat{\mathbf{s}}(t) = \mathbf{W}(q) * \mathbf{x}(t) \tag{10.6}$$

where $\hat{\mathbf{s}}(t)$ denotes the estimated sources. Figure 10.5 is a block diagram for $J = I = 2$.

For nonstationary sources, a moving block is usually performed on received signals

$$\hat{\mathbf{x}}(t,m) = \mathbf{H}(p) * \hat{\mathbf{s}}(t,m), \quad t = 1, \ldots, N, \, m = 1, \ldots, M \tag{10.7}$$

where m is the index for blocks, M is the total number of blocks, and N is the length of each block. After the fast Fourier transform (FFT) is applied on the three components in Equation 10.7, we obtain

$$\hat{\mathbf{X}}(w,m) = \mathbf{H}(w)\hat{\mathbf{S}}(w,m), \quad w = 1, \ldots, N \tag{10.8}$$

where w denotes the frequency bin. Thus, the convolutive mixtures in time-domain turn into the instantaneous mixtures in frequency domain. This is the basic idea of the frequency method for BSCM. Then, the separation is done in each frequency bin for the unmixed filters,

$$\hat{\mathbf{Y}}(w,m) = \mathbf{W}(w)\hat{\mathbf{X}}(w,m) \tag{10.9}$$

where $\mathbf{W}(w)$ and $\hat{\mathbf{Y}}(w,m)$ are the unmixed filter and unmixed signals in frequency bin w, respectively.

Since the envelope of the signal spectrum is correlated, $\hat{\mathbf{X}}_j(w,m)$ must be a colored signal. Under the independence assumption of signals in the time and frequency domains, we apply the cost function as follows:

$$\Phi = \frac{1}{4} \sum_{\tau=0}^{T-1} \|\mathbf{W}(w,r)\mathbf{G}_{\hat{x}}(w,\tau)\mathbf{W}^H(w,r) - \mathbf{I}\|^2 \tag{10.10}$$

The algorithm for the unmixing matrix is given by

$$\mathbf{W}(w,r+1) = \mathbf{W}(w,r) + \mu(w) \sum_{\tau=0}^{T-1} \left[\mathbf{W}(w,r)\mathbf{G}_{\hat{x}}(w,\tau)\mathbf{W}^H(w,r) - \mathbf{I} \right] \mathbf{W}(w,r)\mathbf{G}_{\hat{x}}(w,\tau) \quad (10.11)$$

where μ is the learning rate, r is the iteration time, T is the total number of delayed samples, $[\bullet]^H$ is the conjugate transpose operation, $\mathbf{I}$ is the identity matrix, and with τ samples delayed, $\mathbf{G}_{\hat{x}}(w,\tau)$ is the autocorrelation matrix of $\hat{x}(w,m)$. BSS may introduce two indeterminacies including the permutation and scaling problems [14,17]. We apply the method in Ref. [19] to resolve the permutation. More methods may be found in Refs. [20–32]. Also, the scaling ambiguity is a nearly open problem to deal with. Some methods to deal with this problem are provided in Refs. [33,34]. Generally, to normalize the permuted estimated matrix is a fast and effective way. This is helpful in maintaining a stable convergence too [35]. We do not discuss this much here. After the inverse FFT is applied on $\mathbf{W}(w)$ we get $\mathbf{W}(q)$, and then BSCM is performed according to Equation 10.6. For more information about BSCM, please refer to the literatures mentioned above.

10.4 Experiments and Analysis

10.4.1 Backward Matrix of Active Sonar Data for Approximating the BSCM Model

In Section 10.3, BSCM is introduced with some assumptions. However, in applications like speech enhancement, remote sensing, communication systems, geophysics, and biomedical engineering, conditions may not meet the basic theoretical requirements of BSCM. Interestingly, as long as algorithms are converged, BSCM still performs signal deconvolution effectively after some approximations are made [14,17]. For active sonar, the target echo and the oceanic reverberation are all generated by the same transmitted signal, so they must be somewhat correlated nevertheless. Principal component analysis [36] is the method for decorrelation, and PCI has been applied in separating the target echo and reverberation with some prior knowledge provided [5–13]. As Neumann and Krawczyk have pointed out, the forward matrix model used is not a principal issue of the PCI algorithm, and any other forward model can be used [37]. Approximation implies that the processing result must be discounted under the comparison to the corresponding theory. Fortunately, as we pointed out in Section 10.2 the detection in the presence of reverberation may also be reliable with the matched filter even when the reverberation is not absolutely canceled, but the SRR should be within a certain limit. So, the approximation to BSCM is possible by the one-dimensional active sonar data. Intuitively, we generate the backward data matrix from active sonar data $d(t)$ as the received signals $\mathbf{x}(t)$ to approximate the classical BSCM model:

$$\mathbf{x}(t) = \left[x_1(t), \ldots, x_J(t) \right]^T$$
$$x_j(t) = d[t + j(B-1) - 1] \quad (10.12)$$

where $x_j(t)$ is the so-called jth received signal, J is the total number of received signals, and B is the size of a moving block, and it is an empirical parameter. By adjusting J and B, we may produce a different backward matrix. Also, $x_j(t)$ will not miss the target because the two numbers J and B are relatively small in comparison to the whole length of $d(t)$. The losses of $x_j(t)$ are at the beginning of the received reverberation and are often not useful. These make the approximation entirely reasonable.

10.4.2 Examples of Canceling Real Sea Reverberation

To simplify the discussion in this chapter, we only outline briefly the experimental setup here. Further details are in Ref. [11] about the underwater acoustic data and experiments. The transmitted signal is sinusoid with a frequency of 1750 Hz, the duration is 200 msec, and the sampling frequency is 6000 Hz. The examples are all with different targets simulated and added into the real sea reverberation. Before the targets are added, the reverberation is through the band-pass filter from 1500 to 2000 Hz. The reverberation we use is over 6 sec. For an enlarged plot, we show only the first 4 sec, which has no effect on the result as the reverberation is very light in the last 2 sec. The block size is $N = 200$ for Equation 10.7, and the total number of delayed samples is $T = M$ for Equation 10.9. Since the energy varies in different frequency bins, the learning rate $\mu(w)$ is supposed to correspond to the various frequency bins, and then it is defined as in Ref. [38]:

$$\mu(w) = \frac{0.6}{\sqrt{\sum_{\tau=0}^{T-1} |G_{\tilde{x}}(w,\tau)|^2}} \quad (10.13)$$

10.4.2.1 Example 1: Simulation 1

We take the example of the last plot in Figure 10.3. The corresponding part in Figure 10.4 shows that the detection is no good. A sinusoid target echo of 1750 Hz is simulated and added to the reverberation with SRR $= -14$ dB. The target echo is located between the 9,001st and 11,000th samples. No Doppler effect is produced, and $J = 38$ and $B = 1$. After BSCM is performed on the backward matrix, J deconvolved signals are given, and then, all signals pass through the matched filter. We compute the kurtosis of all the outputs of the matched filter and choose the 18th deconvolved signal that has the highest kurtosis in Figure 10.6 as the BSCM result $\tilde{d}(t)$. In the next simulations, the BSCM results follow this rule. Figure 10.7 is the matched filter on the BSCM result $\tilde{d}(t)$. By contrast to the fourth plot in Figure 10.4, it is apparent that the detection is more reliable after BSCM. Next, we show an example with Doppler effect.

10.4.2.2 Example 2: Simulation 2

A sinusoid target echo of 1745 Hz is simulated and added to the reverberation with SRR $= -19$ dB. The target echo is located between the 12,001st and 14,000th samples. Doppler effect exists, and $J = 20$ and $B = 1$. Figure 10.8 is the matched filter on $d(t)$ and the BSCM result $\tilde{d}(t)$. Figure 10.8 indicates that the BSCM is more effective when the Doppler effect is introduced because the reverberation and the target echo are more different. Figure 10.7 and Figure 10.8 show that the SRR is improved to the degree that the detection is more reliable than before. Since the transmitted signal and the target echo are all sinusoids of a single frequency or with very small bandwidth, obvious changes in frequency domain between the original $d(t)$ and the BSCM result $\tilde{d}(t)$ are not found. So, we perform the third

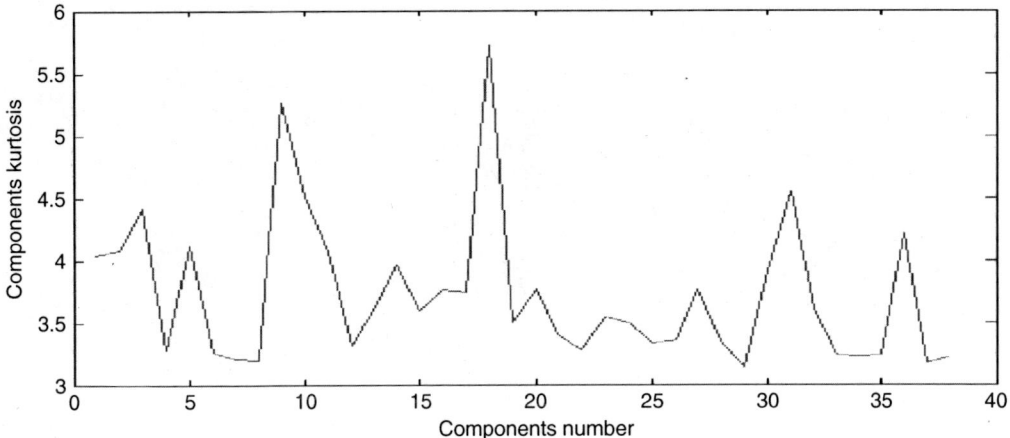

FIGURE 10.6
Kurtosis of outputs from the matched filter on the deconvoluted signals.

simulation. Although it is not very reasonable, it does help to show the validation of BSCM algorithm.

10.4.2.3 Example 3: Simulation 3

A hyperbolic frequency modulated (HFM) target echo is simulated and added to the reverberation with the SRR = −28 dB. The center frequency is 1750 Hz and the bandwidth is 500 Hz. The target echo is located between the 12,001st and 18,000th samples, and $J = 30$ and $B = 2$. Figure 10.9 is the matched filter on $d(t)$ and the BSCM result $\tilde{d}(t)$. In Figure 10.9, it is amazing that the detection is improved greatly after BSCM. To compare the changes in frequency domain, we take out 6000 samples as the target echo from the 12,001st to 18,000th samples $d(t)$ and $\tilde{d}(t)$, respectively, and then, FFT is applied on the two groups of data. Figure 10.10 shows the changes. It is apparent that the SRR has improved a lot after BSCM was applied.

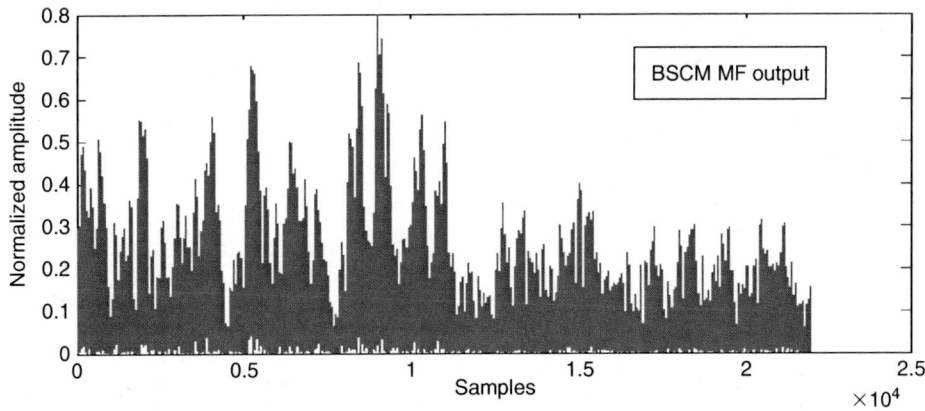

FIGURE 10.7
Output of matched filter on the BSCM result on reverberation with a sinusoid simulated target (SRR = −14 dB, no Doppler effect exists).

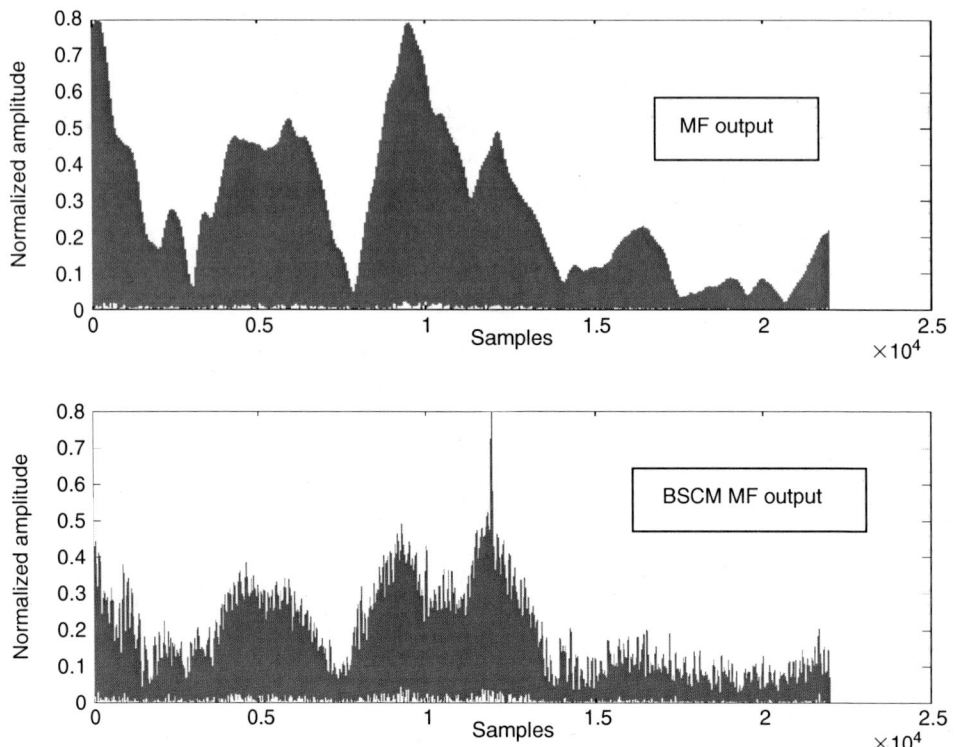

FIGURE 10.8
Outputs of matched filter on a real reverberation with a sinusoid simulated target (SRR = −19 dB, Doppler effect exists).

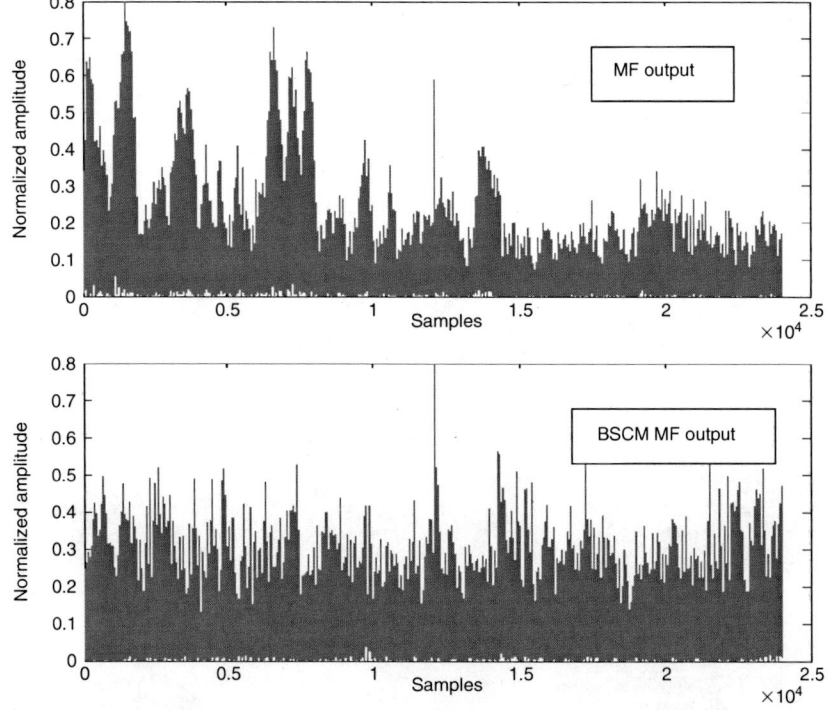

FIGURE 10.9
Outputs of matched filter on reverberation with a HFM simulated target with SRR = −28 dB.

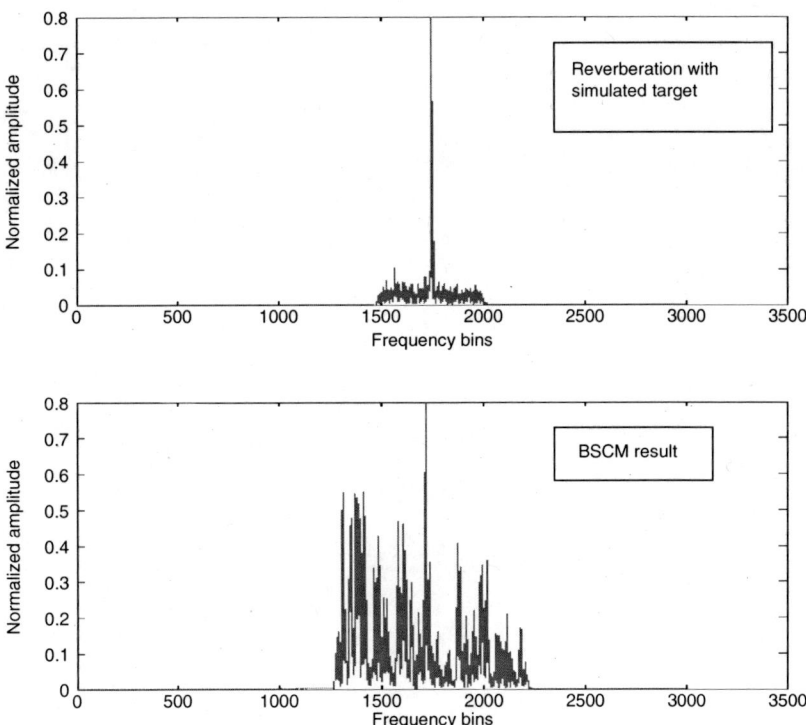

FIGURE 10.10
The changes in frequency domain.

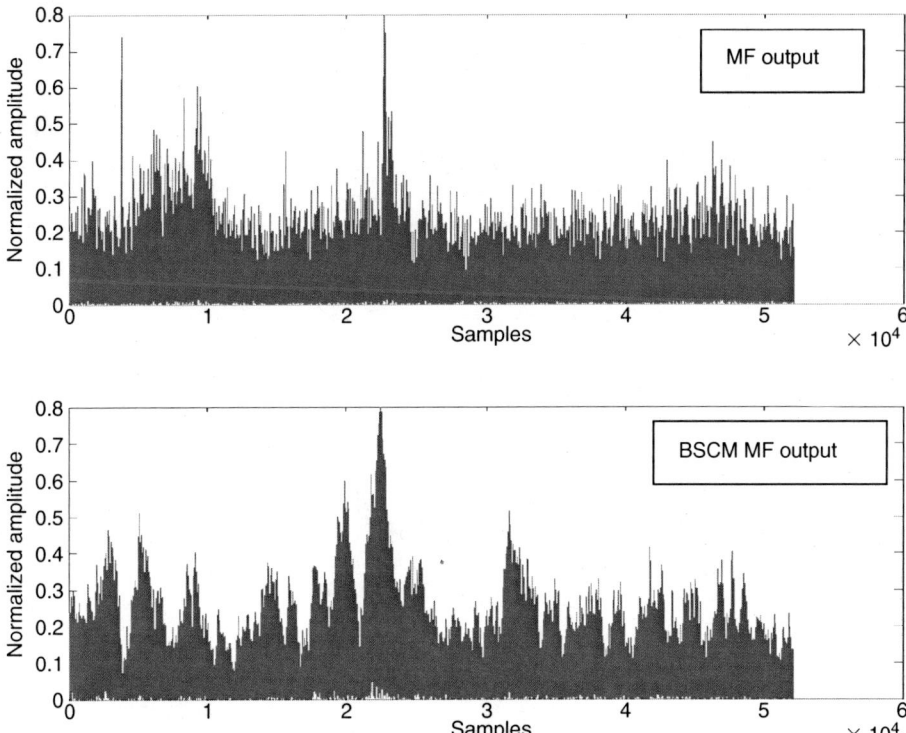

FIGURE 10.11
Outputs of matched filter on the active sonar data and the BSCM output in real target experiment.

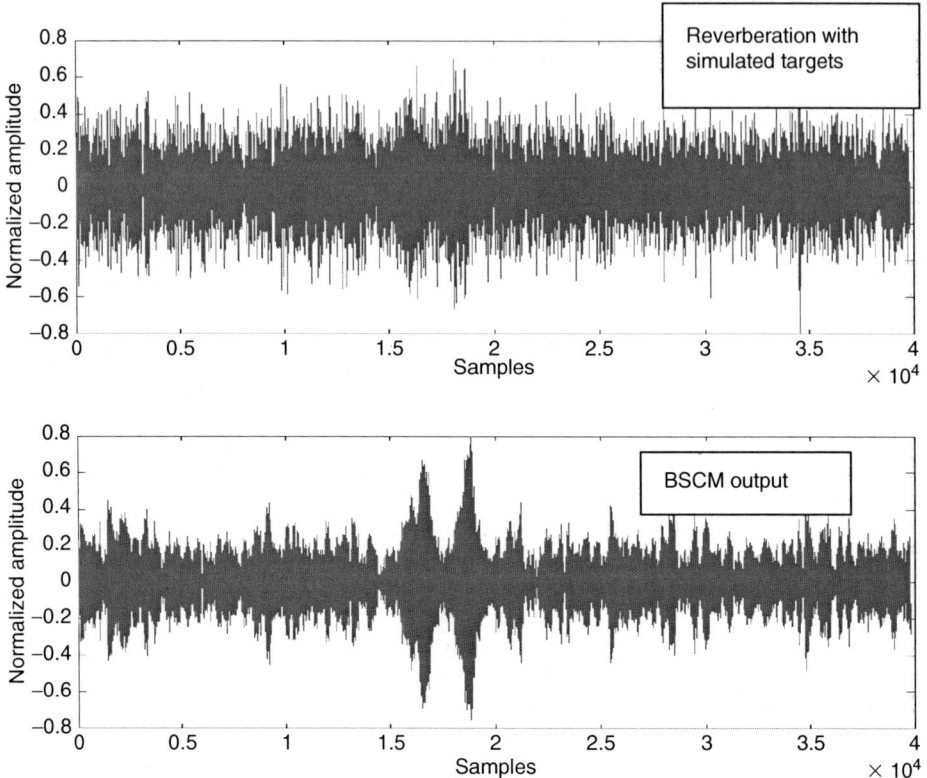

FIGURE 10.12
Waveforms of deep sea reverberation with two simulated targets and BSCM output—no overlap between the target echoes.

10.4.2.4 Example 4: Real Target Experiment

To simplify the discussion in this paper, we only outline briefly the experimental setup here. Further details on the underwater acoustic data and experimental results are in Ref. [11]. The transmitted signal is HFM, the duration is 1 sec, and the bandwidth is 500 Hz. The target speed is about 2.0 m/sec, the range is about 2500 m. The sampling frequency is also 6000 Hz. The data we use is after beamforming. Figure 10.11 shows the matched filter output of the original data and the BSCM output signals. The figure reveals the effect in canceling oceanic reverberation by BSCM without any expensive prior knowledge.

10.4.2.5 Example 5: Simulation 4

The above examples are all reverberations in shallow water. In this simulation, we give an example for canceling deep reverberation. The sea depth is beyond 4000 m. The sampling frequency is also 20,000 Hz. The transmitted signal is HFM and the bandwidth is 100 Hz. In deep sea, more than one target echo may occur within the reverberation issued by one ping; that is, Equation 10.1 is now written as

$$d(t) = \sum_{k=1}^{K_1} \underbrace{h_k(t - \tau_k)}_{\text{target}} \times e(t) + \sum_{k=K_1+1}^{K_2} \underbrace{h_k(t - \tau_k)}_{\text{reverberation}} \times e(t) + \underbrace{n(t)}_{\text{noise}} \qquad (10.14)$$

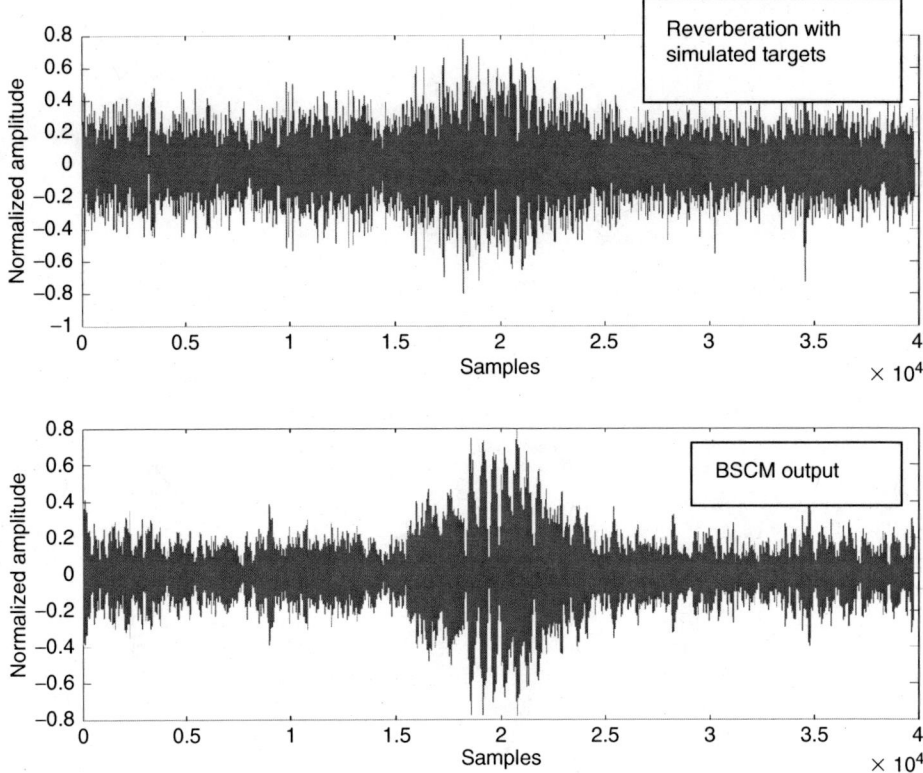

FIGURE 10.13
Waveforms of deep sea reverberation with two simulated targets and BSCM output—overlap between the target echoes.

In this example two target echoes are simulated. The SRR is 0 dB. The first case is shown in Figure 10.12. The two echoes do not overlap, and the duration is 0.05 sec. The second overlaps at the rate of 66.7% as shown in Figure 10.13 and the duration is 0.3 sec. Both Figure 10.12 and Figure 10.13 show that SRR has improved. This simulation implies that BSCM is also effective in canceling deep sea reverberation.

In all simulations, we do not use BSS, but it does not imply that BSS is not useful. Compared to BSCM, BSS needs larger J and B, so the computation is more expensive. Sometimes, BSS cannot provide sound results, but BSCM can be used because it is much closer to the convolutive model of the active sonar data.

10.5 Conclusions

In this chapter we applied three useful and easily obtainable statistical characteristics to remove reverberation by deconvolving and distinguishing the target echoes from the reverberation echoes. They are the nonstationarity and the colored features of reverberation, and the non-Gaussianity of outputs of the matched filter on BSCM results. Except for statistical information, BSCM usually does not need other expensive prior knowledge, such as the target echo energy or the Doppler shift, and so on. This is the advantage of our method. Though the active sonar data used in this paper is one-dimensional convolutive mixtures, the elimination of real sea reverberation proves that the approximation to the BSCM model is effective through backward data matrix. Examples show that BSCM results can improve the SRR to a degree of satisfactory detection.

Finding a good criterion for choosing the appropriate data matrix, as in Ref. [15], is open for exploration.

References

1. T.J. Barnard and F. Khan, Statistical normalization of spherically invariant non-Gaussian clutter, *IEEE Journal of Oceanic Engineering*, 29(2), 303–309, 2004.
2. J.M. Fialkowski, R.C. Gauss, and D.M. Drumheller, Measurements and modeling of low-frequency near-surface scattering statistics, *IEEE Journal of Oceanic Engineering*, 29(2), 197–214, 2004.
3. K.D. LePage, Statistics of broad-band bottom reverberation predictions in shallow-water waveguides, *IEEE Journal of Ocean Engineering*, 29(2), 330–346, 2004.
4. J.R. Preston and D.A. Abraham, Non-Rayleigh reverberation characteristics near 400 Hz observed on the New Jersey shelf, *IEEE Journal of Oceanic Engineering*, 29(2), 215–235, 2004.
5. I.P. Kirsteins and D.W. Tufts, Adaptive detection using low rank approximation to a data matrix, *IEEE Transactions on Aerospace and Electronic Systems*, 30(1), 55–67, 1994.
6. I.P. Kirsteins and D.W. Tufts, Rapidly adaptive nulling of interference, *IEEE International Conference on Systems Engineering*, pp. 269–272, 24–26 August 1989.
7. B.E. Freburger and D.W. Tufts, Case study of principal component inverse and cross spectral metric for low rank interference adaptation, in *Proceedings of ICASSP '98*, Vol. 4, pp. 1977–1980, 12–15 May 1998.
8. B.E. Freburger and D.W. Tufts, Adaptive detection performance of principal components inverse, cross spectral metric and the partially adaptive multistage Wiener filter, *Conference Record of the Thirty-Second Asilomar Conference on Signals, Systems & Computers*, Vol. 2, pp. 1522–1526, 1–4 November 1998.
9. B.E. Freburger and D.W. Tufts, Rapidly adaptive signal detection using the principal component inverse (PCI) method, *Conference Record of the Thirty-First Asilomar Conference on Signals, Systems & Computers*, Vol. 1, pp. 765–769, 2–5 November 1997.
10. T.A. Palka and D.W. Tufts, Reverberation characterization and suppression by means of principal components, in *OCEANS '98 Conference Proceedings*, Vol. 3, pp. 1501–1506, 28 September–1 October 1998.
11. G. Ginolhac and G. Jourdain, Principal component inverse algorithm for detection in the presence of reverberation, *IEEE Journal of Oceanic Engineering*, 27(2), 310–321, 2002.
12. G. Ginolhac and G. Jourdain, Detection in presence of reverberation, *OCEANS 2000 MTS/IEEE Conference and Exhibition*, Vol. 2, pp. 1043–1046, 11–14 September 2000.
13. V. Carmillet, P.O. Amblard, and G. Jourdain, Detection of phase- or frequency-modulated signals in reverberation noise, *JASA*, 105(6), 3375–3389, 1999.
14. A. Cichocki and S. Amari, *Adaptive Blind Signal Image Processing: Learning Algorithm and Application*, John Wiley & Sons, New York, 2002.
15. F. Cong et al., Blind signal separation and reverberation canceling with active sonar data, in *Proceedings of ISSPA*, Australia 2005.
16. G.S. Edelson and I.P. Kirsteins, Modeling and suppression of reverberation components, *IEEE Seventh SP Workshop on Statistical Signal and Array Processing*, pp. 437–440, 26–29 June 1994.
17. A. Hyvärinen, J. Karhunen, and E. Oja, *Independent Component Analysis*, Wiley Interscience, New York, 2001.
18. D.A. Abraham and A.P. Lyons, Simulation of non-Rayleigh reverberation and clutter, *IEEE Journal of Oceanic Engineering*, 29(2), 347–362, 2004.

19. F. Cong et al., Approach based on colored character to blind deconvolution for speech signals, in *Proceedings of ICIMA*, pp. 396–399, 2004.
20. S. Amari, S.C. Douglas, A. Cichocki, and H.H. Yang, Multichannel blind deconvolution and equalization using the natural gradient, in *Proceedings of IEEE Workshop Signal Processing Advances Wireless Communications*, pp. 101–104, April 1997.
21. M. Kawamoto, K. Matsuoka, and N. Ohnishi, A method of blind separation for convolved nonstationary signals, *Neurocomputing*, 22, 157–171, 1998.
22. S.C. Douglas and X. Sun, Convolutive blind separation of speech mixtures using the natural gradient, *Speech Communications*, 39, 65–78, 2003.
23. P. Smaragdis, Blind separation of convolved mixtures in the frequency domain, *Neurocomputing*, 22, 21–34, 1998.
24. H. Sawada, R. Mukai, S. Araki, and S. Makino, Polar coordinate based nonlinear function for frequency domain blind source separation, *IEICE Transactions Fundamentals*, E86-A(3), 590–596, 2003.
25. L. Schobben and W. Sommen, A frequency domain blind signal separation method based on decorrelation, *IEEE Transactions on Signal Processing*, 50, 1855–1865, 2002.
26. H. Sawada, R. Mukai, S. Araki, and S. Makino, A robust and precise method for solving the permutation problem of frequency-domain blind source separation, *IEEE Transactions on Speech and Audio Processing*, 12(5), 530–538, 2004.
27. W. Lu and J.C. Rajapakse, Eliminating indeterminacy in ICA, *Neurocomputing*, 50, 271–290, 2003.
28. S. Araki, R. Mukai, S. Makino, T. Nishikawa, and H. Saruwatari, The fundamental limitation of frequency domain blind source separation for convolutive mixtures of speech, *IEEE Transactions on Speech and Audio Processing*, 11(2), 109–116, 2003.
29. M.Z. Ikram and D.R. Morgan, Permutation inconsistency in blind speech separation: investigation and solutions, *IEEE Transactions on Speech and Audio Processing*, 13(1), 1–13, 2005.
30. A. Dapena and L. Castedo, A novel frequency domain approach for separating convolutive mixtures of temporally white signals, *Digital Signal Processing*, 13(2), 301–316, 2003.
31. A. Dapena, M.F. Bugallo, and L. Catcdo, Separation of convolutive mixtures of temporally-white signals: a novel frequency-domain approach, in *Proceedings of Third ICA*, San Diego, California, pp. 179–184, 2001.
32. C. Mejuto, A. Dapena, and L. Casteda, Frequency-domain infomax for blind source separation of convolutive mixtures, in *Proceedings of ICA 2000*, pp. 315–320, Helsinki, 2000.
33. K. Matsuoka and S. Nakashima, Minimal distortion principle for blind source separation, in *Proceedings of ICA*, pp. 722–727, December 2001.
34. N. Murata, S. Ikeda, and A. Ziehe, An approach to blind source separation based on temporal structure of speech signals, *Neurocomputing*, 41(1), 1–24, 2001.
35. W. Lu and J.C. Rajapakse, Eliminating indeterminacy in ICA, *Neurocomputing*, 50, 271–290, 2003.
36. K.I. Diamantaras and S.Y. Kung, *Principal Component and Neural Network Theory and Application*, John Wiley & Sons, New York, 1996.
37. A. Neumann and H. Krawczyk, *Principal Component Inversion*, Training Course on Remote Sensing of Ocean Color, Ahmedabad, India, February 2001.
38. M.Z. Ikram, *Multichannel Blind Separation of Speech Signals in a Reverberant Environment*, Ph.D thesis, Georgia Institute of Technology, 2001.

11

Neural Network Retrievals of Atmospheric Temperature and Moisture Profiles from High-Resolution Infrared and Microwave Sounding Data

William J. Blackwell

CONTENTS

11.1 Introduction 204
11.2 A Brief Overview of Spaceborne Atmospheric Remote Sensing 205
      11.2.1 Geophysical Parameter Retrieval 207
      11.2.2 The Motivation for Computationally Efficient Algorithms 208
11.3 Principal Components Analysis of Hyperspectral Sounding Data 208
      11.3.1 The PC Transform 209
      11.3.2 The NAPC Transform 209
      11.3.3 The Projected PC Transform 209
      11.3.4 Evaluation of Compression Performance Using
            Two Different Metrics 210
            11.3.4.1 PC Filtering 210
            11.3.4.2 PC Regression 211
      11.3.5 NAPC of Clear and Cloudy Radiance Data 212
      11.3.6 NAPC of Infrared Cloud Perturbations 212
      11.3.7 PPC of Clear and Cloudy Radiance Data 214
11.4 Neural Network Retrieval of Temperature and Moisture Profiles 216
      11.4.1 An Introduction to Multi-Layer Neural Networks 216
      11.4.2 The PPC–NN Algorithm 217
            11.4.2.1 Network Topology 218
            11.4.2.2 Network Training 218
      11.4.3 Error Analyses for Simulated Clear and Cloudy Atmospheres 218
      11.4.4 Validation of the PPC–NN Algorithm with AIRS/AMSU
            Observations of Partially Cloudy Scenes over Land and Ocean 220
            11.4.4.1 Cloud Clearing of AIRS Radiances 220
            11.4.4.2 The AIRS/AMSU/ECMWF Data Set 221
            11.4.4.3 AIRS/AMSU Channel Selection 221
            11.4.4.4 PPC–NN Retrieval Enhancements for Variable Sensor
                    Scan Angle and Surface Pressure 223
            11.4.4.5 Retrieval Performance 223
            11.4.4.6 Retrieval Sensitivity to Cloud Amount 223
      11.4.5 Discussion and Future Work 224

11.5 Summary	225
Acknowledgments	228
References	228

11.1 Introduction

Modern atmospheric sounders measure radiance with unprecedented resolution and accuracy in spatial, spectral, and temporal dimensions. For example, the Atmospheric Infrared Sounder (AIRS), operational on the NASA EOS Aqua satellite since 2002, provides a spatial resolution of ~15 km, a spectral resolution of $\nu/\Delta\nu \approx 1200$ (with 2,378 channels from 650 to 2675 cm^{-1}), and a radiometric accuracy on the order of ± 0.2 K. Typical polar-orbiting atmospheric sounders measure approximately 90% of the Earth's atmosphere (in the horizontal dimension) approximately every 12 h. This wealth of data presents two major challenges in the development of retrieval algorithms, which estimate the geophysical state of the atmosphere as a function of space and time from upwelling spectral radiances measured by the sensor. The first challenge concerns the robustness of the retrieval operator and involves maximal use of the geophysical content of the radiance data with minimal interference from instrument and atmospheric noise. The second is to implement a robust algorithm within a given computational budget. Estimation techniques based on neural networks (NNs) are becoming more common in high-resolution atmospheric remote sensing largely because their simplicity, flexibility, and ability to accurately represent complex multi-dimensional statistical relationships allow both of these challenges to be overcome.

In this chapter, we consider the retrieval of atmospheric temperature and moisture profiles (quantity as a function of altitude) from radiance measurements at microwave and thermal infrared wavelengths. A projected principal components (PPC) transform is used to reduce the dimensionality of and optimally extract geophysical information from the spectral radiance data, and a multi-layer feedforward NN is subsequently used to estimate the desired geophysical profiles. This algorithm is henceforth referred to as the "PPC–NN" algorithm. The PPC–NN algorithm offers the numerical stability and efficiency of statistical methods without sacrificing the accuracy of physical, model-based methods.

The chapter is organized as follows. First, the physics of spaceborne atmospheric remote sensing is reviewed. The application of principal components transforms to hyperspectral sounding data is then presented and a new approach is introduced, where the sensor radiances are projected into a subspace that reduces spectral redundancy and maximizes the resulting correlation to a given parameter. This method is very similar to the concept of canonical correlations introduced by Hotelling over 70 years ago [1], but its application in the hyperspectral sounding context is new. Second, the use of multi-layer feedforward NNs for geophysical parameter retrieval from hyperspectral measurements (first proposed in 1993 [2]) is reviewed, and an overview of the network parameters used in this work is given. The combination of the PPC radiance compression operator with an NN is then discussed, and performance analyses comparing the PPC–NN algorithm to traditional retrieval methods are presented.

11.2 A Brief Overview of Spaceborne Atmospheric Remote Sensing

The typical measurement scenario for spaceborne atmospheric remote sensing is shown in Figure 11.1. A sensor measures upwelling spectral radiance (intensity as a function of frequency) at various incidence angles. The sensor data is usually calibrated to remove measurement artifacts such as gain drift, nonlinearities, and noise. The spectral radiances measured by the sensor are related to geophysical quantities, such as the vertical temperature profile of the atmosphere, and therefore must be converted into a geophysical quantity of interest through the use of an appropriate retrieval algorithm.

The radiative transfer equation describing the radiation intensity observed at altitude L, viewing angle θ, and frequency ν can be formulated by including reflected atmospheric and cosmic contributions and the radiance emitted by the surface as follows [3,4]:

$$R_\nu(L) = \int_0^L \kappa_\nu(z) J_\nu[T(z)] \exp\left(-\int_z^L \sec\theta \kappa_\nu(z') dz'\right) \sec\theta \, dz$$
$$+ \rho_\nu e^{-\tau^* \sec\theta} \int_0^L \kappa_\nu(z) J_\nu[T(z)] \exp\left(-\int_0^z \sec\theta \kappa_\nu(z') dz'\right) \sec\theta \, dz$$
$$+ \rho_\nu e^{-2\tau^* \sec\theta} J_\nu(T_c)$$
$$+ \varepsilon_\nu e^{-\tau^* \sec\theta} J_\nu(T_s) \tag{11.1}$$

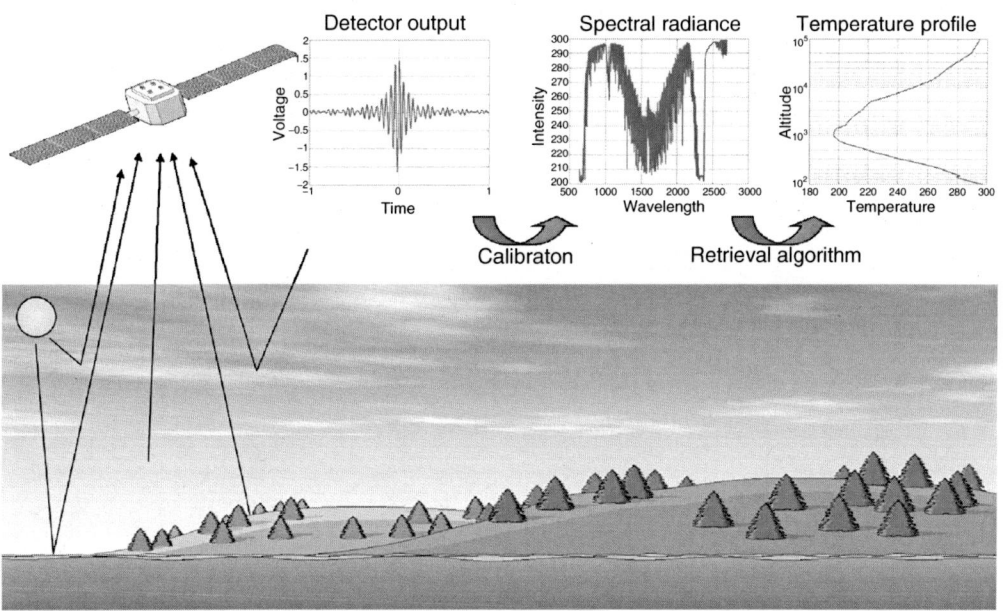

FIGURE 11.1
Typical measurement scenario for spaceborne atmospheric remote sensing. Electromagnetic radiation that reaches the sensor is emitted by the sun, cosmic background, atmosphere, surface, and clouds. This radiation can also be reflected or scattered by the surface, atmosphere, or clouds. The spectral radiances measured by the sensor are related to geophysical quantities, such as the vertical temperature profile of the atmosphere, and therefore must be converted into a geophysical quantity of interest through the use of an appropriate retrieval algorithm.

where ε_ν is the surface emissivity, ρ_ν is the surface reflectivity, T_s is the surface temperature, $\kappa_\nu(z)$ is the atmospheric absorption coefficient, τ^* is the atmospheric zenith opacity, T_c is the cosmic background temperature (2.736 ± 0.017 K), and $J_\nu(T)$ is the radiance intensity emitted by a blackbody at temperature T, which is given by the Planck equation:

$$J_\nu(T) = \frac{h\nu^3}{c^2} \frac{1}{e^{h\nu/kT} - 1} \text{W} \cdot \text{m}^{-2} \cdot \text{ster}^{-1} \cdot \text{Hz}^{-1} \tag{11.2}$$

The first term in Equation 11.1 can be recast in terms of a transmittance function $T_\nu(z)$:

$$R_\nu(L) = \int_0^L J_\nu[T(z)] \left(\frac{dT_\nu(z)}{dz}\right) dz \tag{11.3}$$

The derivative of the transmittance function with respect to altitude is often called the temperature weighting function

$$W_\nu(z) \frac{dT_\nu(z)}{dz} \tag{11.4}$$

and gives the relative contribution of the radiance emanating from each altitude. The temperature weighting functions for the Advanced Microwave Sounding Unit (AMSU) are shown in Figure 11.2.

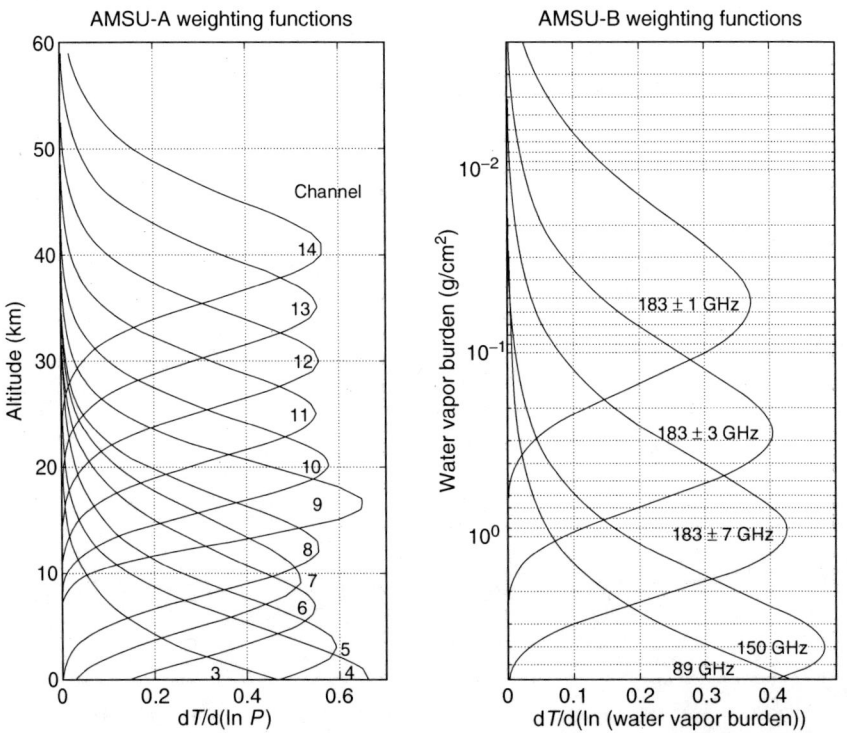

FIGURE 11.2
AMSU-A temperature profile (left) and AMSU-B water vapor profile (right) weighting functions

11.2.1 Geophysical Parameter Retrieval

The objective of the geophysical parameter retrieval algorithm is to estimate the state of the atmosphere (represented by parameter matrix **X**, say), given observations of spectral radiance (represented by radiance matrix **R**, say). There are generally two approaches to this problem, as shown in Figure 11.3. The first approach, referred to here as the variational approach, uses a forward model (for example, the transmittance and radiative transfer models previously discussed) to calculate the sensor radiance that would be measured given a specific atmospheric state. Note that the inverse model typically does not exist, as there are generally an infinite number of atmospheric states that could give rise to a particular radiance measurement. In the variational approach, a "guess" of the atmospheric state is made (this is usually obtained through a forecast model or historical statistics), and this guess is propagated through the forward models thereby producing an estimate of the at-sensor radiance. The measured radiance is compared with this estimated radiance, and the state vector is adjusted so as to reduce the difference between the measured and estimated radiance vectors. Details on this methodology are discussed at length by Rodgers [5], and the interested reader is referred there for a more thorough treatment of the methodology and implementation of variational retrieval methods. The second approach, referred to here as the statistical, or regression-based, approach, does not use the forward model explicitly to derive the estimate of the atmospheric state vector. Instead, an ensemble of radiance–state vector pairs is assembled, and a statistical characterization ($p(\mathbf{X})$, $p(\mathbf{R})$, and $p(\mathbf{X},\mathbf{R})$) is sought. In practice, it is difficult to obtain these probability density functions (PDFs) directly from the data, and alternative methods are often used. Two of these methods are linear least-squares estimation (LLSE), or linear regression, and nonlinear least-squares estimation (NLLSE). NNs are a special class of NLLSEs, and will be discussed later.

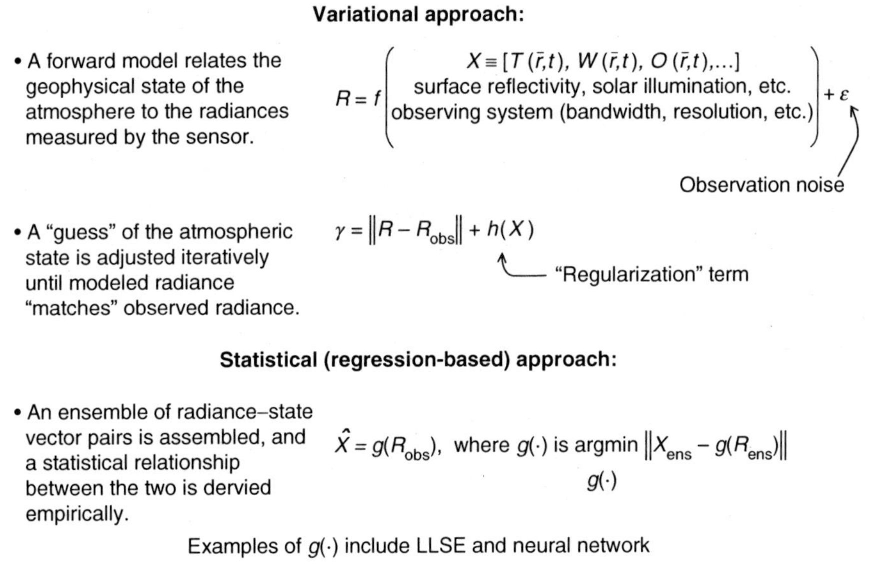

Variational approach:

- A forward model relates the geophysical state of the atmosphere to the radiances measured by the sensor.

$$R = f\left(\begin{array}{c} X \equiv [T(\bar{r},t),\, W(\bar{r},t),\, O(\bar{r},t),\ldots] \\ \text{surface reflectivity, solar illumination, etc.} \\ \text{observing system (bandwidth, resolution, etc.)} \end{array}\right) + \varepsilon$$

Observation noise

- A "guess" of the atmospheric state is adjusted iteratively until modeled radiance "matches" observed radiance.

$$\gamma = \|R - R_{obs}\| + h(X)$$

"Regularization" term

Statistical (regression-based) approach:

- An ensemble of radiance–state vector pairs is assembled, and a statistical relationship between the two is dervied empirically.

$$\hat{X} = g(R_{obs}), \text{ where } g(\cdot) \text{ is } \underset{g(\cdot)}{\operatorname{argmin}} \|X_{ens} - g(R_{ens})\|$$

Examples of $g(\cdot)$ include LLSE and neural network

FIGURE 11.3
Variational and statistical approaches to geophysical parameter retrieval. In the variational approach, a forward model is used to predict at-sensor radiances based on atmospheric state. In the statistical approach, an empirical relationship between at-sensor radiances and atmospheric state is derived using an ensemble of radiance–state vectors.

11.2.2 The Motivation for Computationally Efficient Algorithms

The principal advantage of regression-based methods is their simplicity—once the coefficients are derived from "training" data, the calculation of atmospheric state vectors is relatively easy. The variational approaches require multiple calls to the forward models, which can be computationally prohibitive. The computational complexity of the forward models is usually nonlinearly related (often $O(n^2)$ or more) to the number of spectral channels. As shown in Figure 11.4, the spectral and spatial resolution of infrared sounders has increased dramatically over the last 35 years, and the required computation needed for real-time operation with variational algorithms has outpaced Moore's Law. There is, therefore, a motivation to reduce the computational burden of current and next-generation retrieval algorithms to allow real-time ingestion of satellite-derived geophysical products into numerical weather forecast models.

11.3 Principal Components Analysis of Hyperspectral Sounding Data

Principal components (PC) transforms can be used to represent radiance measurements in a statistically compact form, enabling subsequent retrieval operators to be substantially

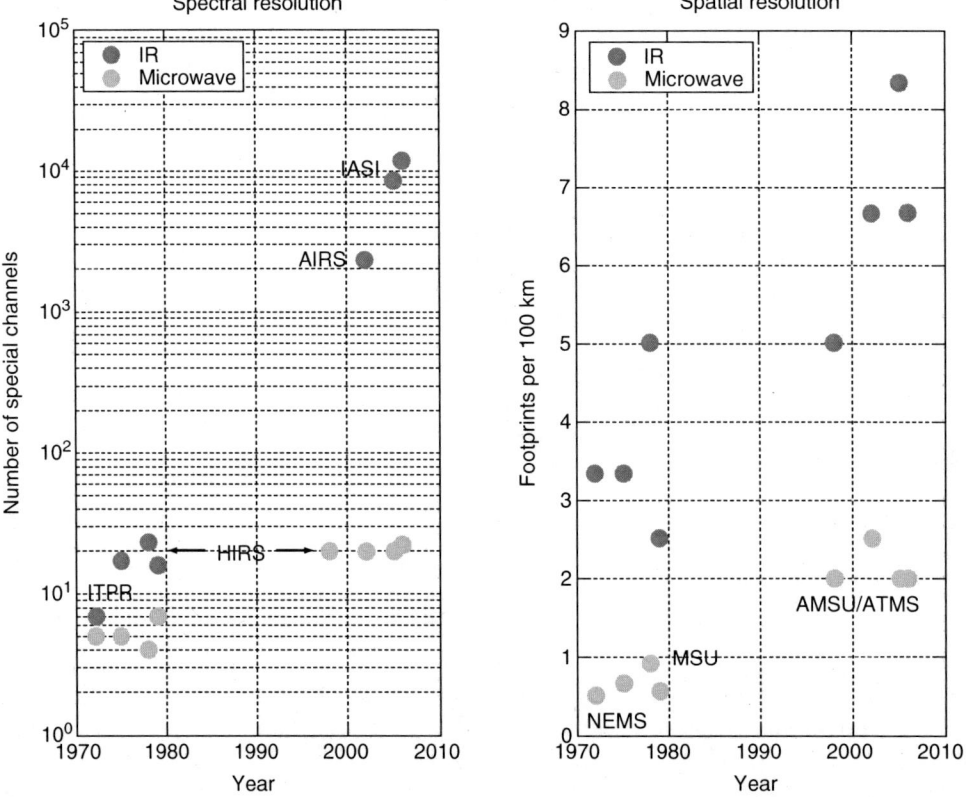

FIGURE 11.4
Improvements in sensor spectral and spatial resolution over the last 35 years is shown. The recent increases in the spectral resolutions afforded by infrared sensors has far surpassed that available from microwave sensors. The trends in spatial resolution are similar for infrared and microwave sensors.

more efficient and robust (see Ref. [6], for example). Furthermore, measurement noise can be dramatically reduced through the use of PC filtering [7,8], and it has also been shown [9] that PC transforms can be used to represent variability in high-spectral-resolution radiances perturbed by clouds. In the following sections, several variants of the PC transform are briefly discussed, with emphasis focused on the ability of each to extract geophysical information from the noisy radiance data.

11.3.1 The PC Transform

The PC transform is a linear, orthonormal operator[1] $\mathbf{Q}_r^T$, which projects a noisy m-dimensional radiance vector, $\tilde{R} = R + \Psi$, into an r-dimensional ($r \leq m$) subspace. The additive noise vector Ψ is assumed to be uncorrelated with the radiance vector R, and is characterized by the noise covariance matrix $\mathbf{C}_{\Psi\Psi}$. The "PC" of $\tilde{R}$, that is, $\tilde{P} = \mathbf{Q}_r^T \tilde{R}$ have two desirable properties: (1) the components are statistically uncorrelated and (2) the reduced-rank reconstruction error.

$$c_1(\cdot) = E[(\hat{\tilde{R}}_r - \tilde{R})^T(\hat{\tilde{R}}_r - \tilde{R})] \tag{11.5}$$

where $\hat{\tilde{R}}_r \triangleq \mathbf{G}_r \tilde{R}$ for some linear operator $\mathbf{G}_r$ with rank r, is minimized when $\mathbf{G}_r = \mathbf{Q}_r \mathbf{Q}_r^T$. The rows of $\mathbf{Q}_r^T$ contain the r most-significant (ordered by descending eigenvalue) eigenvectors of the noisy data covariance matrix $\mathbf{C}_{\tilde{R}\tilde{R}} = \mathbf{C}_{RR} + \mathbf{C}_{\Psi\Psi}$.

11.3.2 The NAPC Transform

Cost criteria other than in Equation 11.5 are often more suitable for typical hyperspectral compression applications. For example, it might be desirable to reconstruct the noise-free radiances and filter the noise. The cost equation thus becomes

$$c_2(\cdot) = E[(\hat{R}_r - R)^T(\hat{R}_r - R)] \tag{11.6}$$

where $\hat{R}_r \triangleq \mathbf{H}_r \tilde{R}$ for some linear operator $\mathbf{H}_r$ with rank r. The noise-adjusted principal components (NAPC) transform [10], where $\mathbf{H}_r = \mathbf{C}_{\Psi\Psi}^{1/2} \mathbf{W}_r \mathbf{W}_r^T \mathbf{C}_{\Psi\Psi}^{-1/2}$ and $\mathbf{W}_r^T$ contains the r most-significant eigenvectors of the whitened noisy covariance matrix $\mathbf{C}_{\tilde{w}\tilde{w}} = \mathbf{C}_{\Psi\Psi}^{-1/2}(\mathbf{C}_{RR} + \mathbf{C}_{\Psi\Psi})\mathbf{C}_{\Psi\Psi}^{-1/2}$, maximizes the signal-to-noise ratio of each component, and is superior to the PC transform for most noise-filtering applications where the noise statistics are known *a priori*.

11.3.3 The Projected PC Transform

It is often unnecessary to require that the PC be uncorrelated, and linear operators can be derived that offer improved performance over PC transforms for minimizing cost functions such as in Equation 11.6. It can be shown [11] that the optimal linear operator with rank r that minimizes Equation 11.6 is

$$\mathbf{L}_r = \mathbf{E}_r \mathbf{E}_r^T \mathbf{C}_{RR}(\mathbf{C}_{RR} + \mathbf{C}_{\Psi\Psi})^{-1} \tag{11.7}$$

where $\mathbf{E}_r = [E_1 \mid E_2 \mid \ldots \mid E_r]$ are the r most-significant eigenvectors of $\mathbf{C}_{RR}(\mathbf{C}_{RR} + \mathbf{C}_{\Psi\Psi})^{-1}\mathbf{C}_{RR}$. Examination of Equation 11.7 reveals that the Wiener-filtered radiances are projected onto the r-dimensional subspace spanned by $\mathbf{E}_r$. It is this projection that

---

[1] The following mathematical notation is used in this chapter: $(\cdot)^T$ denotes the transpose, $(\tilde{\cdot})$ denotes a noisy random vector, and $(\hat{\cdot})$ denotes an estimate of a random vector. Matrices are indicated by bold upper case, vectors by upper case, and scalars by lower case.

motivates the name "PPC." An orthonormal basis for this r-dimensional subspace of the original m-dimensional radiance vector space $\mathcal{R}$ is given by the r most-significant right eigenvectors, $\mathbf{V}_r$, of the reduced-rank linear regression matrix, $\mathbf{L}_r$, given in Equation 11.7. We then define the PPC of $\tilde{R}$ as

$$\tilde{P} = \mathbf{V}_r^T \tilde{R} \qquad (11.8)$$

Note that the elements of $\tilde{P}$ are correlated, as $\mathbf{V}_r^T(\mathbf{C}_{RR}+\mathbf{C}_{\Psi\Psi})\mathbf{V}_r$ is not a diagonal matrix.

Another useful application of the PPC transform is the compression of spectral radiance information that is correlated with a geophysical parameter, such as the temperature profile. The r-rank linear operator that captures the most radiance information, which is correlated to the temperature profile, is similar to Equation 11.7 and is given below:

$$\mathbf{L}_r = \mathbf{E}_r \mathbf{E}_r^T \mathbf{C}_{TR}(\mathbf{C}_{RR} + \mathbf{C}_{\Psi\Psi})^{-1} \qquad (11.9)$$

where $\mathbf{E}_r = [E_1 | E_2 | \cdots | E_r]$ are the r most-significant eigenvectors of $\mathbf{C}_{TR}(\mathbf{C}_{RR} + \mathbf{C}_{\Psi\Psi})^{-1}\mathbf{C}_{RT}$, and $\mathbf{C}_{TR}$ is the cross-covariance of the temperature profile and the spectral radiance.

11.3.4 Evaluation of Compression Performance Using Two Different Metrics

The compression performance of each of the PC transforms discussed previously was evaluated using two performance metrics. First, we seek the transform that yields the best (in the sum-squared sense) reconstruction of the noise-free radiance spectrum given a noisy spectrum. Thus, we seek the optimal reduced-rank linear filter. The second performance metric is quite different and is based on the temperature retrieval performance in the following way. A radiance spectrum is first compressed using each of the PC transforms for a given number of coefficients. The resulting coefficients are then used in a linear regression to estimate the temperature profile.

The results that follow were obtained using simulated, clear-air radiance intensity spectra from an AIRS-like sounder. Approximately, seven thousand and five-hundred 1750-channel radiance vectors were generated with spectral coverage from approximately 4 to 15 μm using the NOAA88b radiosonde set. The simulated intensities were expressed in spectral radiance units (mW m^{-2} sr^{-1}(cm^{-1})$^{-1}$).

11.3.4.1 PC Filtering

Figure 11.5a shows the sum-squared radiance distortion (Equation 11.5) as a function of the number of components used in the various PC decomposition techniques. The *a priori* level indicates the sum-squared error due to sensor noise. Results from two variants of the PC transform are plotted, where the first variant (the "PC" curve) uses eigenvectors of $\mathbf{C}_{\tilde{R}\tilde{R}}$ as the transform basis vectors, and the second variant (the "noise-free PC" curve) uses eigenvectors of $\mathbf{C}_{RR}$ as the transform basis vectors. It is shown in Figure 11.5a that the PPC reconstruction of noise-free radiances (PPC[R]) yields lower distortion than both the PC and NAPC transforms for any number of components (r). It is noteworthy that the "PC" and "noise-free PC" curves never reach the theoretically optimal level, defined by the full-rank Wiener filter. Furthermore, the PPC distortion curves decrease monotonically with coefficient number, while all the PC distortion curves exhibit a local minimum, after which the distortion increases with coefficient number as noisy, high-order terms are

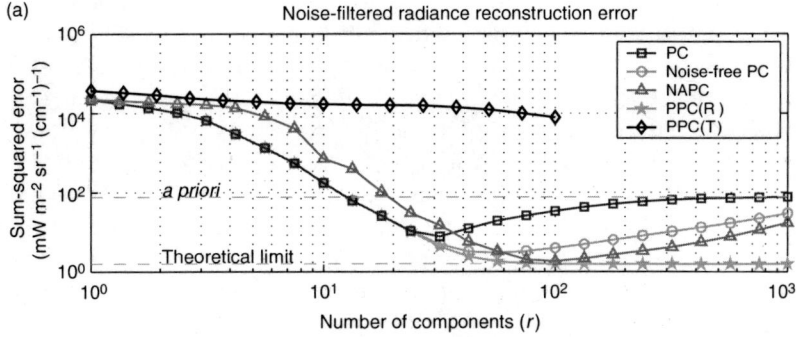

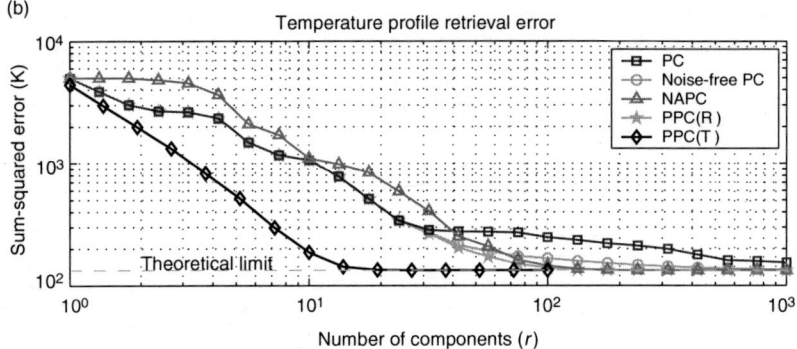

FIGURE 11.5
Performance comparisons of the PC (where the components are derived from both noisy and noise-free radiances), NAPC, and PPC transforms for a hypothetical 1750-channel infrared (4–15 μm) sounder. Two projected principal components transforms were considered, PPC(R) and PPC(T), which are, respectively: (a) maximum representation of noise-free radiance energy, and (b) maximum representation of temperature profile energy. The first plot shows the sum-squared error of the reduced-rank reconstruction of the noise-free spectral radiances. The second plot shows the temperature profile retrieval error (trace of the error covariance matrix) obtained using linear regression with r components.

included. The noise in the high-order PPC terms is effectively zeroed out, because it is uncorrelated with the spectral radiances.

11.3.4.2 PC Regression

The PC coefficients derived in the previous example are now used in a linear regression to estimate the temperature profile. Figure 11.5b shows the temperature profile error (integrated over all altitude levels) as a function of the number of coefficients used in the linear regression, for each of the PC transforms. To reach the theoretically optimal value achieved by linear regression with all channels requires approximately 20 PPC coefficients, 200 NAPC coefficients, and 1000 PC coefficients. Thus, the PPC transform results in a factor of ten improvement over the NAPC transform when compressing temperature-correlated radiances (20 versus 200 coefficients required), and approximately a factor of 100 improvement over the original spectral radiance vector (20 versus 1750). Note that the first guess in the AIRS Science Team Level-2 retrieval uses a linear regression derived from approximately 60 of the most-significant NAPC coefficients of the 2378-channel AIRS spectrum (in units of brightness temperature) [6]. Results for the moisture profile are similar, although more coefficients (typically 35 versus 25 for temperature) are needed because of the higher degree of nonlinearity in the underlying physical relationship between atmospheric moisture and the observed spectral radiance. This substantial compression enables the use of relatively small (and thus very stable and fast) NN estimators to retrieve the desired geophysical parameters.

It is interesting to consider the two variants of the PPC transform shown in Figure 11.5, namely PPC(R), where the basis for the noise-free radiance subspace is desired, and PPC(T), where the basis for only the temperature profile information is desired. As shown in Figure 11.5a, the PPC(T) transform poorly represents the noise-free radiance space, because there is substantial information that is uncorrelated with temperature (and thus ignored by the PPC(T) transform) but correlated with the noise-free radiance. Conversely, the PPC(R) transform offers a significantly less compact representation of temperature profile information (see Figure 11.5b), because the transform is representing information that is not correlated with temperature and thus superfluous when retrieving the temperature profile.

11.3.5 NAPC of Clear and Cloudy Radiance Data

In the following sections we compute the NAPC (and associated eigenvalues) of clear and cloudy radiance data, the NAPC of the infrared radiance perturbations due to clouds, and the projected (temperature) principal components of clear and cloudy radiance data. The 2378 AIRS radiances were converted from spectral intensities to brightness temperatures using Equation 11.2, and were concatenated with the 20 microwave brightness temperatures from AMSU-A and AMSU-B into a single vector R of length 2398. The NAPC were computed as follows:

$$P_{\text{NAPC}} = \mathbf{Q}^T R \quad (11.10)$$

where $\mathbf{Q}$ are the eigenvectors of $\mathbf{C}_{\tilde{W}\tilde{W}}$, sorted in descending order by eigenvalue. $\mathbf{C}_{\tilde{W}\tilde{W}}$ is the prewhitened covariance matrix discussed in Section 11.3. The eigenvalues corresponding to the top 100 NAPC are shown in Figure 11.6 for simulated clear-air and cloudy data. Also shown are scatterplots of the first three NAPC.

The eigenvalues of the 90 lowest order terms are very similar. The principal differences occur in the three highest order terms, which are dominated by channels with weighting function peaks in the lower part of the atmosphere. The eigenvalues associated with the clear-air and cloudy NAPC cluster into roughly five groups: 1, 2–3, 4–9, 10–11, and 12–100. The first 11 NAPC capture 99.96% of the total radiance variance for both the clear-air and cloudy data. The top three NAPCs of both clear-air and cloudy data appear to be jointly Gaussian to a close approximation, with the exception of clear-air NAPC #1 versus NAPC #2.

11.3.6 NAPC of Infrared Cloud Perturbations

We define the infrared cloud perturbation ΔR_{IR} as

$$\Delta R_{\text{IR}} \triangleq R_{\text{IR}}^{\text{clr}} - R_{\text{IR}}^{\text{cld}} \quad (11.11)$$

where $R_{\text{IR}}^{\text{clr}}$ is the clear-air infrared brightness temperature and $R_{\text{IR}}^{\text{cld}}$ is the cloudy infrared brightness temperature. The NAPC of ΔR_{IR} were calculated using the method described above. The results are shown in Figure 11.7.

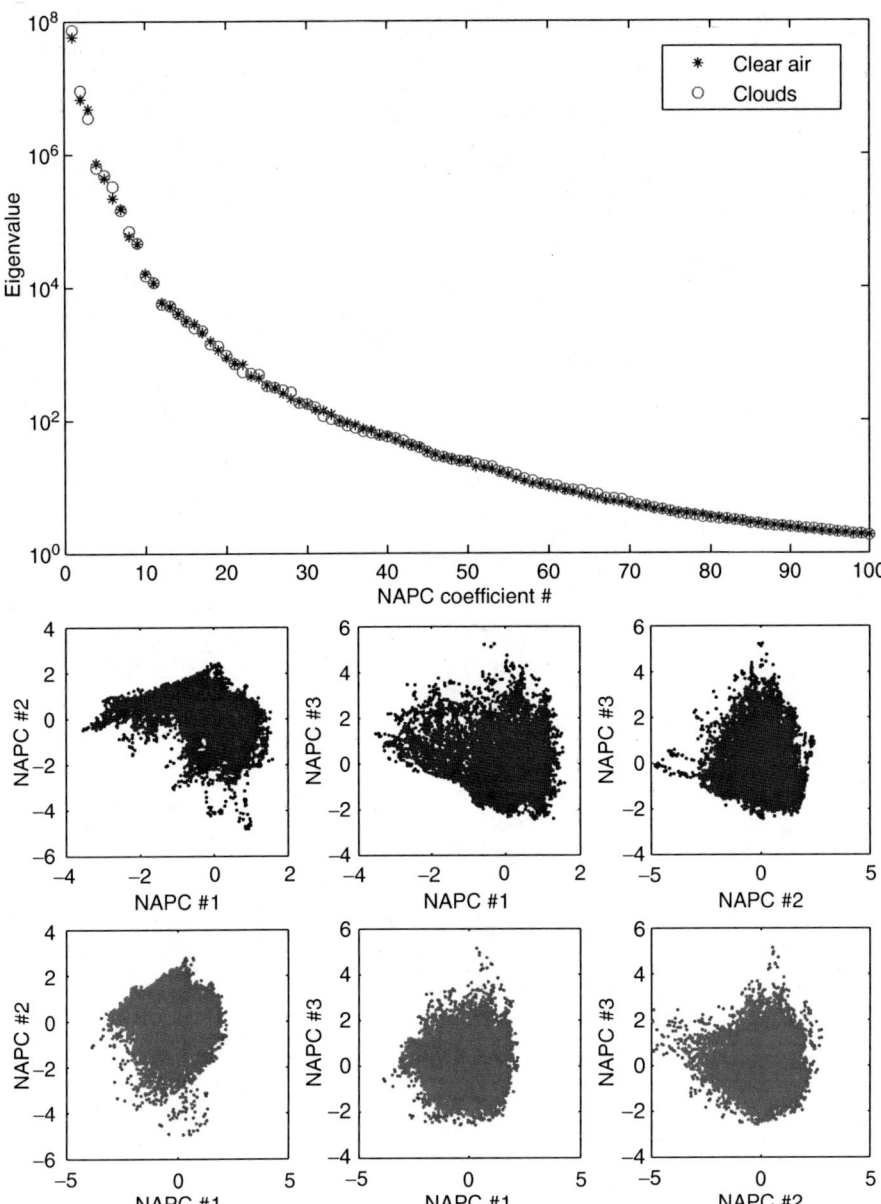

FIGURE 11.6
Noise-adjusted principal components transform analysis of clear and cloudy simulated AIRS/AMSU data. The top plot shows the eigenvalues of each NAPC coefficient for clear and cloudy data. The middle row presents scatterplots of the three clear-air NAPC coefficients with the largest variance (shown normalized to unit variance). The bottom row presents scatterplots of the three cloudy NAPC coefficients with the largest variance (shown normalized to unit variance).

The six highest order NAPC of ΔR_{IR} capture approximately 99.96% of the total cloud perturbation variance, which suggests that there are more degrees of freedom in the atmosphere than there are in the clouds. Furthermore, there is significant crosstalk between the cloud perturbation and the underlying atmosphere, and this crosstalk is highly nonlinear and non-Gaussian. Evidence of this can be seen in the scatterplot of NAPC #1 versus NAPC #2, shown in the lower left corner of Figure 11.7.

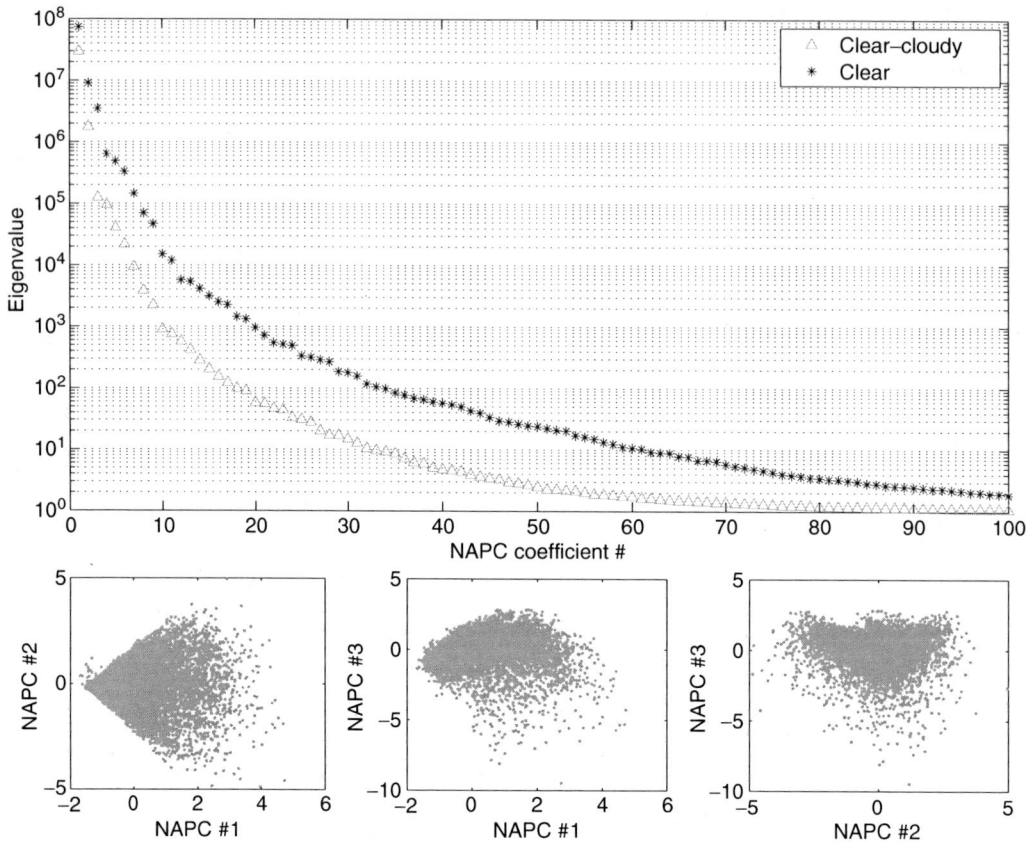

FIGURE 11.7
Noise-adjusted principal components transform analysis of the cloud impact (clear radiance–cloudy radiance) for simulated AIRS data. The top plot shows the eigenvalue of each NAPC coefficient of cloud impact, along with the NAPC coefficients of clear-air data (shown in Figure 11.6). The bottom row presents scatterplots of the three cloud-impact NAPC coefficients with the largest variance (shown normalized to unit variance).

The temperature weighting functions of NAPC #1 and NAPC #2 are shown in Figure 11.8. NAPC #1 consists primarily of surface channels and NAPC #2 consists primarily of channels that peak near 3–6 km and channels that peak near the surface. Therefore, NAPC #1 is sensitive principally to the overall cloud amount, while NAPC #2 is also sensitive to cloud altitude.

11.3.7 PPC of Clear and Cloudy Radiance Data

The PPC transform discussed previously was used to identify temperature information contained in the clear and cloudy radiances. Figure 11.9 shows the mean temperature profile retrieval error for the reduced-rank regression operator given in Equation 11.9 as a function of rank (the number of PPC coefficients retained) for clear-air and cloudy radiance data.

Both curves have asymptotes near 15 coefficients, and clouds degrade the temperature retrieval by an average of approximately 0.3 K RMS.

Retrievals of Atmospheric Temperature and Moisture Profiles

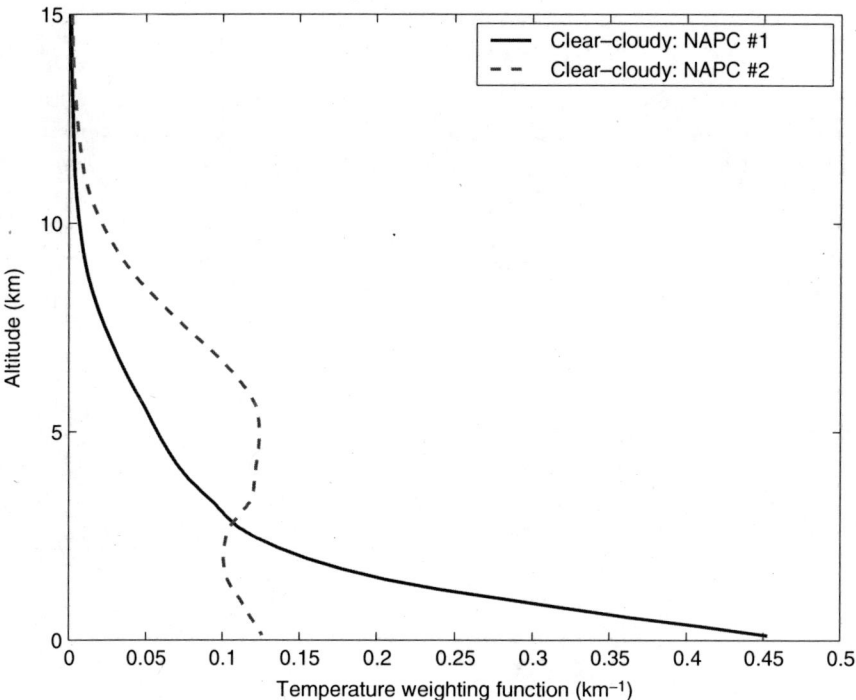

FIGURE 11.8
Temperature weighting functions of the first two noise-adjusted principal components of AIRS cloud perturbations (clear radiance–cloudy radiance).

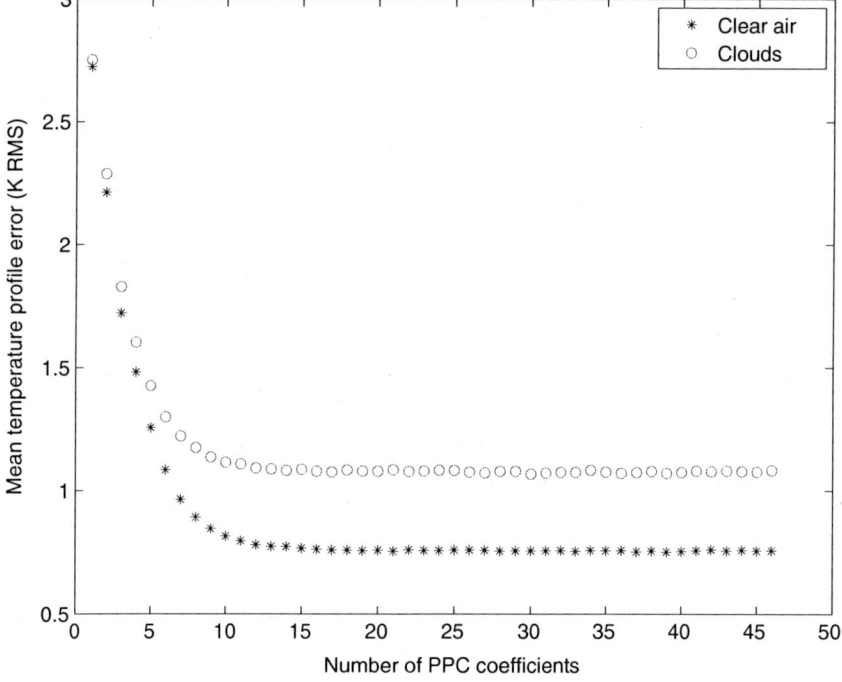

FIGURE 11.9
Projected principal components transform analysis of clear and cloudy simulated AIRS/AMSU data. The mean temperature profile retrieval error (K RMS) is shown as a function of the number of PPC coefficients used in a linear regression for simulated clear and cloudy data.

11.4 Neural Network Retrieval of Temperature and Moisture Profiles

An NN is an interconnection of simple computational elements, or nodes, with activation functions that are usually nonlinear, monotonically increasing, and differentiable. NNs are able to deduce input–output relationships directly from the training ensemble without requiring underlying assumptions about the distribution of the data. Furthermore, an NN with only a single hidden layer of a sufficient number of nodes with nonlinear activation functions is capable of approximating any real-valued continuous scalar function to a given precision over a finite domain [12,13].

11.4.1 An Introduction to Multi-Layer Neural Networks

Consider a multi-layer feedforward NN consisting of an input layer, an arbitrary number of hidden layers (usually one or two), and an output layer (see Figure 11.10). The hidden layers typically contain sigmoidal activation functions of the form $z_j = \tanh(a_j)$, where $a_j = \sum_{i=1}^{d} w_{ji} x_i + b_j$. The output layer is typically linear. The weights (w_{ji}) and biases (b_j) for the jth neuron are chosen to minimize a cost function over a set of P training patterns. A common choice for the cost function is the sum-squared error, defined as

$$E(w) = \frac{1}{2} \sum_p \sum_k \left(t_k^{(p)} - y_k^{(p)} \right)^2 \qquad (11.12)$$

where $y_k^{(p)}$ and $t_k^{(p)}$ denote the network outputs and target responses, respectively, of each output node k given a pattern p, and $\mathbf{w}$ is a vector containing all the weights and biases of the network. The "training" process involves iteratively finding the weights and biases that minimize the cost function through some numerical optimization procedure. Second-order methods are commonly used, where the local approximation of the cost function by a quadratic form is given by

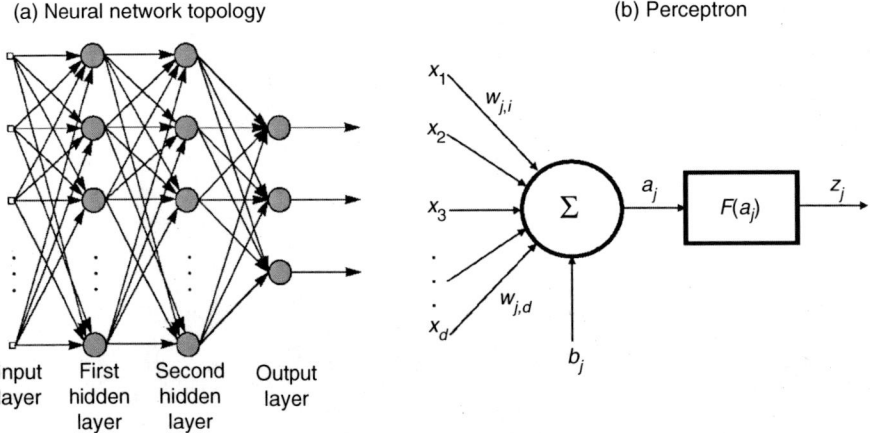

FIGURE 11.10
The structure of the multi-layer feedforward NN (specifically, the multi-layer perceptron) is shown in (a), and the perceptron (or node) is shown in (b).

$$E(\mathbf{w}+\mathbf{dw}) \approx E(\mathbf{w}) + \nabla E(\mathbf{w})^T \mathbf{dw} + \frac{1}{2}\mathbf{dw}^T \nabla^2 E(\mathbf{w})\,\mathbf{dw} \qquad (11.13)$$

where $\nabla E(\mathbf{w})$ and $\nabla^2 E(\mathbf{w})$ are the gradient vector and the Hessian matrix of the cost function, respectively. Setting the derivative of Equation 11.13 to zero and solving for the weight update vector $\mathbf{dw}$ yields

$$\mathbf{dw} = -[\nabla^2 E(\mathbf{w})]^{-1} \nabla E(\mathbf{w}) \qquad (11.14)$$

Direct application of Equation 11.14 is difficult in practice, because computation of the Hessian matrix (and its inverse) is nontrivial, and usually needs to be repeated at each iteration. For sum-squared error cost functions, it can be shown that

$$\nabla E(\mathbf{w}) = \mathbf{J}^T \mathbf{e} \qquad (11.15)$$

$$\nabla^2 E(\mathbf{w}) = \mathbf{J}^T \mathbf{J} + S \qquad (11.16)$$

where $\mathbf{J}$ is the Jacobian matrix that contains first derivatives of the network errors with respect to the weights and biases, $\mathbf{e}$ is a vector of network errors, and $S = \sum_{p=1}^{P} \mathbf{e}_p \nabla \mathbf{e}_p^2$. The Jacobian matrix can be computed using a standard backpropagation technique [14] that is significantly more computationally efficient than direct calculation of the Hessian matrix [15]. However, an inversion of a square matrix with dimensions equal to the total number of weights and biases in the network is required. For the Gauss–Newton method, it is assumed that S is zero (a reasonable assumption only near the solution), and the updated Equation 11.14 becomes

$$\mathbf{dw} = -[\mathbf{J}^T \mathbf{J}]^{-1} \mathbf{J} \mathbf{e} \qquad (11.17)$$

The Levenberg–Marquardt modification [16] to the Gauss–Newton method is

$$\mathbf{dw} = -[\mathbf{J}^T \mathbf{J} + \mu \mathbf{I}]^{-1} \mathbf{J} \mathbf{e} \qquad (11.18)$$

As μ varies between zero and ∞, $\mathbf{dw}$ varies continuously between the Gauss–Newton step and the steepest descent. The Levenberg–Marquardt method is thus an example of a model-trust-region approach in which the model (in this case the linearized approximation of the error function) is trusted only within some region around the current search point [17]. The size of this region is governed by the value μ.

The use of multi-layer feedforward NNs, such as the multi-layer perceptron (MLP), to retrieve temperature profiles from hyperspectral radiance measurements has been addressed by several investigators (see Refs. [18,19], for example). NN retrieval of moisture profiles from hyperspectral data is relatively new [20], but follows the same methodology used to retrieve temperature.

11.4.2 The PPC–NN Algorithm

A first attempt to combine the properties of both NN estimators and PC transforms for the inversion of microwave radiometric data to retrieve atmospheric temperature and moisture profiles is reported in Ref. [21], and a more recent study with hyperspectral data is presented in Ref. [20]. A conceptually similar approach is taken in this work by combining the PPC compression technique described in Section 11.3.3 with the NN estimator discussed in the previous section. PPC compression offers substantial performance advantages over traditional principal components analysis (PCA) and is the cornerstone of the present work.

11.4.2.1 Network Topology

All MLPs used in the PPC–NN algorithm are composed of one or two hidden layers of nonlinear (hyperbolic tangent) nodes and an output layer of linear nodes. For the temperature retrieval, 25 PPC coefficients are input to six NNs, each with a single hidden layer of 15 nodes. Separate NNs are used for different vertical regions of the atmosphere; a total of six networks are used to estimate the temperature profile at 65 points from the surface to 50 mbar. For the water vapor retrieval, 35 PPC coefficients are input to nine NNs, each with a single hidden layer of 25 nodes. The water vapor profile (mass mixing ratio) is estimated at 58 points from the surface to 75 mbar. These network parameters were determined largely through empirical analyses. Work is underway to dynamically optimize these parameters as the NN is trained. Separate training and testing data sets are used and are discussed in more detail in Section 11.4.2.2.

11.4.2.2 Network Training

The weights and biases were initialized using the Nguyen–Widrow method [22]. This method reduces the training time by initializing the weights so that each node is "active" (in the linear region of the activation function) over the input range of interest. The NN was trained using the Levenberg–Marquardt backpropagation algorithm discussed in Section 11.4.1. For each epoch, the μ parameter was initialized to 0.001. If a successful step was taken (i.e., $E(\mathbf{w} + \mathbf{dw}) < E(\mathbf{w})$), then μ was decreased by a factor of ten. If the current step was unsuccessful, the value of μ was increased by a factor of ten until a successful step could be found (or until μ reached 10^{10}). The network training was stopped when the error on a separate data set did not decrease for 10 consecutive epochs. The sensor noise was changed on each training epoch to desensitize the network to radiance measurement errors.

11.4.3 Error Analyses for Simulated Clear and Cloudy Atmospheres

AIRS and AMSU-A/B clear and cloudy radiances were simulated for an ensemble of approximately 10,000 profiles. These profiles were produced using a numerical weather prediction (NWP) model, and are substantially smoother (vertically) than the NOAA88b profile set that is used in the following sections. The profiles were separated into mutually exclusive sets for training and testing of the PPC–NN algorithm. The RMS errors for the LLSE and NN temperature retrievals are shown in Figure 11.11 for clear and cloudy atmospheres. The NN estimator significantly outperforms the LLSE in both cases.

The sensitivity of the retrieval to instrument noise is examined by repeating the retrieval with instrument noise set to zero. The difference in retrieval errors (with and without noise) is shown in the first panel of Figure 11.12. The atmospheric contribution to the retrieval error (i.e., the noise-free retrieval error) is shown in the second panel of Figure 11.12. Finally, the differences (net minus LLSE) in error contributions for the two methods are shown in the third panel of Figure 11.12. It is noteworthy that the NN is a much better filter of instrument noise than is the LLSE.

As a final test of sensitivity to instrument noise, the LLSE and NN retrievals were repeated while varying the instrument noise between 10% and 1000% of its nominal value. The resulting retrieval errors are shown in Figure 11.13. The NN retrieval is significantly less sensitive to instrument noise than is the LLSE retrieval.

Retrievals of Atmospheric Temperature and Moisture Profiles 219

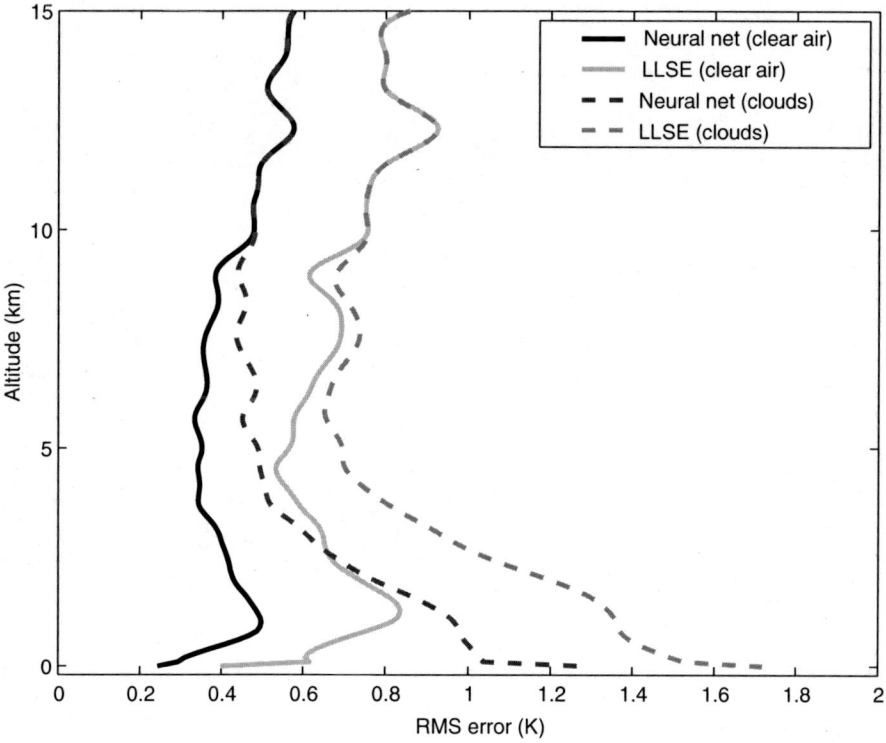

FIGURE 11.11
RMS temperature profile retrieval error for the NN and LLSE estimators. Results for clear air and clouds are shown. Surface emissivities were modeled as random variables with clipped-Gaussian PDFs; mean = 0.975, standard deviation = 0.025 (AIRS); mean = 0.95, standard deviation = 0.05 (AMSU).

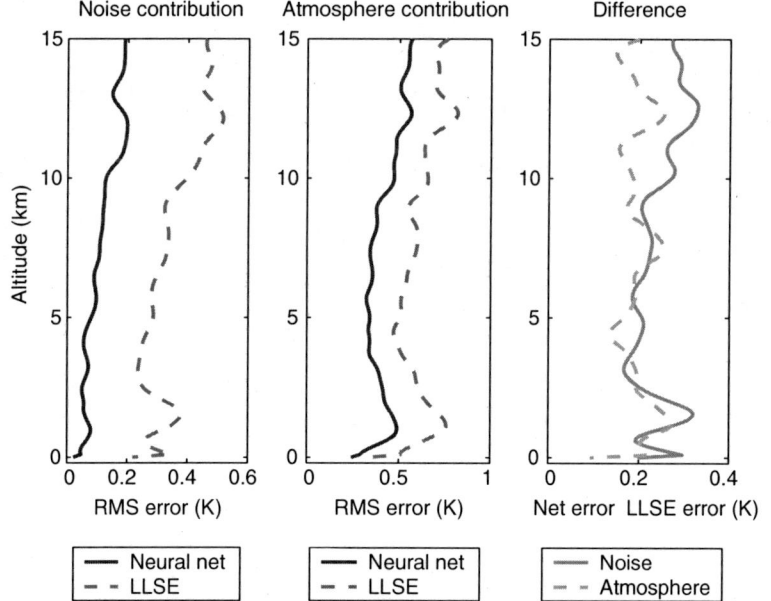

FIGURE 11.12
Contributions to temperature profile retrieval error due to measurement noise and atmospheric noise for the NN and LLSE estimators.

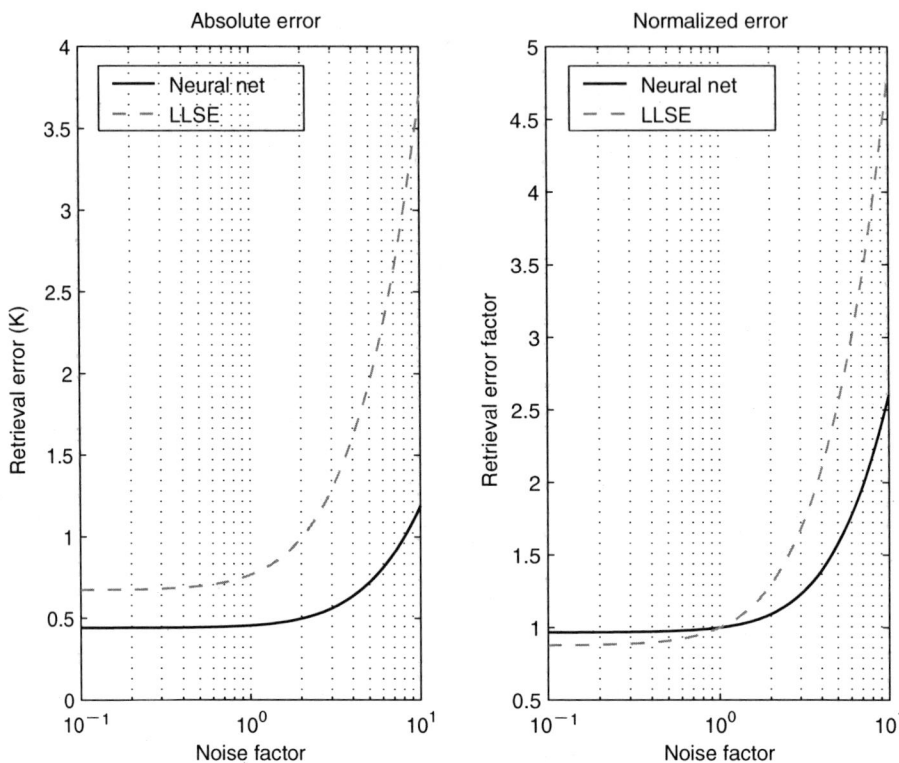

FIGURE 11.13
Sensitivity of temperature profile retrieval to measurement noise. The mean RMS error over 15 km is shown as a function of the noise amplification factor.

11.4.4 Validation of the PPC–NN Algorithm with AIRS/AMSU Observations of Partially Cloudy Scenes over Land and Ocean

In this section, the performance of the PPC–NN algorithm is evaluated using cloud-cleared AIRS observations (not simulations, as were used in the previous section) and colocated ECMWF (European Center for Medium-range Weather Forecasting) forecast fields. The cloud clearing is performed using both AIRS and AMSU data. The PPC–NN retrieval performance is compared with that obtained using the AIRS Level-2 algorithm. Both ocean and land cases are considered, including elevated surface terrain, and retrievals at all sensor scan angles (out to $\pm 48°$) are derived. Finally, sensitivity analyses of PPC–NN retrieval performance are presented with respect to scan angle, orbit type (ascending or descending), cloud amount, and training set comprehensiveness.

11.4.4.1 Cloud Clearing of AIRS Radiances

The cloud-clearing approach discussed in Susskind et al. [23] was applied to the AIRS data by the AIRS Science Team. Version 3.x of the algorithm was used in this work (see Table 11.1 for a detailed listing of the software version numbers). The algorithm seeks to estimate a clear-column radiance (the radiance that would have been measured if the scene were cloud-free) from a number of adjacent cloud-impacted fields of view.

TABLE 11.1

AIRS Software Version Numbers for the Seven Days Used in the Match-Up Data Set

	6 Sep 2002	25 Jan 2003	8 Jun 2003	21 Aug 2003	3 Sep 2003	12 Oct 2003	5 Dec 2003
Cloud clearing	3.7.0	3.7.0	3.7.0	3.1.9	3.1.9	3.1.9	3.7.0
Level-2	3.0.8	3.0.8	3.0.8	3.0.8	3.0.8	3.0.8	3.0.10

11.4.4.2 The AIRS/AMSU/ECMWF Data Set

The performance of the PPC–NN algorithm was evaluated using 352,903 AIRS/AMSU observations and colocated ECMWF atmospheric fields collected on seven days throughout 2002 and 2003 (see Table 11.1). Various software version changes were made over the course of this work (see Table 11.1 for details), but these changes were primarily with regard to quality control and do not significantly affect the results presented here. However, the version 4.x release of the AIRS software, which was not available in time to be included in this work, should offer many enhancements over version 3.x, including improved cloud clearing, retrieval accuracies, quality control, and retrieval yield [24]. Reanalyses of the results presented in this section are therefore planned with the new AIRS software release.

The 352,903 observations were randomly divided into a training set of 302,903 observations (206,061 of which were over ocean) and a separate validation set of 50,000 observations (40,000 of which were over ocean). The *a priori* RMS variation of the temperature and water vapor (mass mixing ratio) profiles contained in the validation set are shown in Figure 11.14. The observations in the validation set were matched with AIRS Level-2 retrievals obtained from the Earth Observing System (EOS) Data Gateway (EDG). As advised in the AIRS Version 3.0 L2 Data Release Documentation, only retrievals that met certain quality standards (specifically, RetQAFlag = 0 for ocean and RetQAFlag = 256 for land) were included in the analyses. There were 17,856 AIRS Level-2 retrievals (all within $\pm 40°$ latitude) that met this criterion. Re-analysis with AIRS Level-2 version 4.x software is planned, as the version 4.x products have been validated over both ocean and land at near-polar latitudes.

To facilitate comparison with results published in the AIRS v3.0 Validation Report [25], layer error statistics are calculated as follows. First, layer averages are calculated in layers of approximately (but not exactly) 1-km width—the exact layer widths can be found in Appendix III in the AIRS v3.0 Validation Report. Second, weighted water vapor errors in each layer are calculated by dividing the RMS mass mixing ratio error by the RMS variation of the true mass mixing ratio (as opposed to dividing the mass mixing ratio error of each profile by the true mass mixing ratio for that profile and computing the RMS of the resulting ensemble).

11.4.4.3 AIRS/AMSU Channel Selection

Thirty-seven percent (888 of the 2378) of the AIRS channels were discarded for the analysis, as the radiance values for these channels frequently were flagged as invalid by the AIRS calibration software. A simulated AIRS brightness temperature spectrum is shown in Figure 11.15, which shows the original 2378 AIRS channels and the 1490 channels that were selected for use with the PPC–NN algorithm. All 15 AMSU-A channels were used. The algorithm automatically discounts channels that are excessively corrupted by sensor noise (for example, AMSU-A channel 7 on EOS Aqua) or other interfering

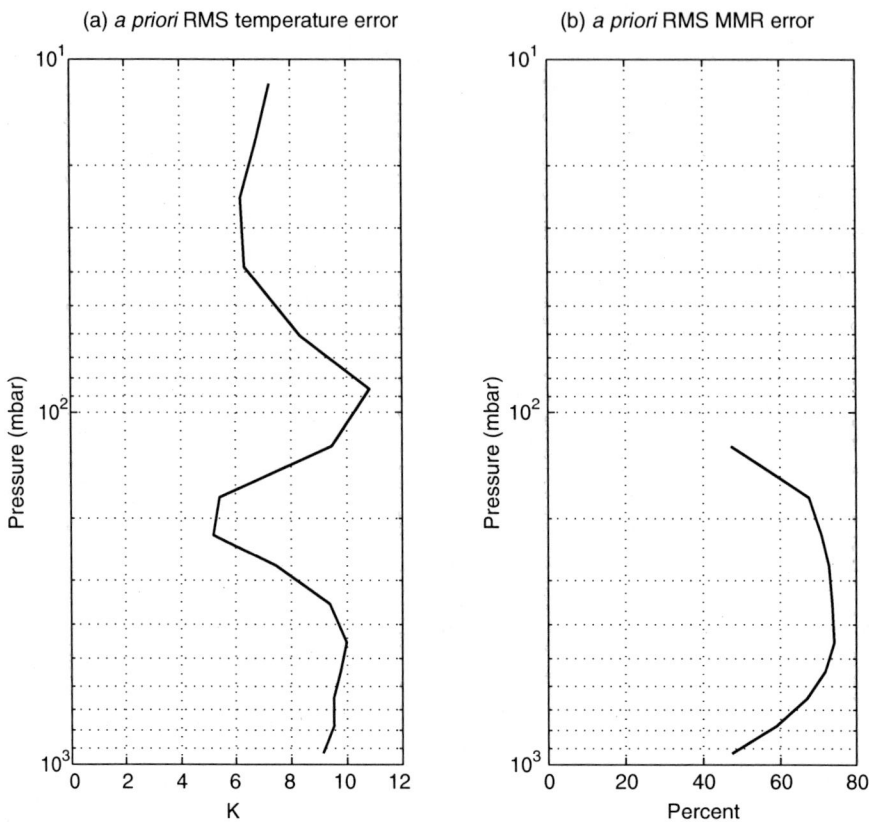

FIGURE 11.14
Temperature and water vapor profile statistics for the validation data set used in the analysis. See the text for details on how the statistics are computed at each layer.

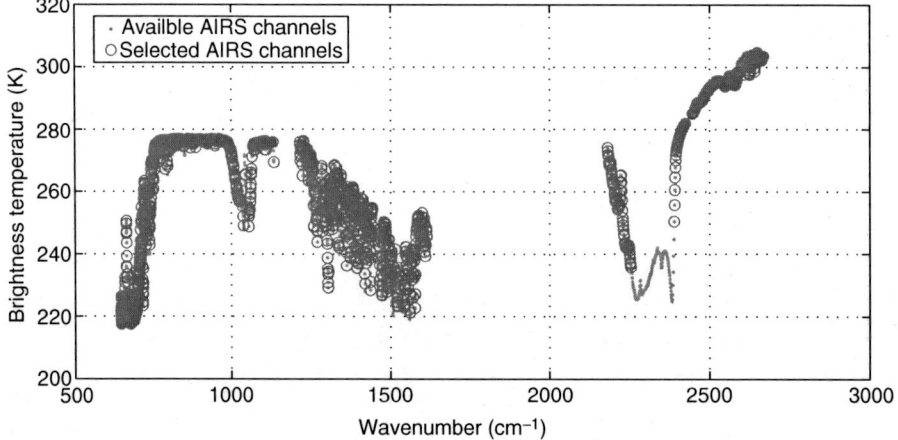

FIGURE 11.15
A typical AIRS spectrum (simulated) is shown. 1490 out of 2378 AIRS channels were selected.

signals (for example, the effects of nonlocal thermodynamic equilibrium) because the corruptive signals are largely uncorrelated with the geophysical parameters that are to be estimated.

11.4.4.4 PPC–NN Retrieval Enhancements for Variable Sensor Scan Angle and Surface Pressure

When dealing with global AIRS/AMSU data, a variety of scan angles and surface pressures must be accommodated. Therefore, two additional inputs were added to the NNs discussed previously: (1) the secant of the scan angle and (2) the forecast surface pressure (in mbar) divided by 1013.25. The resulting temperature and water vapor profile estimates were reported on a variable pressure grid anchored by the forecast surface pressure.

Because the number of inputs to the NNs increased, the number of hidden nodes in NNs used for temperature retrievals was increased from 15 to 20. For water vapor retrievals, the number of hidden nodes in the first hidden layer was maintained at 25, but a second hidden layer of 15 hidden nodes was added.

11.4.4.5 Retrieval Performance

We now compare the retrieval performance of the PPC–NN, linear regression, and AIRS Level-2 methods. For both the ocean and land cases, the PPC–NN and linear regression retrievals were derived using the same training set, and the same validation set was used for all methods.

The temperature profile retrieval performance over ocean for the linear regression retrieval, the PPC–NN retrieval, and the AIRS Level-2 retrieval is shown in Figure 11.16, and the water vapor retrieval performance is shown in Figure 11.17. The error statistics were calculated using the 13,156 (out of 40,000) AIRS Level-2 retrievals that converged successfully. A bias of approximately 1 K near 100 mbar was found between the AIRS Level-2 temperature retrievals and the ECMWF data (ECMWF was colder). This bias was removed prior to computation of the AIRS Level-2 retrieval error statistics, which are shown in Figure 11.16.

The temperature profile retrieval performance over land for the linear regression retrieval, the PPC–NN retrieval, and the AIRS Level-2 retrieval is shown in Figure 11.18, and the water vapor retrieval performance is shown in Figure 11.19. The error statistics were calculated using the 4,700 (out of 10,000) AIRS Level-2 retrievals that converged successfully.

There are several features in Figure 11.16 through Figure 11.19 that are worthy of note. First, for all retrieval methods, the performance over land is worse than that over ocean, as expected. The cloud-clearing problem is significantly more difficult over land, as variations in surface emissivity can be mistaken for cloud perturbations, thus resulting in improper radiance corrections. Second, the magnitude of the temperature profile error degradation for land versus ocean is larger for the PPC–NN algorithm than for the AIRS Level-2 algorithm. In fact, the temperature profile retrieval performance of the AIRS Level-2 algorithm is superior to that of the PPC–NN algorithm throughout most of the lower troposphere over land. Further analyses of this discrepancy suggest that the performance of the PPC–NN method over elevated terrain is suboptimal, and could be improved. This work is currently underway.

11.4.4.6 Retrieval Sensitivity to Cloud Amount

A plot of the temperature retrieval error in the layer closest to the surface as a function of the cloud fraction retrieved by the AIRS Level-2 algorithm is shown in Figure 11.20.

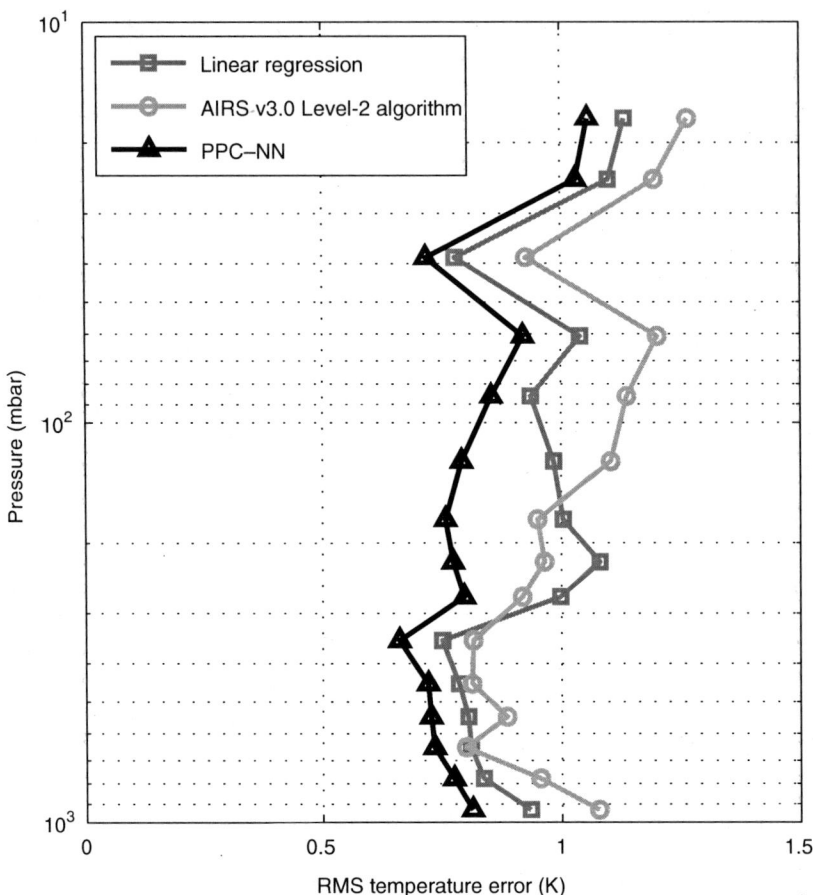

FIGURE 11.16
Temperature retrieval performance of the PPC–NN, linear regression, and AIRS Level-2 methods over ocean. Statistics were calculated over 13,156 fields of regard.

Similar curves for the water vapor retrieval performance are shown in Figure 11.21. Both methods produce temperature and moisture retrievals with RMS errors near 1 K and 15%, respectively, even in cases with large cloud fractions. The figures show that the PPC–NN temperature and moisture retrievals are less sensitive than the AIRS Level-2 retrievals to cloud amount. Furthermore, it has been shown that the PPC–NN retrieval technique is relatively insensitive to sensor scan angle, orbit type, and training set comprehensiveness [26].

11.4.5 Discussion and Future Work

Although the PPC–NN performance results presented in the previous section are very encouraging, several caveats must be mentioned. The ECMWF fields used for "ground truth" contain errors, and the NN will tune to these errors as part of its training process. Therefore, the PPC–NN RMS errors shown in the previous section may be underestimated, and the AIRS Level-2 RMS errors may be overestimated, as the ECMWF data is not an accurate representation of the true state of the atmosphere. Therefore, the "true" spread between the performance of the PPC–NN and AIRS Level-2 algorithms is

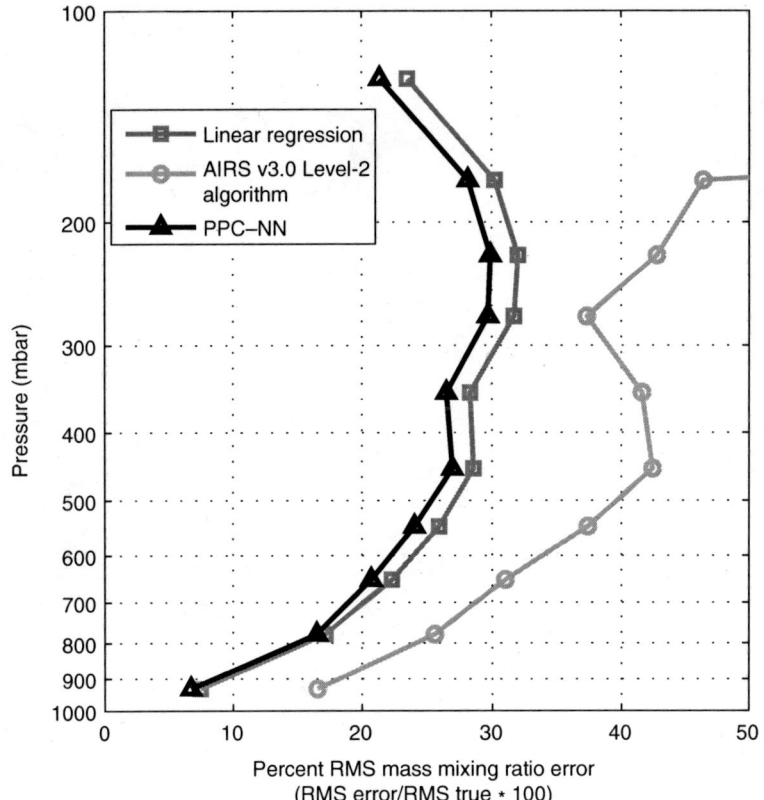

FIGURE 11.17
Water vapor (mass mixing ratio) retrieval performance of the PPC–NN, linear regression, and AIRS Level-2 methods over ocean. Statistics were calculated over 13,156 fields of regard.

almost certainly smaller than that shown here. Work is currently underway to test the performance of both the PPC–NN and AIRS Level-2 algorithms with additional ground-truth data, including radiosonde data, and ground- and aircraft-based measurements. It should be noted that the PPC–NN algorithm as implemented in this work is currently not a stand-alone system, as both AIRS cloud-cleared radiances and quality flags produced by the AIRS Level-2 algorithm are required. Future work is planned to adapt the PPC–NN algorithm for use directly on cloudy AIRS/AMSU radiances and to produce quality assessments of the retrieved products. Finally, assimilation of PPC–NN-derived atmospheric parameters into NWP models is planned, and the resulting impact on forecast accuracy will be an excellent indicator of retrieval quality.

11.5 Summary

The PPC–NN temperature and moisture profile retrieval technique combines a linear radiance compression operator with an NN estimator. The PPC transform was shown to be well suited for this application because information correlated to the geophysical

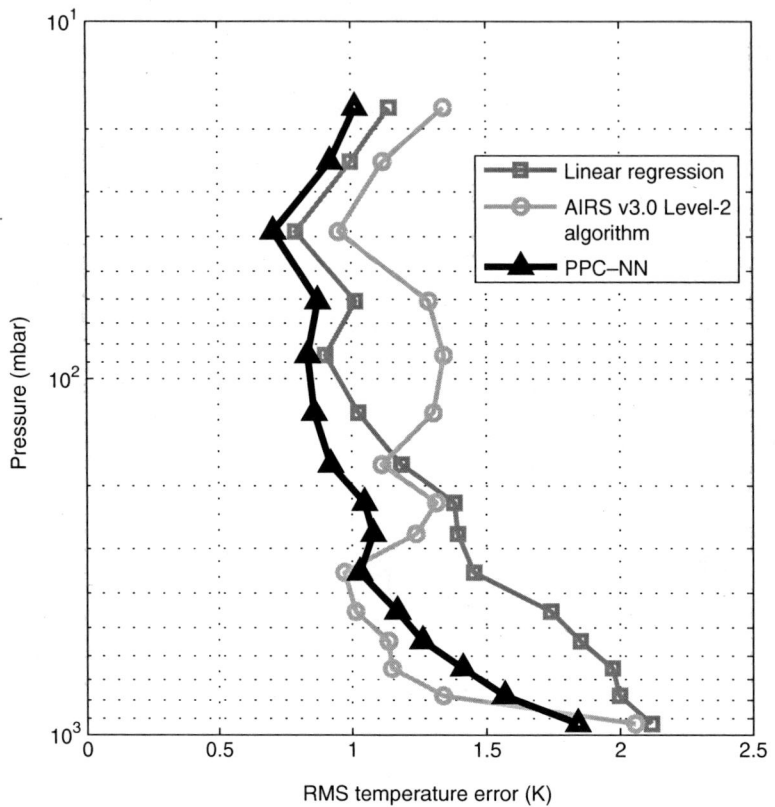

FIGURE 11.18
Temperature retrieval performance of the PPC–NN, linear regression, and AIRS Level-2 methods over land. Statistics were calculated over 4,700 fields of regard.

quantity of interest is optimally represented with only a few dozen components. This substantial amount of radiance compression (approximately a factor of 100) allows relatively small NNs to be used thereby improving both the stability and computational efficiency of the algorithm. Test cases with observed partially cloudy AIRS/AMSU data demonstrate that the PPC–NN temperature and moisture retrievals yield accuracies commensurate with those of physical methods at a substantially reduced computational burden. Retrieval accuracies (defined as agreement with ECMWF fields) near 1 K for temperature and 25% for water vapor mass mixing ratio in layers of approximately 1-km thickness were obtained using the PPC–NN retrieval method with AIRS/AMSU data in partially cloudy areas. PPC–NN retrievals with partially cloudy AIRS/AMSU data over land were also performed. The PPC–NN retrieval technique is relatively insensitive to cloud amount, sensor scan angle, orbit type, and training set comprehensiveness. These results further suggest the AIRS Level-2 algorithm that produced the cloud-cleared radiances and quality flags used by the PPC–NN retrieval is performing well.

The high level of performance achieved by the PPC–NN algorithm suggests it would be a suitable candidate for the retrieval of geophysical parameters other than temperature and moisture from high resolution spectral data. Potential applications include the retrieval of ozone profiles and trace gas amounts. Future work will involve further evaluation of the algorithm with simulated and observed partially cloudy data, including global radiosonde data and ground- and aircraft-based observations.

Retrievals of Atmospheric Temperature and Moisture Profiles

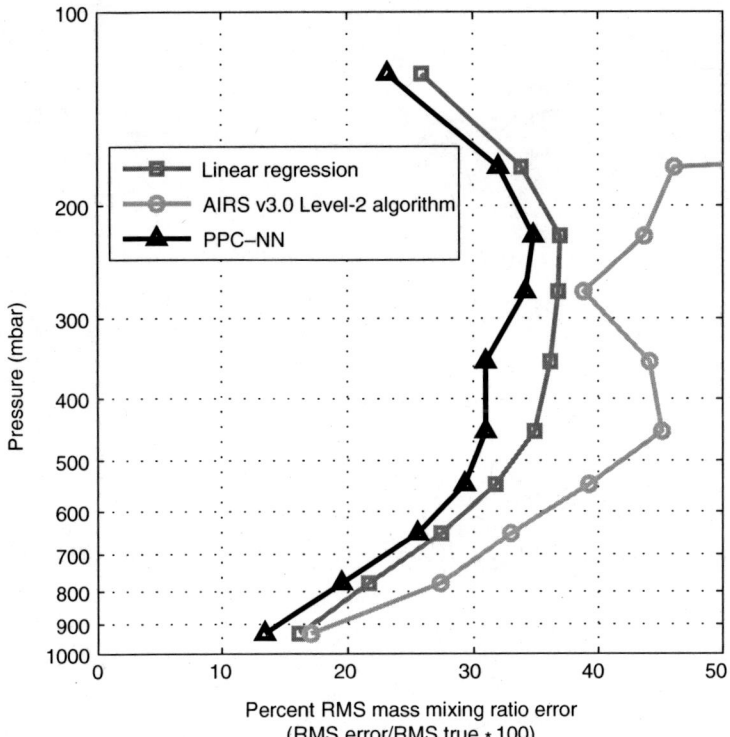

FIGURE 11.19
Water vapor (mass mixing ratio) retrieval performance of the PPC–NN, linear regression, and AIRS Level-2 methods over land. Statistics were calculated over 4,700 fields of regard.

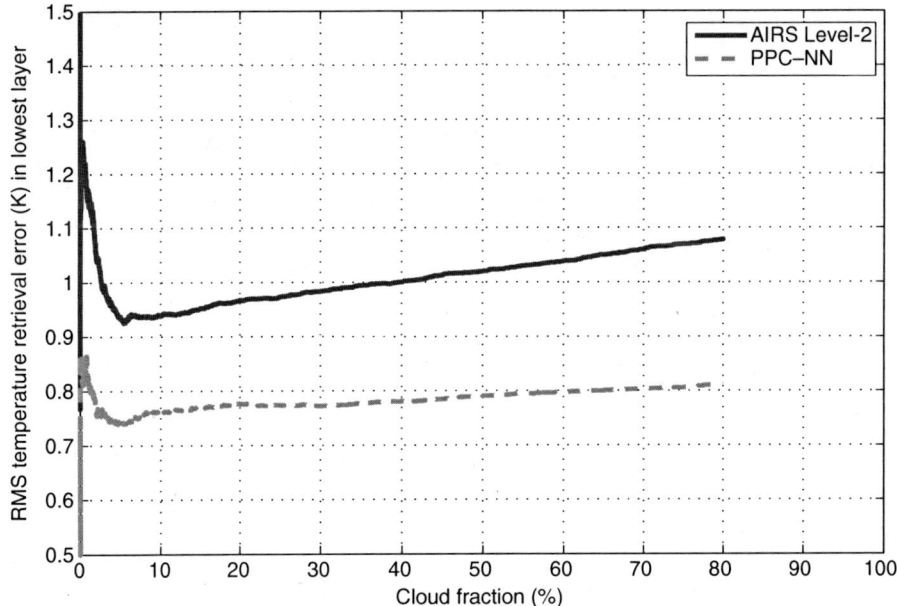

FIGURE 11.20
Cumulative RMS temperature error in the layer closest to the surface. Pixels were ranked in order of increasing cloudiness according to the retrieved cloud fraction from the AIRS Level-2 algorithm. No retrievals were attempted by the AIRS Level-2 algorithm if the retrieved cloud fraction exceeded 80%.

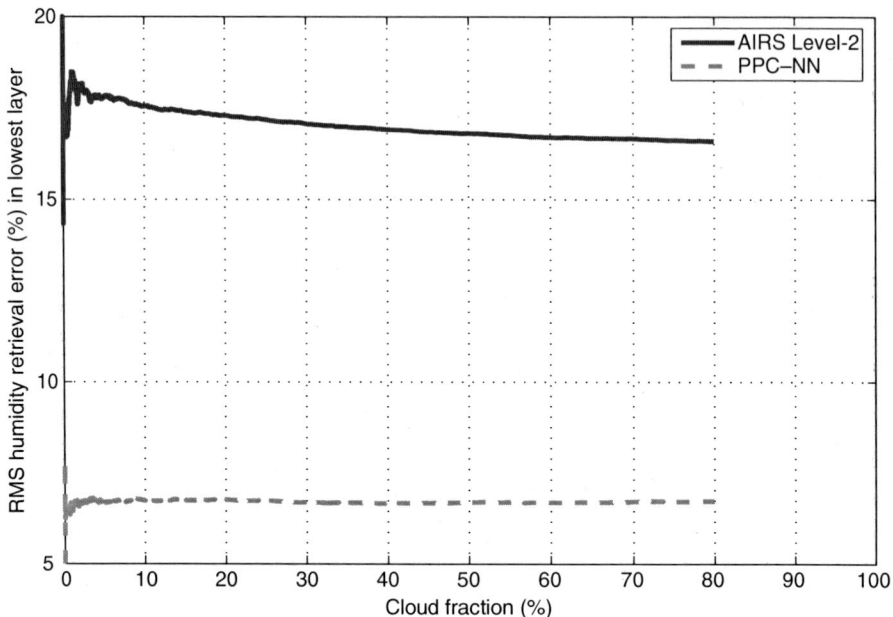

FIGURE 11.21
Cumulative RMS water vapor error in the layer closest to the surface. Pixels were ranked in order of increasing cloudiness, according to the retrieved cloud fraction from the AIRS Level-2 algorithm. No retrievals were attempted by the AIRS Level-2 algorithm if the retrieved cloud fraction exceeded 80%.

Acknowledgments

The author would like to thank NOAA NESDIS, Office of Systems Development, and the National Aeronautics and Space Administration for financial support of this work. The colocated AIRS/AMSU/ECMWF data sets were provided by the AIRS Science Team, with assistance from M. Goldberg (NOAA NESDIS, Office of Research Applications). The author would also like to thank L. Zhou for help in obtaining and interpreting the AIRS/AMSU/ECMWF data sets, and D. Staelin and P. Rosenkranz for many useful discussions on all facets of this work.

This work was supported by the National Oceanic and Atmospheric administration under Air Force Contract F19628-00-C-0002. Opinions, interpretations, conclusions, and recommendations are those of the author and are not necessarily endorsed by the United States Government.

References

1. H. Hotelling, The most predictable criterion, *J. Educ. Psychol.*, vol. 26, pp. 139–142, 1935.
2. J. Escobar-Munoz, A. Chédin, F. Cheruy, and N. Scott, Réseaux de neurons multi-couches pour la restitution de variables thermodynamiques atmosphériques à l'aide de sondeurs verticaux satellitaires, *Comptes-Rendus de L'Académie des Sciences; Série II*, vol. 317, no. 7, pp. 911–918, 1993.

3. D.H. Staelin, Passive remote sensing at microwave wavelengths, *Proc. IEEE*, vol. 57, no. 4, pp. 427–439, 1969.
4. M.A. Janssen, *Atmospheric Remote Sensing by Microwave Radiometry*, New York, John Wiley & Sons, Inc., 1993.
5. C.D. Rodgers, Retrieval of atmospheric temperature and composition from remote measurements of thermal radiation, *J. Geophys. Res.*, vol. 41, no. 7, pp. 609–624, 1976.
6. M.D. Goldberg, Y. Qu, L.M. McMillin, W. Wolff, L. Zhou, and M. Divakarla, AIRS near-real-time products and algorithms in support of operational numerical weather prediction, *IEEE Trans. Geosci. Rem. Sens.*, vol. 41, no. 2, pp. 379–389, 2003.
7. H. Huang and P. Antonelli, Application of principal component analysis to high-resolution infrared measurement compression and retrieval, *J. Appl. Meteorol.*, vol. 40, pp. 365–388, Mar. 2001.
8. F. Aires, W.B. Rossow, N.A. Scott, and A. Chédin, Remote sensing from the infrared atmospheric sounding interferometer instrument: 1. compression, denoising, first-guess retrieval inversion algorithms, *J. Geophys. Res.*, vol. 107, Nov. 2002.
9. W.J. Blackwell and D.H. Staelin, Cloud flagging and clearing using high-resolution infrared and microwave sounding data, *IEEE Int. Geosci. Rem. Sens. Symp. Proc.*, June 2002.
10. J.B. Lee, A.S. Woodyatt, and M. Berman, Enhancement of high spectral resolution remote-sensing data by a noise-adjusted principal components transform, *IEEE Trans. Geosci. Rem. Sens.*, vol. 28, pp. 295–304, May 1990.
11. W.J. Blackwell, Retrieval of cloud-cleared atmospheric temperature profiles from hyperspectral infrared and microwave observations, Ph.D. dissertation, Massachusetts Institute of Technology, Department of Electrical Engineering and Computer Science, June 2002.
12. K.M. Hornik, M. Stinchcombe, and H. White, Multilayer feedforward networks are universal approximators, *Neural Netw.*, vol. 4, no. 5, pp. 359–366, 1989.
13. S. Haykin, *Neural Networks: A Comprehensive Foundation*, New York, Macmillan College Publishing Company, 1994.
14. D.E. Rumelhart, G. Hinton, and R. Williams, *Parallel Distributed Processing: Explorations in the Microstructure of Cognition*, Vol. 1: Foundations., ser. D.E. Rumelhart and J.L. McClelland, Eds., Cambridge, MA, MIT Press, 1986.
15. M.T. Hagan and M.B. Menhaj, Training feedforward networks with the Marquardt algorithm, *IEEE Trans. Neural Netw.*, vol. 5, pp. 989–993, Nov. 1994.
16. P.E. Gill, W. Murray, and M.H. Wright, The Levenberg–Marquardt method, in *Practical Optimization*, London, Academic Press, 1981.
17. C.M. Bishop, *Neural Networks for Pattern Recognition*, Oxford, Oxford University Press, 1995.
18. H.E. Motteler, L.L. Strow, L. McMillin, and J.A. Gualtieri, Comparison of neural networks and regression based methods for temperature retrievals, *Appl. Opt.*, vol. 34, no. 24, pp. 5390–5397, 1995.
19. F. Aires, A. Chédin, N.A. Scott, and W.B. Rossow, A regularized neural net approach for retrieval of atmospheric and surface temperatures with the IASI instrument, *J. Appl. Meteorol.*, vol. 41, pp. 144–159, Feb. 2002.
20. F. Aires, W.B. Rossow, N.A. Scott, and A. Chédin, Remote sensing from the infrared atmospheric sounding interferometer instrument: 2. simultaneous retrieval of temperature, water vapor, and ozone atmospheric profiles, *J. Geophys. Res.*, vol. 107, Nov. 2002.
21. F.D. Frate and G. Schiavon, A combined natural orthogonal functions/neural network technique for the radiometric estimation of atmospheric profiles, *Radio Sci.*, vol. 33, no. 2, pp. 405–410, 1998.
22. D. Nguyen and B. Widrow, Improving the learning speed of two-layer neural networks by choosing initial values of the adaptive weights, *IJCNN*, vol. 3, pp. 21–26, 1990.
23. J. Susskind, C.D. Barnet, and J.M. Blaisdell, Retrieval of atmospheric and surface parameters from AIRS/AMSU/HSB data in the presence of clouds, *IEEE Trans. Geosci. Rem. Sens.*, vol. 41, no. 2, pp. 390–409, 2003.
24. J. Susskind et al., Accuracy of geophysical parameters derived from AIRS/AMSU as a function of fractional cloud cover, *J. Geophys. Res.*, III; May 2006.

25. H.H. Aumann et al., Validation of AIRS/AMSU/HSB core products for data release version 3.0, *NASA JPL Tech. Rep. D-26538*, Aug 2003.
26. W.J. Blackwell, A neural-network technique for the retrieval of atmospheric temperature and moisture profiles from high spectral resolution sounding data, *IEEE Trans. Geosci. Rem. Sens.*, vol. 43, no. 11, pp. 2535–2546, 2005.

12

Satellite-Based Precipitation Retrieval Using Neural Networks, Principal Component Analysis, and Image Sharpening*

Frederick W. Chen

CONTENTS

12.1 Introduction 231
12.2 Physical Basis of Passive Microwave Precipitation Sensing 232
12.3 Description of AMSU-A/B and AMSU/HSB.. 238
12.4 Signal Processing 240
    12.4.1 Regional Laplacian Filtering 240
    12.4.2 Principal Component Analysis... 241
        12.4.2.1 Basic PCA.. 242
        12.4.2.2 Constrained PCA 242
    12.4.3 Data Fusion 243
    12.4.4 Neural Nets.. 243
12.5 The Chen–Staelin Algorithm 245
    12.5.1 Limb-Correction of Temperature Profile Channels 247
    12.5.2 Detection of Precipitation 249
    12.5.3 Cloud-Clearing by Regional Laplacian Interpolation............... 250
    12.5.4 Image Sharpening 250
    12.5.5 Temperature and Water Vapor Profile Principal Components 252
    12.5.6 The Neural Net.. 252
12.6 Retrieval Performance Evaluation 253
    12.6.1 Image Comparisons of NEXRAD and AMSU-A/B Retrievals 253
    12.6.2 Numerical Comparisons of NEXRAD and AMSU-A/B Retrievals 254
    12.6.3 Global Retrievals of Rain and Snow 257
12.7 Conclusions... 258
Acknowledgments 259
References 259

12.1 Introduction

Global estimation of precipitation is important to studies in areas such as atmospheric dynamics, hydrology, climatology and meteorology. Improvements in methods for

---
*The material in this chapter is taken from Refs. [66] and [67].

satellite-based estimation of precipitation can lead to improvements in weather forecasting, climate studies, and climate models.

Precipitation presents many challenges because of its complicated physics and statistics. Existing models of clouds do not adequately account for all of the possible variations in clouds and precipitation that can be encountered. Instead of a physics-based approach, Chen and Staelin developed a method using a statistics-based approach. This method was developed for the Advanced Microwave Sounding Unit (AMSU) instruments AMSU-A/B on the NOAA-15, NOAA-16, and NOAA-17 satellites. The development process made use of the fact that although satellite-based passive microwave data cannot completely characterize the observed clouds and precipitation, the data can still yield useful information relevant to precipitation. This chapter discusses the methods used by Chen and Staelin to extract from the data information related to atmospheric state variables that are highly correlated with precipitation. Principal component analysis for signal separation and data compression, data fusion for resolution sharpening, neural nets, and regional filtering are among the signal and image processing methods used.

AMSU-A/B and the corresponding instruments AMSU/HSB (Humidity Sounder for Brazil) collect data near 23.8, 31.4, 54, 89, and 183 GHz. These frequency bands provide useful information about precipitation. In this chapter, the precipitation estimation algorithm developed by Chen and Staelin is discussed with an emphasis on the signal processing employed to sense important parameters about precipitation. Section 12.2 explains why it is possible to use the data on AMSU-A/B to estimate precipitation. Section 12.3 provides a more detailed description of AMSU-A/B and AMSU/HSB. Section 12.4 describes the Chen–Staelin algorithm. Section 12.7 provides concluding remarks.

12.2 Physical Basis of Passive Microwave Precipitation Sensing

Matter radiates thermal energy depending on its physical temperature and characteristic properties. When a spaceborne radiometer observes a location on the Earth, the amount of energy it receives depends on contributions from the various atmospheric and topographical constituents within its field of view (Figure 12.1). One useful quantity describing the amount of thermal radiation emitted by a body is spectral brightness. Spectral brightness is a measure of how much energy a body radiates at a specified frequency per unit receiving area, per transmitting solid angle, and per unit frequency. The spectral brightness of a blackbody (W·ster^{-1}·m^{-2}·Hz^{-1}) is a function of its physical temperature T (K) and frequency f (Hz) and is given by the following formula

$$B(f,T) = \frac{2hf^3}{c^2(e^{hf/kT} - 1)} \qquad (12.1)$$

where h is Planck's constant (J·s), c is the speed of light (m/s), and k is Boltzmann's constant (J/K). For this chapter, f never exceeds 200 GHz, and T never falls below 100 K, so $hf/kT < 0.1$. Then, the Taylor series expansion for exponential functions is used to simplify Equation 12.1

$$e^{hf/kT} - 1 = \left[1 + \left(\frac{hf}{kT}\right) + \frac{1}{2!}\left(\frac{hf}{kT}\right)^2 + \cdots\right] - 1 \qquad (12.2)$$

$$\approx \frac{hf}{kT} \qquad (12.3)$$

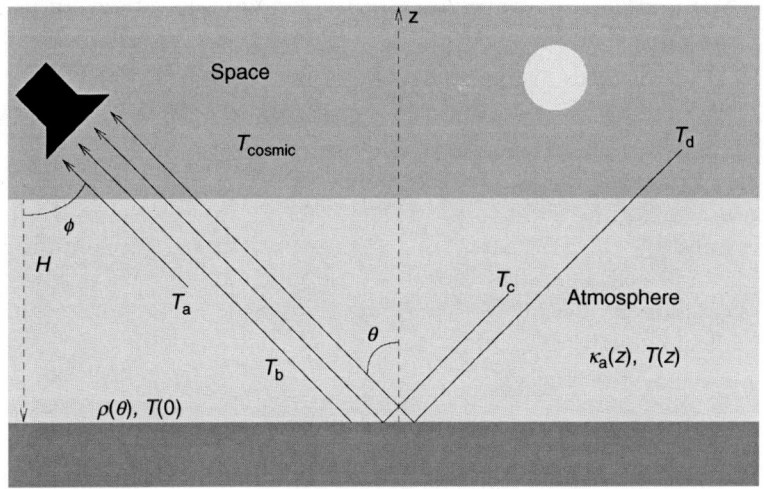

FIGURE 12.1
The major components of the radiative transfer equation (*note*: $\theta \neq \varphi$ since the surface of the Earth is spherical).

$$B(f,T) = \frac{2kT}{\lambda^2} \qquad (12.4)$$

where λ is the wavelength (m) associated with the frequency f.

For blackbodies, given an observation of spectral brightness, one can calculate the physical temperature of a blackbody as follows:

$$T = \frac{B(f,t)\lambda^2}{2k} \qquad (12.5)$$

Unlike blackbodies, gray bodies reflect some of the energy incident on them, so the intrinsic spectral brightness of a gray body is not equal to that of a blackbody. For a gray body,

$$B(f,T) = \varepsilon \frac{2kT}{\lambda^2} \qquad (12.6)$$

where ε is the emissivity of the gray body. This equation is rewritten as follows:

$$B(f,T) = \frac{2k}{\lambda^2} \varepsilon T \qquad (12.7)$$

The quantity εT is called the *brightness temperature* of an unilluminated gray body; that is, the temperature of a blackbody radiating the same brightness. Another useful property of a gray body is its reflectivity ρ, the fraction of incident energy that is reflected. A body in thermal equilibrium emits the same amount of energy that it absorbs. Therefore, $\rho + \varepsilon = 1$. For a blackbody, $\varepsilon = 1$ and $\rho = 0$.

In an atmosphere without hydrometeors, the brightness temperature observed by an Earth-observing spaceborne radiometer can be divided into four components (Figure 12.1):

1. T_a, the brightness temperature due to radiation emitted by atmospheric gases that is not reflected off the surface

2. T_b, the brightness temperature due to radiation emitted by the surface
3. T_c, the brightness temperature due to radiation that is emitted by atmospheric gases that is reflected off the surface
4. T_d, the brightness temperature due to cosmic background radiation that passes through the atmosphere twice and is reflected off the surface

Thermal radiation can be attenuated through absorption or reflected off the surface before being received by a radiometer. The observed brightness temperature is, therefore, a function of several variables such as the atmospheric temperature profile, water vapor profile, the emissivity of the surface (as a function of satellite zenith angle), and the absorption coefficients of atmospheric gases (as a function of altitude z). Given the atmospheric absorption coefficients $\kappa_a(z)$ (in Np per unit length), the reflectivity of the surface $\rho(\theta)$, the altitude of the satellite H, the cosmic background temperature T_{cosmic} (in K), and the satellite zenith angle θ, the brightness temperature components can be computed as follows (assuming specular surface reflection and the absence of hydrometeors):

$$T_a = \sec\theta \int_{z_0}^{H} T(z')\kappa_a(z') e^{-\tau(z',H)\sec\theta}\, dz' \tag{12.8}$$

$$T_b = [1 - \rho(\theta)] T(z_0) e^{-\tau(z_0,H)\sec\theta} \tag{12.9}$$

$$T_c = \rho(\theta)\sec\theta \cdot e^{-\tau(z_0,H)\sec\theta} \int_{z_0}^{H} T(z')\kappa_a(z') e^{-\tau(z_0,z')\sec\theta}\, dz' \tag{12.10}$$

$$T_d = \rho(\theta) T_{\text{cosmic}} e^{-2\tau(z_0,H)\sec\theta} \tag{12.11}$$

Then, the brightness temperature T_B measured by the radiometer is the sum of T_a, T_b, T_c, and T_d.

$$\begin{aligned} T_B = &\sec\theta \int_{z_0}^{H} T(z')\kappa_a(z') e^{-\tau(z',H)\sec\theta}\, dz' + [1 - \rho(\theta)] T(z_0) e^{-\tau(z_0,H)\sec\theta} \\ &+ \rho(\theta)\sec\theta \cdot e^{-\tau(z_0,H)\sec\theta} \int_{z_0}^{H} T(z')\kappa_a(z') e^{-\tau(z_0,z')\sec\theta}\, dz' + \rho(\theta) T_{\text{cosmic}} e^{-2\tau(z_0,H)\sec\theta} \end{aligned} \tag{12.12}$$

where z_0 is the altitude of the surface and $\tau(z_1,z_2)$ is the integrated atmospheric absorption of the atmosphere between altitudes z_1 and z_2.

$$\tau(z_1,z_2) = \int_{z_1}^{z_2} \kappa_a(z)\, dz \tag{12.13}$$

Equation 12.12 can be rewritten as follows (assuming that the contribution of thermal radiation from altitudes above H to T_c is negligible):

$$T_B \approx \int_{z_0}^{H} T(z') W(z')\, dz' + [1 - \rho(\theta)] T(z_0) e^{-\tau(z_0,H)\sec\theta} + \rho(\theta) T_{\text{cosmic}} e^{-2\tau(z_0,H)\sec\theta} \tag{12.14}$$

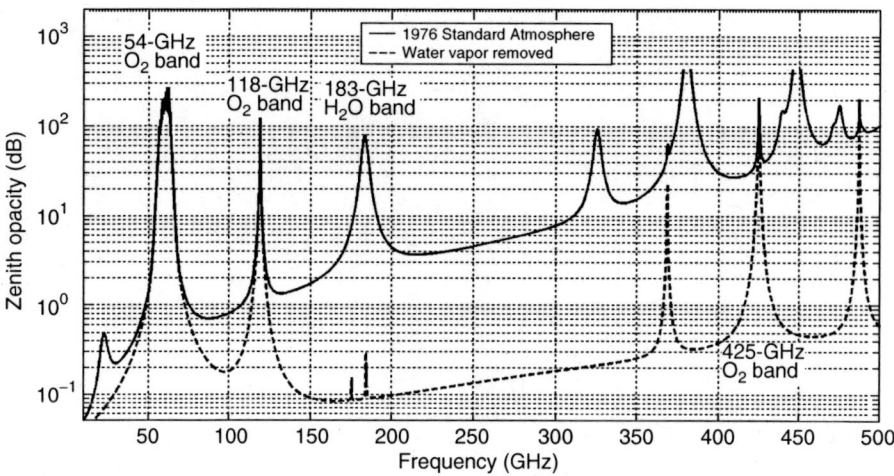

FIGURE 12.2
Zenith opacity for the microwave spectrum.

where

$$W(z) = \sec\theta \cdot \kappa_a(z')e^{-\tau(z,H)\sec\theta} + \rho(\theta)\sec\theta \cdot e^{-\tau(z_0,H)\sec\theta}\kappa_a(z')e^{-\tau(z_0,z)\sec\theta} \quad (12.15)$$

$W(z)$ is called the *weighting function* or *Jacobian*.

Equation 12.12 shows that the relationship between the thermal energy radiated and the physical temperature of a body is linear; that is, the brightness temperature can be expressed as a linear integral of physical temperatures in the field of view, where these radiated signals are uncorrelated and superimposed [1].

The previous section shows that the brightness temperatures seen by a satellite-borne radiometer depend on many variables in a highly complex and nonlinear fashion, so retrievals of temperature and water vapor profiles by direct inversion would be very difficult. However, the physics of the atmosphere still allows the extraction of useful information about the atmosphere from microwave frequency bands.

Two of the most important determinants of precipitation rate are the temperature and water vapor profiles. The presence of oxygen and water vapor resonance frequencies in the microwave spectrum and the presence of oxygen and water vapor in the atmosphere result in frequency bands that are sensitive primarily to a specific range of altitudes. Figure 12.2 shows the zenith opacity as a function of frequency for the range from 10 to 500 GHz for a ground-based zenith-observing radiometer. There are absorption spikes around oxygen resonance frequencies such as those in the neighborhood of 54, 118.75, and 424.76 GHz and at water vapor resonance frequencies such as 183.31 GHz. A satellite-borne radiometer observing at these frequencies would sense only the highest layers of the atmosphere. On the other hand, a satellite-borne radiometer observing at frequencies away from the spikes would be sensitive to the surface. By observing at frequency bands that are near the resonance frequencies, but still on the sides of the spikes, one can capture information on conditions at specific layers of the atmosphere.

The AMSU focuses primarily on the 54-GHz oxygen band and the 183-GHz water vapor band. Figure 12.3 shows the weighting functions of the channels on AMSU.

Opaque microwave channels were used with great success to retrieve atmospheric conditions. Rosenkranz used AMSU-A and AMSU-B data from NOAA satellites to estimate temperature and water vapor profiles [2,3]. Shi used AMSU-A to estimate temperature

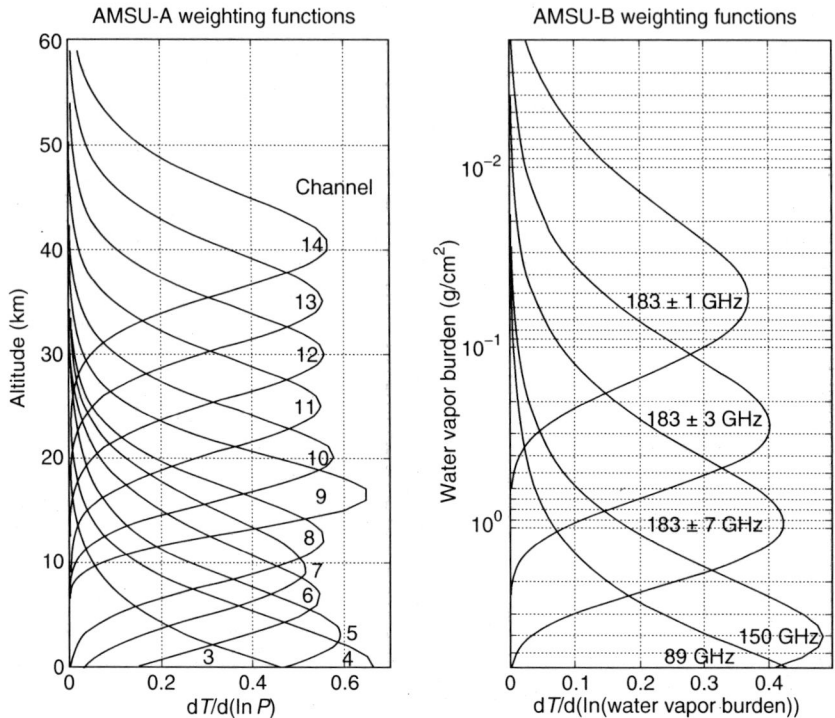

FIGURE 12.3
Weighting functions of AMSU-A and AMSU-B channels.

profiles [4]. Blackwell et al. used the 54-GHz and 118-GHz bands aboard the National Polar-Orbiting Environmental Satellite System (NPOESS) Aircraft Sounder Testbed-Microwave (NAST-M) [5]. Leslie et al. added the 183-GHz and 425-GHz bands to NAST-M [6,7,8].

Clouds and precipitation result from humid air that rises, cools, and condenses. Precipitation typically occurs in two forms: convective and stratiform. Stratiform precipitation occurs when one air mass slides under or over another, as in a cold or warm front, causing the upper air mass to cool and the water vapor within it to condense. Such precipitation is spread out. Convective precipitation is initiated by instabilities in the atmosphere caused by cold, dense air supported by warm, humid, and less dense air. Such instabilities result in the warm, humid air escaping upward and cooling. Water vapor in this ascending air mass condenses and releases latent heat. The latent heat warms the surrounding air, which pushes the air mass and the water and ice particles that have formed further upward. This cycle continues until the original warm humid air mass has cooled to the temperature of the surrounding cooler air [9,10]. The tops of convective clouds spew forth ice particles and can reach more than 10 km above sea level [11]. Convective precipitation often occurs within stratiform precipitation.

Several factors affect the precipitation rate. Higher degrees of instability caused by large vertical temperature gradients force ice particles higher in the atmosphere, causing such particles to pick up more moisture and grow. Precipitation amounts are limited by the water vapor available in the atmosphere; therefore, higher concentrations of water vapor result in higher precipitation rates. Warmer surface air contributes to higher precipitation rates because warmer air holds more water vapor.

Channels in the 54-GHz and 183-GHz bands provide important clues about factors such as cloud-top altitude, temperature profile, water vapor profile, and particle size distribution. Because each channel is sensitive to a specific layer of the atmosphere

(Figure 12.3), it is possible to extract information about the cloud-top altitude. Precipitating clouds exhibit perturbations in channels whose weighting functions have significant values in the range of altitudes occupied by the cloud. For example, one would not expect to detect low-lying clouds below 3 km in the AMSU-A channel 14 because the weighting function of channel 14 peaks at ~40 km, far above the tops of nearly all precipitating clouds.

The 54-GHz and 183-GHz bands together provide information about particle size distribution through their sensitivities to different ranges of ice particle diameters. The scattering of electromagnetic waves by spherical ice particles is described by Mie scattering coefficients. For a single spherical particle with radius r, given the power scattered by the particle P_s and the power density of the incident plane wave S_i, the scattering cross-sectional area Q_s of the particle is defined as follows:

$$Q_s = \frac{P_s}{S_i} \quad (12.16)$$

Then, the Mie scattering coefficient is defined as the ratio of Q_s of the particle to the geometric cross-sectional area of the particle

$$\xi_s = \frac{Q_s}{\pi r^2} \quad (12.17)$$

where r is the radius of the particle. Figure 12.4 shows the Mie scattering coefficients for fresh-water ice spheres at 54 GHz and 183.31 GHz as a function of diameter. The permittivities of ice were calculated using formulas developed by Hufford [12], and the Mie scattering coefficients were calculated using an iterative computational procedure developed by Deirmendjian [1,13]. At 183.31 GHz, the diameter of a particle can increase to about 0.7 mm before its diameter can no longer be uniquely determined by the Mie scattering coefficient curve. At 54 GHz, this limit is about 2.4 mm. For diameters less than 0.7 mm, ξ_s for 54 GHz is smaller than that for 183.31 GHz by a factor of at least

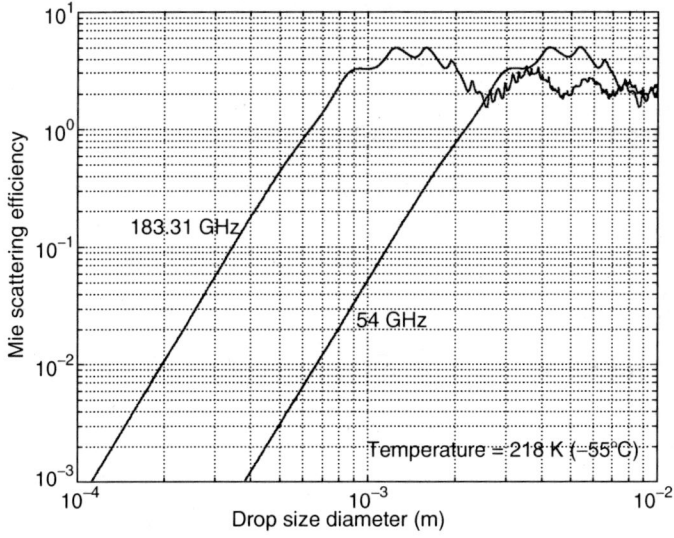

FIGURE 12.4
Mie scattering coefficients for 54 GHz and 183.31 GHz at a temperature of −55°C.

100. These differences make it possible to use both bands together to extract information about particle size distributions. The Mie scattering coefficients in Figure 12.4 were calculated for a temperature of −55°C (218 K), near the temperature at an altitude of 10 km for the 1962 U.S. Standard Atmosphere [1]. Corresponding values for −10°C (263 K) were also calculated, but did not differ from those for −55°C by more than 7.2%.

In addition to particle size distribution, the 54-GHz and 183-GHz bands can also provide information about particle abundance. While the particle size distribution can be sensed from a comparison of the Mie scattering efficiencies in both bands, particle abundance can be sensed from absolute scattering over volumes.

The cloud-top altitude is another important variable in precipitation. This is correlated with particle size density because only higher updraft velocities are able to support larger particles and reach higher altitudes. The sensitivities of the 54-GHz and 183-GHz channels to specific layers of the atmosphere suggest that these channels are able to provide information about cloud-altitude. Spina et al. used data from the opaque 118-GHz band to estimate cloud-top altitudes [14].

Gasiewski showed that the 54-GHz and 118-GHz bands together could be used to estimate cell-top altitude and hydrometeor density (in terms of mass per volume) [15,16].

Window channels contribute information about precipitation through their sensitivity to the warm emission signatures of precipitating clouds against a sea background and their sensitivity to scattering. In addition to opaque channels, AMSU also includes window channels at 23.8 GHz, 31.4 GHz, 89.0 GHz, and 150 GHz. Weng et al. and Grody et al. have developed a precipitation-retrieval algorithm for AMSU that relies primarily on these channels [17,18]. Some of the recent passive microwave instruments that have focused on windows channels and have been used to study precipitation include the following:

1. The Special Sensor Microwave Imager (SSM/I) for the Defense Meteorological Satellite Program (DMSP) [19–21]
2. The Advanced Microwave Sounding Radiometer (AMSR-E) for the Earth Observing System aboard the NASA Aqua satellite [22,23]
3. The Tropical Rainfall Measurement Mission (TRMM) microwave imager (TMI) aboard the TRMM satellite [24,25]

One weakness of window channels is that they tend to be sensitive to the surface. Over land, surface signatures can obscure the emission signatures of precipitation. Also, window-channel-based precipitation-rate retrieval algorithms tend to use one method over ocean and another over land [26,27]. The opaque channels aboard AMSU enable the development of an algorithm that uses the same method over both land and sea.

12.3 Description of AMSU-A/B and AMSU/HSB

The AMSU has been aboard the NOAA-15, NOAA-16, and NOAA-17 satellites launched in May 1998, September 2000, and June 2002, respectively. They are each equipped with the instruments AMSU-A and AMSU-B. AMSU-A has 15 channels: one each at 23.8 GHz, 31.4 GHz, and 89.0 GHz, and 12 channels in the 54-GHz oxygen absorption band (Table 12.1). AMSU-B has 5 channels at the following frequencies: 89.0 GHz, 150 GHz, 183.31 ± 1 GHz, 183.31 ± 3 GHz, and 183.31 ± 7 GHz (Table 12.2). AMSU-A and AMSU-B measure

TABLE 12.1

AMSU-A Channel Frequencies

Channel	Center Frequencies (MHz)	Bandwidth (MHz)
1	23,800 ± 72.5	2×125
2	31,400 ± 50	2×80
3	50,300 ± 50	2×80
4	52,800 ± 105	2×190
5	53,596 ± 115	2×168
6	54,400 ± 105	2×190
7	54,940 ± 105	2×190
8	55,500 ± 87.5	2×155
9	57,290.344 ± 87.5	2×155
10	57,290.344 ± 217	2×77
11	57,290.344 ± 322.2 ± 48	4×35
12	57,290.344 ± 322.2 ± 22	4×15
13	57,290.344 ± 322.2 ± 10	4×8
14	57,290.344 ± 322.2 ± 4.5	4×3
15	89,000 ± 900	2×1000

brightness temperatures at 50 and 15 km nominal resolutions at nadir, respectively. AMSU-A has a 3.33°-diameter 3-dB beamwidth and observes at 30° angles spaced at 3.33° intervals up to 48.33° from nadir every 8.00 sec. AMSU-B has a 1.1°-diameter beamwidth and observes at 90° angles spaced at 1.1° intervals up to 48.95° from nadir every 2.67 sec (Figure 12.5b) [28–30]. AMSU covers a swath width of ~2200 km. NOAA-15, NOAA-16, and NOAA-17 are sun-synchronous polar-orbiting satellites with equatorial crossing times of about 7 A.M./P.M., 2 A.M./P.M., and 10 A.M./P.M., respectively, so together they observe most locations approximately six times a day (Figure 12.6). The ascending local equatorial crossing times are 7 P.M., 2 P.M. 10 P.M. for NOAA–16 and NOAA–17, respectively.

The 15-km, 89.0-GHz channel on AMSU-B was not used in this research in order that the algorithm developed for AMSU–A/B could also be used with AMSU/HSB with minimal adjustment.

We also use data from the NASA Aqua satellite that was launched in May 2002. It is equipped with AMSU and is identical to AMSU-A aboard the NOAA satellites, and the HSB, which is identical to AMSU-B, but without the 89.0-GHz channel [31]. The nominal resolutions of AMSU and HSB are 40.5 and 13.5 km, respectively. Aqua has an equatorial crossing time of about 1:30 A.M./P.M. (Figure 12.6) [31–33]. The scan pattern of AMSU/HSB is slightly different from that of AMSU-A/B on the NOAA satellites in that the path traced during one AMSU scan is more parallel to that traced by a nearly coincident HSB scan (Figure 12.5).

TABLE 12.2

AMSU-B Channel Frequencies

Channel	Center Frequencies (GHz)	Bandwidth (GHz)
1	150 ± 0.9	2×1
2	183.31 ± 1	2×0.5
3	183.31 ± 3	2×1
4	183.31 ± 7	2×2

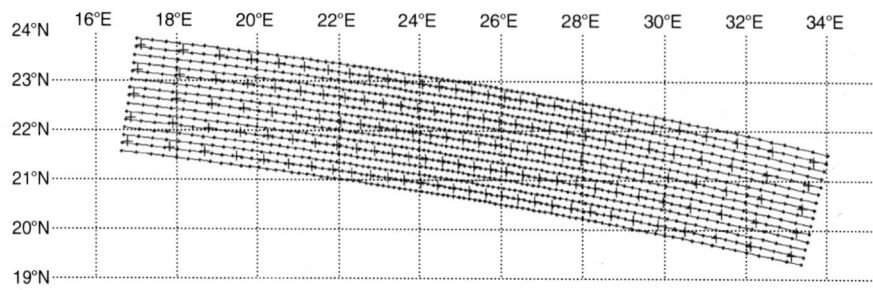

(a) Aqua AMSU/HSB, Southbound

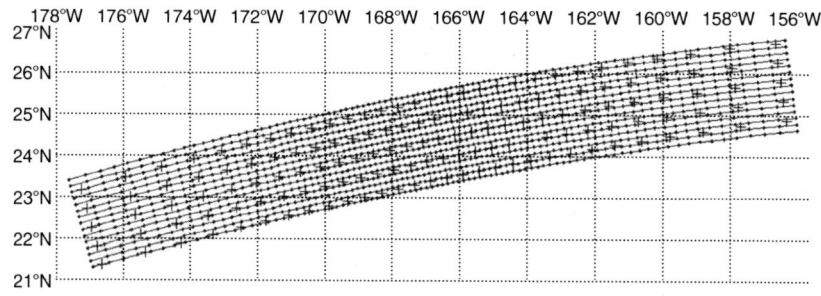

(b) Aqua AMSU/HSB, Northbound

FIGURE 12.5
Scan patterns of (a) Aqua AMSU/HSB and (b) AMSU-A/B on NOAA-15, NOAA-16, and NOAA-17. AMSU and AMSU-A spots are labeled with +'s and HSB and AMSU-B spots with dots.

12.4 Signal Processing

The preceding section provides an overview of the types of information available in data from AMSU-A/B. While developing an algorithm for estimating precipitation, it is important to process the data in a way that makes the information relevant to precipitation stand out as much as possible. This section provides an overview of the signal-processing methods used in the Chen and Staelin algorithm.

12.4.1 Regional Laplacian Filtering

Laplacian filtering is useful for clearing the effects of clouds over regions that are identified as cloudy. This enables computation of not only cloud-cleared brightness temperatures but also the perturbation due to precipitation.

Laplacian filtering gives the Chen and Staelin algorithm a spatial filtering component not found in other algorithms that process data in a manner that treats each pixel independently of any other pixel.

Laplacian filtering of a rectangular field Φ is done by first determining the region over which filtering is to be done. Using this region, a set of boundary conditions is determined. Then, using these boundary conditions, values for pixels in the region of interest are computed so that the discrete Laplace's equation is satisfied. In a continuous two-dimensional (2D) domain, Laplace's equation is as follows:

$$\nabla^2 \Phi = 0 \qquad (12.18)$$

Satellite-Based Precipitation Retrieval Using Neural Networks 241

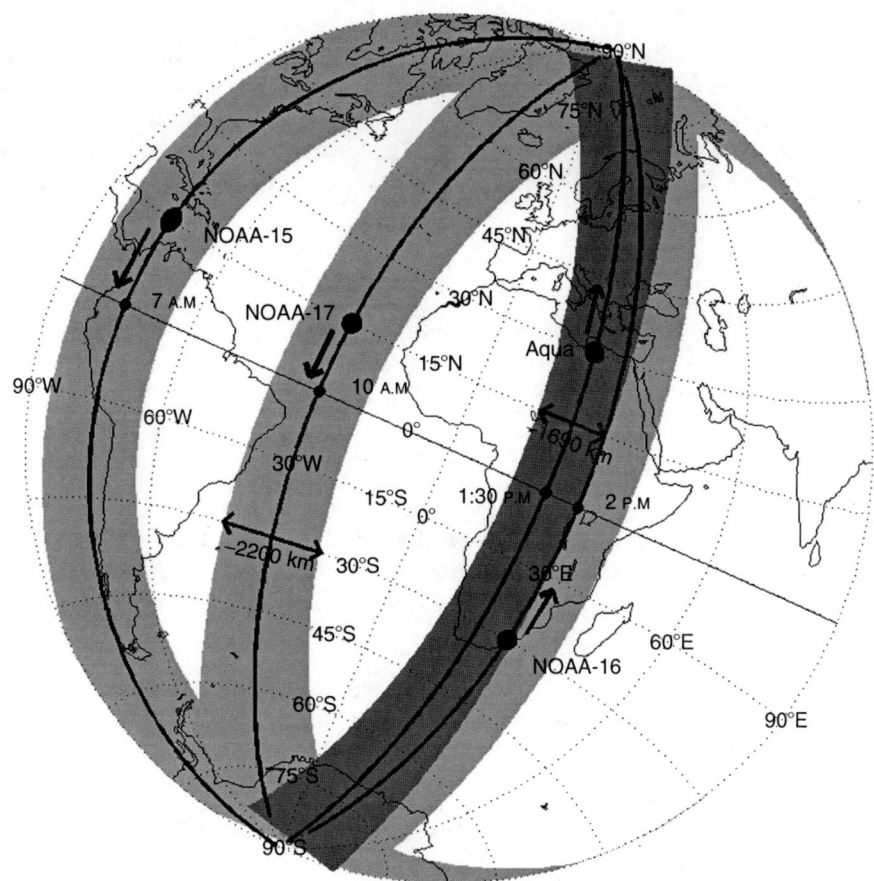

FIGURE 12.6
Orbital patterns of the NOAA-15, NOAA-16, NOAA-17, and Aqua satellites.

12.4.2 Principal Component Analysis

Signal separation is an important concept in this chapter. Although precipitation is very complex, nonstationary, and sporadic, one can process the data in a way that adequately separates the different degrees of freedom that affect precipitation.

Principal component analysis (PCA) is a linear method for reducing the dimensionality of a data set of interrelated variables. PCA transforms the data into a set of uncorrelated random variables that capture all of the variance of the original data set and assign as much variance as possible to the fewest number of variables. PCA is also known by other names such as singular value decomposition (SVD) and Karhunen–Loève transform (KLT).

PCA is useful for data compression. This feature can be critical in problems related to the compression of satellite data where a large amount of data must be downloaded from a satellite using a communications link with limited bandwidth and in situations where computational resources are limited. Blackwell used projected principal component transform, a variant of PCA, to compress data for use in estimating atmospheric temperature profiles. This reduced the number of inputs and, as a result, simplified the neural net, created a more stable neural net, and reduced the training time [34,35]. Cabrera-Mercader used noise-adjusted principal components (NAPC) to compress simulated NASA Atmospheric Infrared Sounder (AIRS) data [36].

PCA is also used to filter out noise in data. Several different versions or extensions of PCA have been used to eliminate various types of noise. A variant of PCA has been used to remove signatures of the surface from passive microwave data for the purpose of detecting and characterizing precipitating clouds [37]. PCA is also an important part of blind multivariate noise estimation and filtering algorithms such as iterative order and noise (ION) estimation and an extension of ION that is capable of estimating mixing matrices [38–40].

12.4.2.1 Basic PCA

Here, a definition of basic PCA is presented. For a random vector $\mathbf{x}$ of p random variables, the *principal components* of $\mathbf{x}$ can be defined inductively. The first principal component is the product $\alpha_1^T \mathbf{x}$, where α_1 is a unit-length vector that maximizes the variance of $\alpha_1^T \mathbf{x}$.

$$\alpha_1 = \arg\max_{\|\alpha\|=1} \mathrm{Var}(\alpha^T \mathbf{x}) \tag{12.19}$$

where $\mathrm{Var}(\cdot)$ denotes the variance of a random variable. Each of the other principal components are defined as follows: the nth principal component is the product $\alpha_n^T \mathbf{x}$, where α_n is a unit-length vector that is orthogonal to $\alpha_1, \alpha_2, \ldots$, and α_{n-1} and maximizes the variance of $\alpha_n^T \mathbf{x}$.

$$\alpha_n = \arg\max_{\|\alpha\|=1,\ \alpha \perp \alpha_i,\ \forall i \in \{1, 2, \ldots, n-1\}} \mathrm{Var}(\alpha^T \mathbf{x}) \tag{12.20}$$

$\alpha_1, \alpha_2, \ldots$, and α_p are derived in Ref. [41]. The nth principal component is the product $\alpha_n^T \mathbf{x}$ where α_n is the eigenvector associated with the nth highest eigenvalue of the covariance matrix $\Lambda_\mathbf{x}$ of $\mathbf{x}$.

In this chapter, the definitions of PCA and the term *principal component* follow the convention of Ref. [41].

12.4.2.2 Constrained PCA

In addition to the basic PCA, the algorithm developed in this chapter uses a variation of PCA known as *constrained PCA*. This form of PCA finds principal components that are constrained to be orthogonal to a given subspace [41]. It can be used to filter out noise in remote-sensing data. Constrained PCA has been used to filter out signatures of surface variations from microwave remote-sensing data [42].

Filtering out unwanted signatures involves the following steps:

1. Select a set of data that captures a good representation of the type of noise to be filtered without capturing too much of the variation of the signal of interest
2. Apply basic PCA to the resulting subset of data
3. Examine the resulting principal components for sensitivity to the type of noise to be filtered out
4. Project the data onto the subspace orthogonal to the noise-sensitive principal components
5. Apply basic PCA to the data resulting from the projection

The principal components resulting from steps (2) and (5) are called *preconstraint principal components* and *postconstraint principal components*, respectively, as in Ref. [37].

12.4.3 Data Fusion

Data fusion is a very broad area involving the combination of information from different sources. A working group set up by the European Association of Remote Sensing Laboratories and the French Society for Electricity and Electronics has adopted the following definition of data fusion [42]:

> Data fusion is a formal framework in which are expressed means and tools for the alliance of data originating from different sources. It aims at obtaining information of greater quality; the exact definition of 'greater quality' will depend upon the application.

Review papers have referred to three levels of data fusion: measurement, feature, and decision [43–45]. The measurement level is sometimes called the *pixel level* [44]. The algorithm described in Section 12.5 involves the measurement and decision levels. In this chapter, nontrivial uses of data fusion occur only at the measurement level.

Some of the applications of image fusion (or data fusion applied to 2D data) include image sharpening or enhancement [45], feature enhancement, and replacement of missing or faulty data [44]. For this research, nonlinear data fusion is applied to sharpen images. Rosenkranz developed a method for nonlinear geophysical parameter estimation through multi-resolution data fusion [46–48].

12.4.4 Neural Nets

Neural nets are computational structures that were developed to mimic the way biological neural nets learn from their environment and are useful for pattern recognition and classification. Neural nets can be used to learn and compute functions for which the relationships between inputs and outputs are unknown or computationally complex.

There are a variety of neural nets such as feedforward neural nets (sometimes called *multilayer perceptrons* [49]), Kohonen self-organizing feature maps, and Hopfield nets [50,51]. The feedforward neural net is used in this chapter.

The basic structural element of feedforward neural nets is called a *perceptron*. It computes a function of the weighted sum of inputs and a bias, as shown in Figure 12.7.

$$y = f\left(\sum_{i=1}^{n} w_i x_i + b\right) \qquad (12.21)$$

where x_i is the ith input, w_i is the weight associated with the ith input, b is the bias, f is the transfer function of the perceptron, and y is the output.

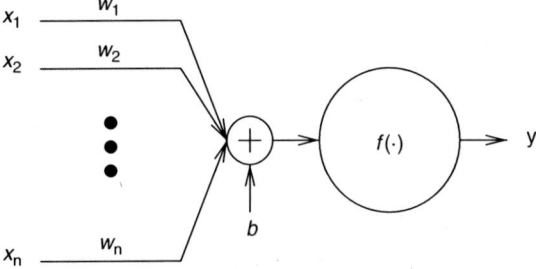

FIGURE 12.7
The structure of a perceptron.

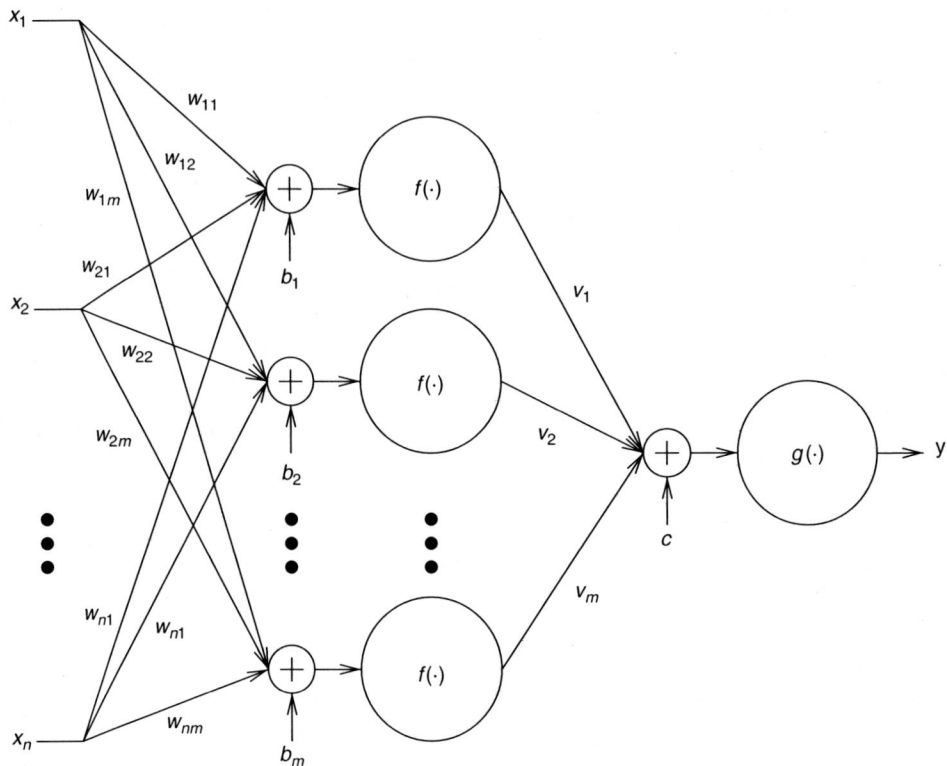

FIGURE 12.8
A two-layer feedforward neural net with one output node.

Perceptrons can be combined to form a multi-layer network, as shown in Figure 12.8. In Figure 12.8, x_i is the ith input, n is the number of inputs, w_{ij} is the weight associated with the connection from the ith input to the jth node in the hidden layer, b_i is the bias of the ith node, m is the number of nodes in the hidden layer, f is the transfer function of the perceptrons in the hidden layer, v_i is the weight between the ith node and the output node, c is the bias of the output node, g is the transfer function of the output node, and y is the output. Then,

$$y = g\left(\sum_{j=1}^{m} v_j f\left(\sum_{i=1}^{n} w_{ij} x_i + b_j \right) + c \right) \quad (12.22)$$

In this chapter, f and g are defined as follows:

$$f(x) = \tanh x = \frac{e^x - e^{-x}}{e^x + e^{-x}} \quad (12.23)$$

$$g(x) = x \quad (12.24)$$

The function $\tanh x$ is approximately linear in the range $-0.6 \leq x \leq 0.6$, and approaches 1 as x tends to 1 and -1 as x tends to -1, so it has a nonlinearity that is not too complex (Figure 12.9). This neural net topology is good for situations in which one wants to develop a simple nonlinear estimator whose output depends approximately monotonically on each input.

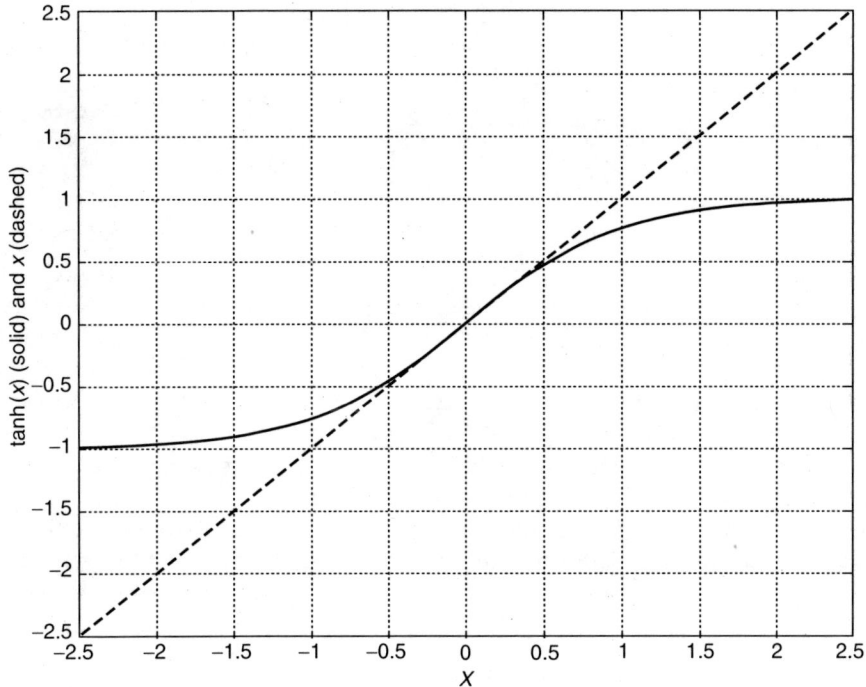

FIGURE 12.9
Neural net transfer functions.

The neural nets for this chapter were trained using the Levenberg–Marquardt training algorithm. Marquardt developed an efficient algorithm (called the *Marquardt–Levenberg algorithm* in Ref. [51]) for nonlinear least-square parameter estimation [52]. Hagan and Menhaj incorporated this algorithm into a backpropagation training algorithm for feedforward neural nets [52]. The weights of the neural net were initialized using the Nguyen–Widrow method to facilitate convergence of the neural net weights during training [53]. The vectors used to train and evaluate the neural nets were divided into three disjoint sets:

1. The *training set*, the set used to determine how the weights of the neural net should be adjusted during the training
2. The *validation set*, the set used to determine when the training should stop
3. The *testing set*, the set used to evaluate the resulting neural net

These definitions are from Ref. [54].

One of the challenges encountered in the course of developing an estimator involved dealing with an output range that covered several orders of magnitude. Chapter 4 in this volume describes how this was accomplished.

12.5 The Chen–Staelin Algorithm

The basic structure of the algorithm includes some signal-processing components and a neural net, as shown in Figure 12.10. The signal processing components process the data

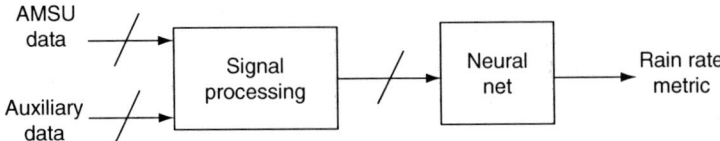

FIGURE 12.10
Basic structure of the algorithm.

into forms that characterize the most important degrees of freedom related to the precipitation rate such as atmospheric temperature profile, water vapor profile, cloud-top altitude, particle size distribution, and vertical updraft velocity. The neural net is trained to learn the nonlinear dependencies of precipitation rate on these variables. The dependence of the precipitation rate on these variables should be monotonic, so the neural net does not need to be complicated. A feedforward neural net with one hidden layer of tangent sigmoid nodes (with transfer function $f(x) = \tanh x$) and one linear output node should be sufficient (Figure 12.8) [49].

The Chen–Staelin algorithm uses the signal-processing methods in the previous section to extract the most relevant information from AMSU data. Figure 12.11 shows a block diagram of the first part of the algorithm and Figure 12.12 shows the final part of the algorithm with a neural net. A neural net that takes the following sets of inputs is at the heart of the algorithm:

1. Inferred 15-km-resolution perturbations at 52.8 GHz, 53.6 GHz, 54.4 GHz, 54.9 GHz, and 55.5 GHz.
2. 183 ± 1-, 183 ± 3-, and 183 ± 7-GHz 15-km HSB data.

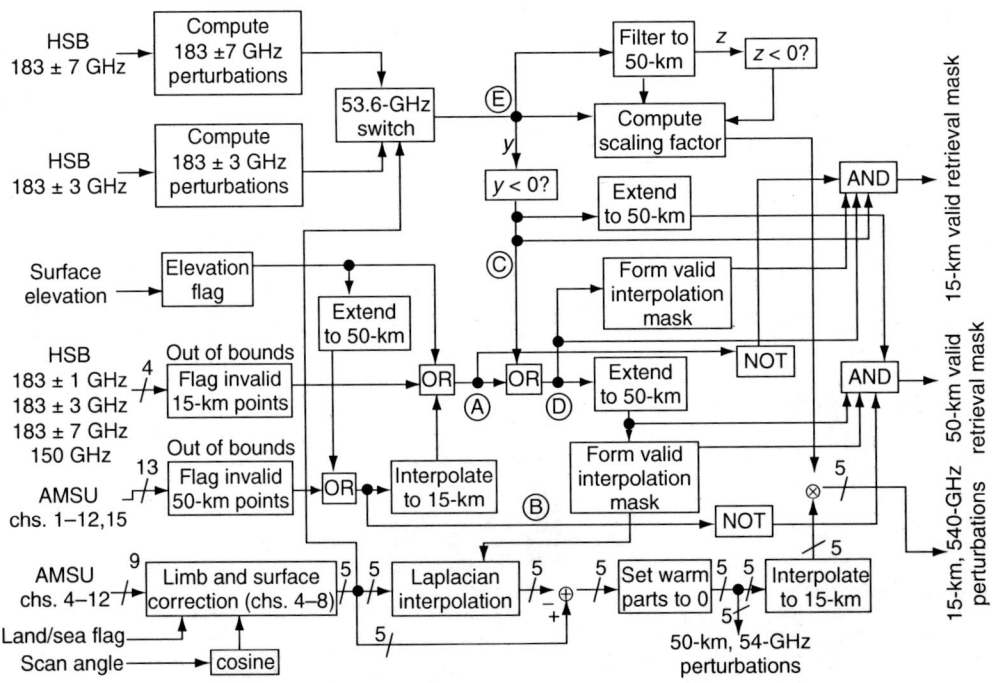

FIGURE 12.11
Block diagram of the algorithm, part 1.

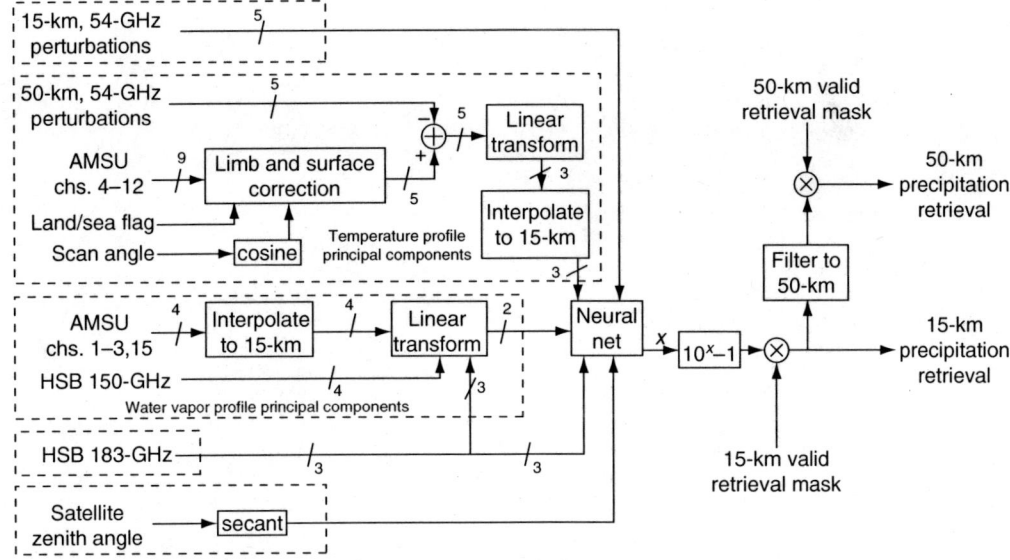

FIGURE 12.12
Block diagram of the algorithm, part 2.

3. The leading three principal components characterizing the original five corrected 50-km AMSU-A temperature radiances.
4. Two surface-insensitive principal components that characterize the window channels at 23.8 GHz, 31.4 GHz, 50 GHz, and 89 GHz, along with the four HSB channels.
5. The secant of the satellite zenith angle θ.

Each of these sets provides the neural net with information that is relevant to precipitation. The three principal components characterizing AMSU-A temperature radiances provide information on the atmospheric temperature profile, which is important because it determines how much water vapor can be precipitated. The secant of the satellite zenith angle θ allows the neural net to account for variations in the data due to the scan angle. The 15-km cloud-induced perturbations provide information on the cloud-top altitude.

The current AMSU/HSB precipitation retrieval algorithm is based on NOAA-15 AMSU comparisons with NEXRAD over the eastern United States during 38 orbits that exhibited significant precipitation and were distributed throughout the year. These orbits are listed in Table 12.3. The primary precipitation-rate retrieval products of AMSU/HSB are ~15- and ~50-km-resolution contiguous retrievals over the viewing positions of HSB and AMSU, respectively, within 43° of nadir. The two outermost 50-km and six outermost 15-km viewing positions on each side of the swath are omitted due to their grazing angles. The algorithm architectures for these two retrieval methods and the derivation of the numerical coefficients characterizing the neural network are described and presented below.

12.5.1 Limb-Correction of Temperature Profile Channels

AMSU observes at angles up to 49° away from the nadir. For angles further away from nadir, the electromagnetic energy originating from a given altitude and atmospheric state

TABLE 12.3

List of Rainy Orbits Used for Training, Validation, and Testing

October 16, 1999, 0030 UTC	April 30, 2000, 1430 UTC
October 31, 1999, 0130 UTC	May 14, 2000, 0030 UTC
November 2, 1999, 0045 UTC	May 19, 2000, 0015 UTC
December 4, 1999, 1445 UTC	May 19, 2000, 0145 UTC
December 12, 1999, 0100 UTC	May 20, 2000, 0130 UTC
January 28, 2000, 0200 UTC	May 25, 2000, 0115 UTC
January 31, 2000, 0045 UTC	June 10, 2000, 0200 UTC
February 14, 2000, 0045 UTC	June 16, 2000, 0130 UTC
February 27, 2000, 0045 UTC	June 30, 2000, 0115 UTC
March 11, 2000, 0100 UTC	July 4, 2000, 0115 UTC
March 17, 2000, 0015 UTC	July 15, 2000, 0030 UTC
March 17, 2000, 0200 UTC	August 1, 2000, 0045 UTC
March 19, 2000, 0115 UTC	August 8, 2000, 0145 UTC
April 2, 2000, 0100 UTC	August 18, 2000, 0115 UTC
April 4, 2000, 0015 UTC	August 23, 2000, 1315 UTC
April 8, 2000, 0030 UTC	September 23, 2000, 1315 UTC
April 12, 2000, 0045 UTC	October 5, 2000, 0130 UTC
April 12, 2000, 0215 UTC	October 6, 2000, 0100 UTC
April 20, 2000, 0100 UTC	October 14, 2000, 0130 UTC

has to travel longer paths before reaching the radiometer and, therefore, is subject to more absorption and scattering effects. This results in scan-angle-dependent effects in brightness temperature images, as shown in Figure 12.13a. A limb and surface correction method for AMSU-A channels 4–8 brightness temperatures is needed to make precipitation-induced perturbations more apparent and for extracting information about atmospheric conditions. AMSU-A channels 4 and 5 are corrected for surface variations, as they are sensitive to the surface. For these two channels, the brightness temperature for pixels over ocean is corrected to what might be observed for the same atmospheric conditions over land. AMSU-A channels 9–14 brightness temperatures are not corrected because they are not significantly perturbed by clouds and therefore are not used for anything other than limb correction.

Limb and surface correction was done by training a neural net of the type shown in Figure 12.8 to estimate nadir-viewing brightness temperatures. For each pixel, the neural net used brightness temperatures from several channels at that pixel to estimate the brightness temperature seen at the pixel closest to nadir at a nearly identical latitude and at nearly the same time. It is assumed that the temperature field does not vary significantly over one scan. Limb and surface correction was done for AMSU-A channels 4–8. The data used to correct each of these channels are listed in Table 12.4. No attempt has been made to correct for the scan-angle-dependent asymmetry in the brightness temperatures. These neural nets were trained using data between 55°N and 55°S from seven orbits spaced over 1 year. Channels 4 and 5 are surface sensitive, so they were trained to estimate brightness temperatures that would be seen over land.

Figure 12.13 shows a sample of (a) uncorrected and (b) limb and surface corrected 54.4-GHz brightness temperatures. The shapes of precipitation systems over Texas and the Mexico–Guatemala border are more apparent after the limb correction. In Figure 12.13(a), the difference between brightness temperatures at nadir and the swath edge is as high as 18 K. In Figure 12.13(b), the angle-dependent variation is less than 3 K.

One limitation of the training is that one cannot really know what the nadir-viewing brightness temperature is supposed to be when there is precipitation.

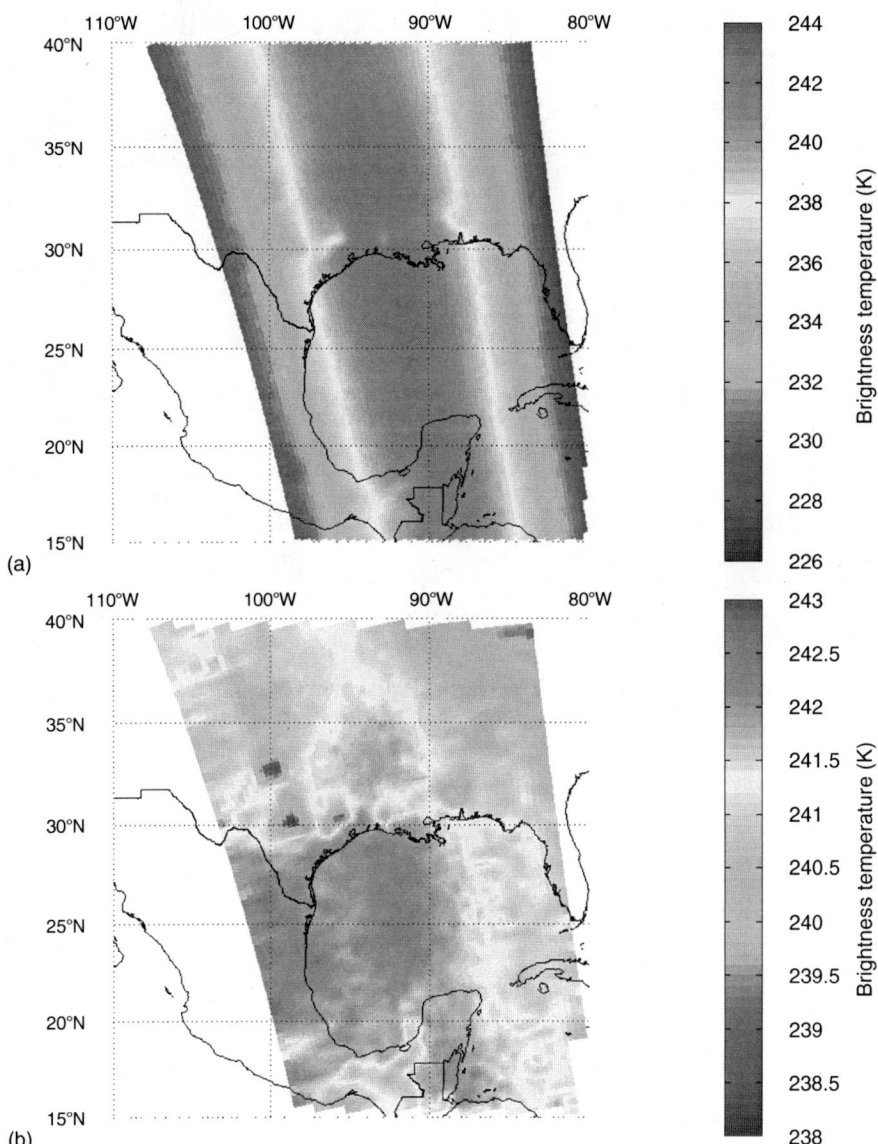

FIGURE 12.13 (See color insert following page 178.)
NOAA-15 AMSU-A 54.4-GHz brightness temperatures for a northbound track on September 13, 2000. (a) Uncorrected and (b) limb and surface corrected.

12.5.2 Detection of Precipitation

The 15-km-resolution precipitation-rate retrieval algorithm, summarized in Figure 5.2, and Figure 5.3 begins with the identification of potentially precipitating pixels. The neural net operates on data from only FOVs labeled as potentially precipitating. This choice eliminates the need to exhaustively learn all of the conditions where the precipitation rate is exactly zero. The neural net is also likely to have difficulty forcing precipitation rates in nonprecipitating FOVs to be exactly zero. This choice also reduces the time needed to train the neural net since, at any given time, precipitation falls over less than 10% of the Earth's surface. All 15-km pixels with brightness temperatures at 183 ± 7 GHz that are below a threshold T_7 are flagged as potentially precipitating, where

TABLE 12.4
Data Used in Limb and Surface Correction of AMSU-A Channels

AMSU-A Channel	Inputs Used for Limb and Surface Correction
4	AMSU-A channels 4–12, land/sea flag, cos φ
5	AMSU-A channels 5–12, land/sea flag, cos φ
6	AMSU-A channels 6–12, cos φ
7	AMSU-A channels 6–12, cos φ
8	AMSU-A channels 6–12, cos φ

$$T_7 = 0.667(T_{53.6} - 248) + 262 + 6\cos\theta \tag{12.25}$$

and where θ is the satellite zenith angle and $T_{53.6}$ is the spatially filtered 53.6-GHz brightness temperature obtained by selecting the warmest brightness temperature within a 7×7 array of AMSU-B pixels. If, however, $T_{53.6}$ is below 248 K, then the brightness temperature at 183 ± 3 GHz is compared, instead, to a different threshold T_3, where

$$T_3 = 242.5 + 5\cos\theta \tag{12.26}$$

The 183 ± 3-GHz band is used to flag potential precipitation when the 183 ± 7-GHz flag could be erroneously set by low-surface emissivity in very cold and dry atmospheres, as indicated by $T_{53.6}$. These thresholds T_7 and T_3 are slightly colder than a saturated atmosphere would be, implying the presence of a microwave-absorbing cloud. If the locally filtered $T_{53.6}$ is less than 242 K, then the pixel is assumed not to be precipitating.

12.5.3 Cloud-Clearing by Regional Laplacian Interpolation

Within regions flagged as potentially precipitating, strong precipitation is generally characterized by cold, cloud-induced perturbations of the AMSU-A tropospheric temperature sounding channels in the range of 52.5–55.6 GHz. Brightness temperature images approximately satisfy Laplace's equation in the absence of precipitation. When the potentially precipitating FOVs have been identified, Laplacian interpolation can be performed to clear the brightness temperature image of the effects of precipitation, and the perturbations due to precipitation can be computed. Examples of 183 ± 7-GHz data and the corresponding 50-km cold perturbations at 52.8 GHz are illustrated in Figure 12.14a and Figure 12.14c. Physical considerations and aircraft data show that convective cells near 54 GHz typically appear slightly off-center and less extended relative to the 183-GHz images [55,56].

12.5.4 Image Sharpening

The small interpolation errors in converting 54-GHz perturbations to 15-km contribute to the total errors and discrepancies discussed in Section 12.3. These 50-km-resolution 52.8-GHz perturbations $\Delta T_{50,52.8}$ are then used to infer the perturbations $\Delta T_{15,52.8}$ [Figure 12.14d]. These might have been observed at 52.8 GHz with a 15-km resolution had those perturbations been distributed spatially in the same way as the cold perturbations observed at either 183 ± 7 or 183 ± 3 GHz, the choice between these two channels being the same as described above. This requires the bilinearly interpolated 50-km AMSU data

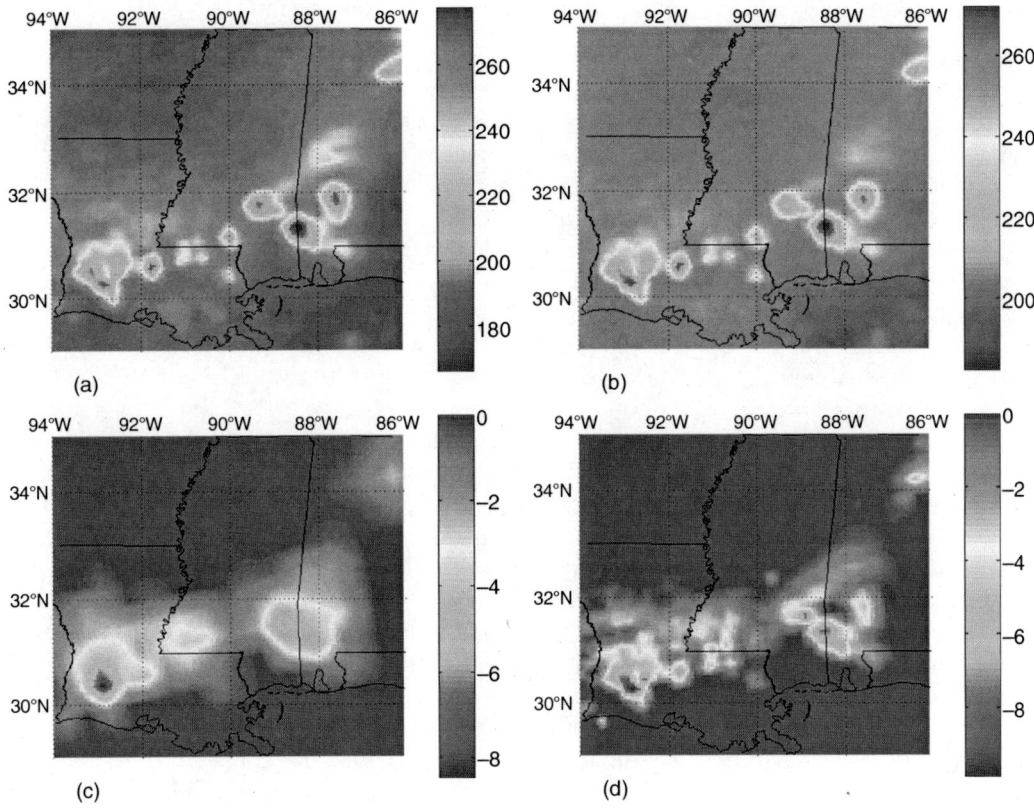

FIGURE 12.14 (See color insert following page 178.)
Frontal system on September 13, 2000, 0130 UTC. (a) Brightness temperatures (K) near 183 ± 7 GHz. (b) Brightness temperatures (K) near 183 ± 3 GHz. (c) Brightness temperature perturbations (K) near 52.8 GHz. (d) Inferred 15-km-resolution brightness temperature perturbations (K) near 52.8 GHz.

to be resampled at the HSB beam positions. These inferred 15-km perturbations are computed for five AMSU-A channels using

$$\Delta T_{15,54} = 20 \tan h\left(\frac{\Delta T_{15,183}}{\Delta T_{50,183}}\right)\Delta T_{50,54} \tag{12.27}$$

The perturbation $\Delta T_{15,183}$ near 183 GHz is defined to be the difference between the observed radiance and the appropriate threshold given by (12.25) or (12.26). The perturbation $\Delta T_{50,54}$ near 54 GHz is defined to be the difference between the observed radiance and the Laplacian-interpolated radiance based on those pixels surrounding the flagged region [58]. Any warm perturbations in the images of $\Delta T_{15,183}$ and $\Delta T_{50,54}$ are set to zero. Limb and surface-emissivity corrections to nadir for the five 54-GHz channels are produced by neural networks for each channel; they operate on nine AMSU-A channels above 52 GHz, the cosine of the viewing angle φ from nadir, and a land–sea flag (Figure 12.12). They were trained on seven orbits spaced over 1 year for latitudes up to ±55°. Inferred 50- and 15-km precipitation-induced perturbations at 52.8 GHz are shown in Figure 12.14c and Figure 12.14d for a frontal system. Such estimates of 15-km perturbations near 54 GHz help characterize heavily precipitating small cells.

12.5.5 Temperature and Water Vapor Profile Principal Components

One important determinant of precipitation is the temperature profile. Warmer atmospheres can hold more water vapor and result in higher vertical updraft velocities. Therefore, inputs to the neural net in Figure 12.10 should include some that have information about the temperature profile. For each of AMSU-A channels 4–8, the brightness temperatures were corrected for limb and surface effects and then processed to eliminate precipitation signatures with the methods described in Section 12.4.1 through Section 12.4.3. The corrected brightness temperatures from all five of these channels could have been inputs to the neural net in Figure 12.10, but it was determined that a more compact representation of these channels was sufficient. PCA was applied to these five channels, and the first three principal components were found to be sufficient for characterizing the temperature profile. Adding the fourth and fifth principal components did not significantly improve the training of the neural net.

The water vapor profile is another important determinant of precipitation. Higher concentrations of water vapor can result in higher precipitation rates. The water vapor principal components are computed using AMSU-A channels 1–3, and 15, and the AMSU-B 150-, 183 ± 7-, 183 ± 3-, and 183 ± 1-GHz channels. Some of these channels are sensitive to surface variations. Therefore, it is necessary to project the vector of these observations onto a subspace that is not significantly sensitive to surface variations. Constrained PCA, which was described in Section 12.4.2.2, was used to compute the water vapor principal components. A set of pixels without precipitation and with different types of surfaces was selected to compute surface-sensitive eigenvectors using PCA. The surface-sensitive eigenvectors were determined by visual inspection of the preconstraint principal components for correlation with surface features (e.g., land and sea boundaries). Then, a set of data that also included precipitation was selected. The observations over this set were projected onto a linear subspace that was orthogonal to the subspace spanned by the surface-sensitive eigenvectors. Then, PCA was done on the resulting data set to determine the water vapor principal components. It was found that two water vapor principal components were adequate for characterizing the eight channels.

12.5.6 The Neural Net

All 13 of the variables listed at the beginning of this section are fed into the neural net used for 15-km precipitation-rate retrievals, as shown in Figure 12.12. The relative insensitivity of these inputs to surface emissivity is important to the success of this technique over land, ice, and snow.

This network was trained to minimize the rms value of the difference between the logarithms of the (AMSU+1 mm/h) and (NEXRAD+1 mm/h) retrievals; the use of logarithms reduced the emphasis on the heaviest rain rates, which were roughly three orders of magnitude greater than the lightest rates. Adding 1 mm/h reduced the emphasis on the lightest rain rates, which were more noise-dominated. These intuitive choices clearly impact the retrieval error distribution, and therefore further studies should enable algorithm improvements. However, retrievals with training optimized for low rain rates did not markedly improve that regime. NEXRAD precipitation retrievals with a 2-km resolution were smoothed to approximate Gaussian spatial averages that were centered on and approximated the view-angle-distorted 15- or 50-km antenna beam patterns. The accuracy of NEXRAD precipitation observations is known to vary with distance; therefore, only points beyond 30 km, but within 110 km, of each NEXRAD radar site were included in the data used to train and test the neural nets. Eighty different networks were trained using the Levenberg–Marquardt algorithm, each with different numbers of nodes and water vapor principal components. A network with nearly the best performance over the testing dataset was chosen; it used two surface-blind water vapor principal components, and a slightly better performance was achieved with five water vapor principal components with increased surface sensitivity. The final network had one hidden layer with five nodes that used the tanh sigmoid function. These neural networks were similar to those described in Ref. [57]. The resulting 15-km-resolution precipitation retrievals were then smoothed to yield 50-km retrievals.

The 15-km retrieval neural network was trained using precipitation data from the 38 orbits listed in Table 12.3. During this period, the radio interference to AMSU-B was negligible relative to other sources of retrieval error. Each 15-km pixel flagged as potentially precipitating using 183 ± 7- or 183 ± 3-GHz radiances (see Figure 12.11 and Figure 12.12) was used for training, validation, or testing of the neural network. For these 38 orbits over the United States, 15 one-hundred and sixty 15-km pixels were flagged and considered suitable for training, validation, and testing; half were used for training and one quarter were used for each of validation and testing, where the validation pixels were used to determine when the training of the neural network should cease. On the basis of the final AMSU and NEXRAD 15-km retrievals, approximately 14 and 38%, respectively, of the flagged 15-km pixels appeared to have been precipitating less than 0.1 mm/h for the test set.

12.6 Retrieval Performance Evaluation

This section presents three forms of evaluation for this initial precipitation-rate retrieval algorithm: (1) representative qualitative comparisons of AMSU and NEXRAD precipitation rate images, (2) quantitative comparisons of AMSU and NEXRAD retrievals stratified by NEXRAD rain rate, and (3) representative precipitation images at more extreme latitudes beyond the NEXRAD training zone.

12.6.1 Image Comparisons of NEXRAD and AMSU-A/B Retrievals

Each NEXRAD comparison at 15-km resolution occurred within 8 min of satellite overpass; such coincidence is needed to characterize single-pixel retrievals because convective precipitation evolves rapidly on this spatial scale. Although comparison with instruments such as TRMM and SSM/I would be useful, their orbits unfortunately overlap those of AMSU within 8 min so infrequently (if ever) that comparisons over precipitation are too rare to be useful until several years of data have been analyzed. This challenge of simultaneity and the sporadic character of rain have restricted most prior instrument comparisons (passive microwave satellites, radar, rain gauges) to dimensions over 100 km and to periods of an hour to a month [58–60]. The uniformity and extent of the NEXRAD network offer a unique degree of simultaneity on 15- and 50-km scales and also the ability to match the Gaussian shape of the AMSU antenna beams.

Although these AMSU/HSB–NEXRAD comparisons are encouraging because they involve single pixels and independent physics and facilities, further extensive analyses are required for real validation. For example, comparisons of precipitation averages and differences over the same time/space units used to validate other precipitation measurement systems (e.g., SSM/I [61], ATOVS, TRMM, rain gauges) are needed to characterize variances and systematic biases based on the precipitation rate, type, location, or season.

These biases include any present in the NEXRAD data used to train the AMSU/HSB algorithm; once characterized, they can be diminished. Any excess variance experienced for rain cells too small to be resolved by AMSU/HSB can also eventually be better characterized, although it is believed to be modest for cells with microwave signatures larger than 10 km. Smaller cells contribute little to the total rainfall.

12.6.2 Numerical Comparisons of NEXRAD and AMSU-A/B Retrievals

Figure 12.15a and Figure 12.15b present 15-km-resolution precipitation retrieval images for September 13, 2000, obtained from NEXRAD and AMSU, respectively. On this occasion, both sensors yielded rain rates over 50 mm/h at similar locations and lower rain rates down to 0.5 mm/h over comparable areas. The revealed morphology is thus very similar, even though AMSU observes 6 min before NEXRAD, and it senses altitudes that are separated by several kilometers; rain falling at a nominal rate of 10 m/s takes 10 min to fall 6 km.

Figure 12.16 shows the scatter between the 15-km AMSU and NEXRAD rain-rate retrievals for the test pixels not used for training or validation. Figure 12.17 shows the scatter between the 50-km AMSU and NEXRAD rain-rate retrievals over all points flagged as precipitating.

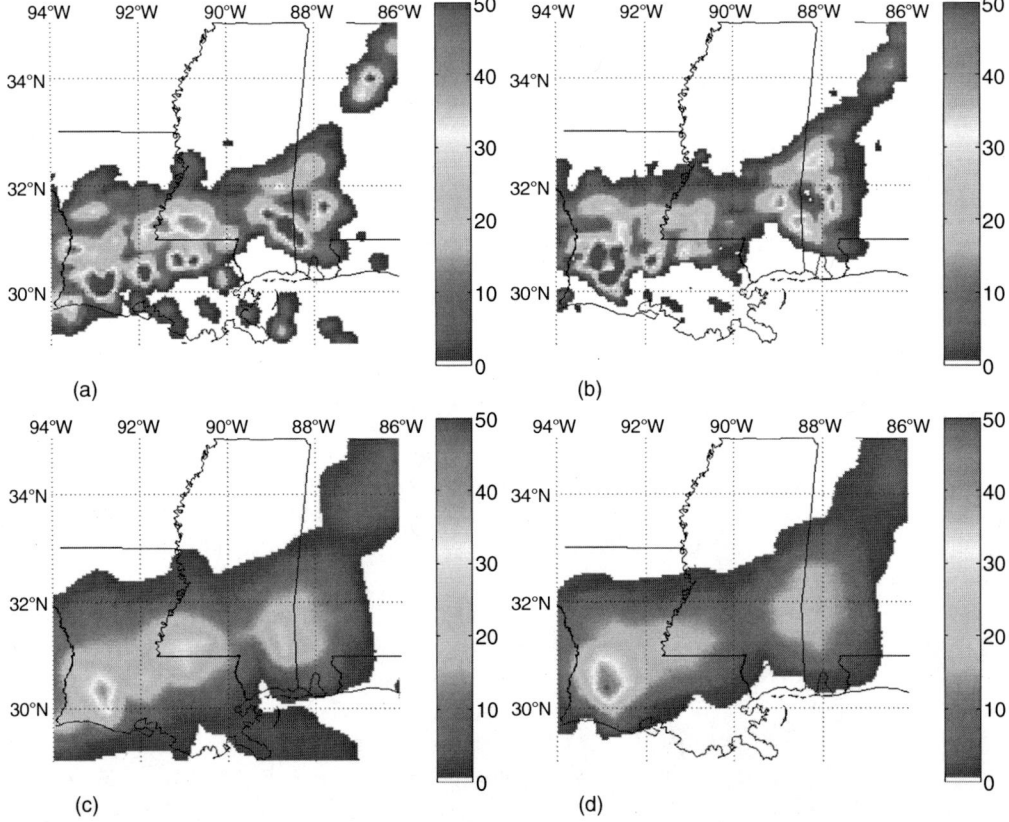

FIGURE 12.15 (See color insert following page 178.)
Precipitation rates (mm/h) above 0.5 mm/h observed on September 13, 2000, 0130 UTC. (a) 15-km-resolution NEXRAD retrievals, (b) 15-km-resolution AMSU retrievals, (c) 50-km-resolution NEXRAD retrievals, and (d) 50-km-resolution AMSU retrievals.

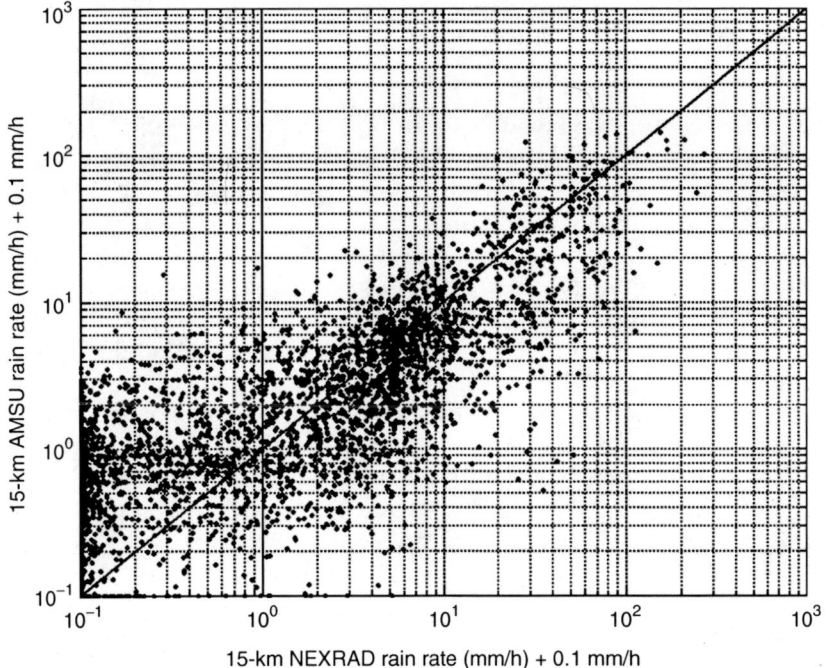

FIGURE 12.16
Comparison of AMSU and NEXRAD estimates of rain rate at 15-km resolution.

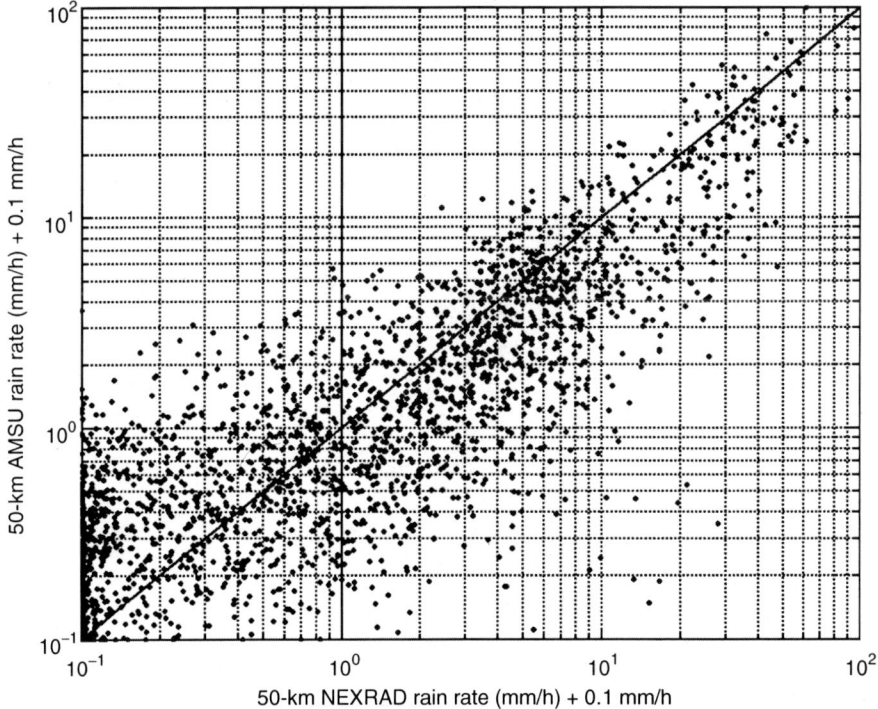

FIGURE 12.17
Comparison of AMSU and NEXRAD estimates of rain rate at 50-km resolution.

The relative sensitivity of AMSU and NEXRAD to light and heavy rain can be seen in Figure 12.17. In general, these figures suggest that AMSU responds less to the highest radar rain rates, perhaps because AMSU is less sensitive to the bright-band or hail anomalies that affect the radar. They also suggest that the risk of false rain detections increases for AMSU retrievals below 0.5 mm/h at a 50-km resolution, although further study is required. Greater accuracy at these low rates requires more space-time averaging and careful calibration. The risk of overestimating rain rate also appears to be limited. Only 3.3% of the total AMSU-derived rainfall was in areas where AMSU saw more than 1 mm/h and NEXRAD saw less than 1 mm/h. Only 7.6% of the total NEXRAD-derived rainfall was in areas where NEXRAD saw more than 1 mm/h and AMSU saw less than 1 mm/h. These percentages were compared with the total percentages of AMSU and NEXRAD rain that fell at rates above 1 mm/h, which were 94 and 97%, respectively. It is also interesting to see to what degree does each sensor retrieve rain when the other does not, and how much rain does each sensor miss. For example, of the 73 NEXRAD 15-km rain-rate retrievals in Figure 12.16 above 54 mm/h, none were found by AMSU to be below 3 mm/h, and of the 61 AMSU 15-km retrievals above 45 mm/h, none were found by NEXRAD to be below 16 mm/h. Also, of the 69 NEXRAD 50-km rain-rate retrievals in Figure 12.17 above 30 mm/h, none were found by AMSU to be below 5 mm/h, and of the 102 AMSU 50-km retrievals above 16 mm/h, none were found by NEXRAD to be below 10 mm/h.

Perhaps the most significant AMSU precipitation performance metric is the rms difference between the NEXRAD and AMSU rain-rate retrievals; these are grouped by retrieved NEXRAD rain rates in octaves. The central 26 AMSU-A scan angles and central 78 AMSU-B scan angles were included in these evaluations; only the outermost two AMSU-A angles on each side were omitted. These comparisons used all 50-km pixels and only the 15-km pixels were not used for training or validation. The results are listed in Table 12.5. The smoothing of the 15-km NEXRAD and AMSU results to a nominal 50-km resolution was consistent with an AMSU-A Gaussian beamwidth of 3.3°.

The rms agreement between these two very different precipitation-rate sensors appears surprisingly good, particularly since a single AMSU neural network is used over all angles, seasons, and latitudes. The 3-GHz radar retrievals respond most strongly to the largest hydrometeors, especially those below the bright band near the freezing level, while AMSU interacts with the general population of hydrometeors in the top few kilometers of the precipitation cell, which may lie several kilometers above the freezing level. Much of the agreement between AMSU and NEXRAD rain-rate retrievals must therefore result from the statistical consistency of the relations between rain rate and its various electromagnetic signatures. It is difficult to say how much of the observed

TABLE 12.5

RMS AMSU/NEXRAD Discrepancies (mm/h)

NEXRAD Range	15-km Resolution	50-km Resolution
<0.5 mm/h	1.0	0.5
0.5–1 mm/h	2.0	0.9
1–2 mm/h	2.3	1.1
2–4 mm/h	2.7	1.8
4–8 mm/h	3.5	3.2
8–16 mm/h	6.9	6.6
16–32 mm/h	19.0	12.9
>32 mm/h	42.9	22.1

discrepancy is due to each sensor or how well each correlates with precipitation reaching the ground.

Furthermore, this study provided an opportunity for evaluation of radar data. The rms discrepancies between AMSU and NEXRAD retrievals were separately calculated over all points at ranges from 110 to 230 km from any radar. For NEXRAD precipitation rates below 16 mm/h, these rms discrepancies were approximately 40% greater than those computed for test points at the 30- to 110-km range. At rain rates greater than 16 mm/h, the accuracies beyond 110 km were more comparable. Most points in the eastern United States are more than 110 km from any NEXRAD radar site.

12.6.3 Global Retrievals of Rain and Snow

Figure 12.18 illustrates precipitation-rate retrievals at points around the globe where radar confirmation data are scarce. Figure 12.18a shows precipitation retrievals in the tropics over a mix of land and sea, while Figure 12.18b shows a more intense tropical

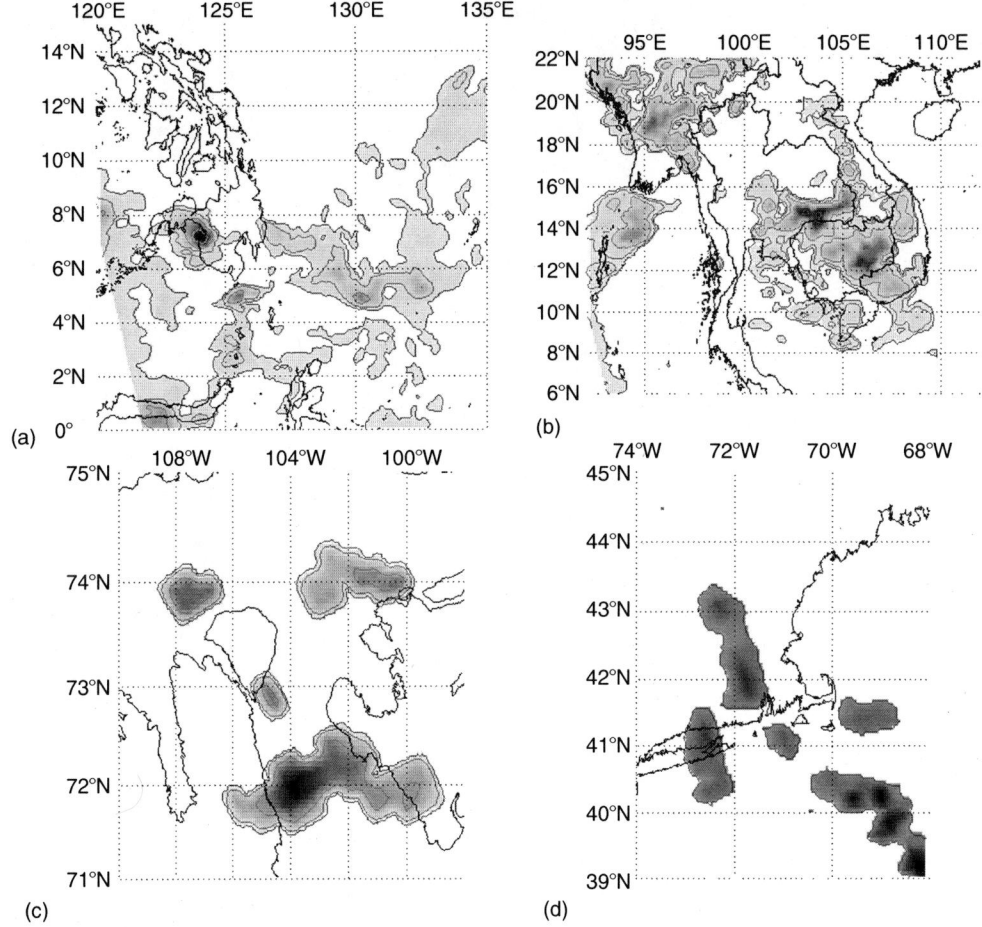

FIGURE 12.18
AMSU precipitation-rate retrievals (mm/h) with 15-km resolution. (a) Philippines, April 16, 2000; (b) Indochina, July 5, 2000; (c) Canada, August 2, 2000; and (d) New England snowstorm, March 5, 2001. Precipitation-rate retrievals exceed 0.5 mm/h in the shaded regions, and contours are drawn for 0.5 mm/h, 2 mm/h, 8 mm/h, 32 mm/h, and 128 mm/h. The peak retrieved values are 47 mm/h, 143 mm/h, 30 mm/h, and 1.5 mm/h in (a), (b), (c), and (d), respectively

event. Figure 12.18c illustrates strong precipitation near 72°N to 74°N, again over both land and sea. Finally, Figure 12.18d illustrates the March 5, 2001, New England snowstorm that deposited roughly a foot of snow within a few hours: an accumulation somewhat greater than is indicated by the retrieved rain rates of ~1.2 mm/h. This applicability of the algorithm to snowfall rate should be expected because the observed radio emission originates exclusively at high altitudes. Whether the hydrometeors are rain or snow on impact depends only on air temperatures near the surface—far below those altitudes being probed. For essentially all of the pixels shown in Figure 12.18, the adjacent clear air exhibited temperature and humidity profiles (inferred from AMSU) within the range of the training set. Nonetheless, regional biases are expected and will require evaluation. For example, polar stratiform precipitation is expected to exhibit relatively weaker radiometric signatures in winter when the temperature lapse rates are lower, and snow-covered mountains in cold polar air can produce false detections.

12.7 Conclusions

In this chapter, the precipitation estimation method for microwave radiometric data from Chen and Staelin and the role of signal processing methods were described. The development of the Chen–Staelin algorithm shows that signal processing can play a useful role in satellite-based precipitation estimation. In this algorithm, which was developed for AMSU-A/B and AMSU/HSB, PCA was used to reduce the dimensionality of selected sets of channels and to separate the effects of surface variations from atmospheric variations. Data fusion was used to sharpen 50-km data from AMSU-A so that 15-km precipitation retrievals could be done. Laplacian filtering was applied to data from the 54-GHz band to quantify the effects of clouds, and neural nets were trained to learn the mathematical relationships between precipitation and the information resulting from the signal processing. The signal processing components of the algorithm were designed to process the brightness temperature measurements in a way that extracts the most relevant information, and the neural net was trained to learn the relationship between precipitation rate and the inputs.

This Chen–Staelin algorithm represents a step in the ongoing development of microwave precipitation retrieval algorithms. The algorithm of Chen and Staelin likely can be improved by choosing more general signal processing methods or fine-tuning the ones already being used. For example, variations or extensions of PCA such as independent component analysis (ICA) could be used [63,64]. Additionally, methods like PCA and ICA can be improved by incorporating components from physics-based methods. Additional improvements can be made by making better use of the data from window channels.

Future instruments also present opportunities for better precipitation retrievals. The Advanced Technology Microwave Sounder (ATMS) to be launched aboard the NPOESS preparatory project (NPP) and NPOESS satellite series could be considered a more advanced version of AMSU-A/B and AMSU/HSB because it has a set of channels very similar to that of AMSU-A/B and offers finer resolution and sampling for most channels [65,66]. Because of the similarity of channel sets, the algorithm of Chen and Staelin can be a starting point for the development of an algorithm for ATMS. The finer resolution and sampling of ATMS will likely lead to better image-sharpening methods and temperature and water vapor profile characterization.

Acknowledgments

The author wishes to thank P.W Rosenkranz, E. Williams, and M.M. Wolfson for helpful discussions, and S.P.L. Maloney and C. Lebell for assistance with the NEXRAD data. This work was supported by the National Oceanographic and Atmospheric Administration under Air Force Contract FA8721-05-C-0002. Opinions, interpretations, conclusions, and recommendations are those of the author and not necessarily endorsed by the United States Government.

References

1. F.T. Ulaby, R.K. Moore, and A.K. Fung, *Microwave Remote Sensing: Active and Passive*, Addison-Wesley, Reading, MA, 1981.
2. P.W. Rosenkranz, Retrieval of temperature of moisture profiles from AMSU-A and AMSU-B measurements, *IEEE Transactions on Geoscience and Remote Sensing*, 39(11), 2429–2435, 2001.
3. P.W. Rosenkranz, Rapid radiative transfer model for AMSU/HSB channels, *IEEE Transactions on Geoscience and Remote Sensing*, 41(2), 362–368, 2003.
4. L. Shi, Retrieval of atmospheric temperature profiles from AMSU-A measurement using a neural network approach, *Journal of Atmospheric and Oceanic Technology*, 18, 340–347, 2001.
5. W.J. Blackwell, J.W. Barrett, F.W. Chen, R.V. Leslie, P.W. Rosenkranz, M.J. Schwartz, and D.H. Staelin, NPOESS Aircraft Sounder Testbed-Microwave (NAST-M), instrument description and initial flight results, *IEEE Transactions on Geoscience and Remote Sensing*, 39(11), 2444–2453, 2001.
6. R.V. Leslie, W.J. Blackwell, P.W. Rosenkranz, and D.H. Staelin, 183-GHz and 425-GHz passive microwave sounders on the NPOESS aircraft sounder testbed-microwave (NAST-M), *Proceedings of the 2003 IEEE International Geoscience and Remote Sensing Symposium*, 1, 506–508, 2003.
7. R.V. Leslie, J.A. Loparo, P.W. Rosenkranz, and D.H. Staelin, Cloud and precipitation observations with the NPOESS Aircraft Sounder Testbed-Microwave (NAST-M) spectrometer suite at 54/118/183/425 GHz, *Proceedings of the 2003 IEEE International Geoscience and Remote Sensing Symposium*, 2, 1212–1214, 2003.
8. R.V. Leslie and D.H. Staelin, NPOESS Aircraft Sounder Testbed-Microwave (NAST-M) observations of clouds and precipitation at 54, 118, 183, and 425 GHz, *IEEE Transactions on Geoscience and Remote Sensing*, 42(10), 2240–2247, 2004.
9. J. Houghton, *The Physics of the Atmospheres*, Cambridge University Press, New York, 2002.
10. Tropical Rainfall Measuring Mission, http://trmm.gsfc.nasa.gov/
11. R.A. Houze, *Cloud Dynamics*, Academic Press, New York, 1993.
12. G. Hufford, A model for the complex permittivity of ice at frequencies below 1 THz, *International Journal of Infrared and Millimeter Waves*, 12(7), 677–682, 1991.
13. D. Deirmendjian, *Electromagnetic Scattering on Spherical Polydispersions*, Elsevier, New York, 1969.
14. M.S. Spina, M.J. Schwartz, D.H. Staelin, and A.J. Gasiewski, Application of multilayer feedforward neural networks to precipitation cell-top altitude estimation, *IEEE Transactions on Geoscience and Remote Sensing*, 36(1), 154–162, 1998.
15. A.J. Gasiewski, *Atmospheric Temperature Sounding and Precipitation Cell Parameter Estimation Using Passive 118-GHz O_2 Observations*, Ph.D. thesis, MIT Department of Electrical Engineering and Computer Science, December 1988.
16. A.J. Gasiewski, Microwave radiative transfer in hydrometeors, *Atmospheric Remote Sensing by Microwave Radiometry*, Ed. M.A. Janssen, John Wiley & Sons, New York, 1993.
17. N. Grody, F. Weng, and R. Ferraro, Application of AMSU for obtaining hydrological parameters, *Microwave Radiometry and Remote Sensing of the Earth's Surface and Atmosphere*, Eds. P. Pampaloni and S. Paloscia, pp. 339–351, 2000.

18. F. Weng, L. Zhao, R.R. Ferraro, G. Poe, X. Li, and N.C. Grody, Advanced microwave sounding unit cloud and precipitation algorithms, *Radio Science*, 38(4), 8068, 2003; doi: 10.1029/2002RS002679.
19. J.P. Hollinger, SSM/I instrument evaluation, *IEEE Transactions on Geoscience and Remote Sensing*, 28(5), 781–790, 1990.
20. C. Kummerow and L. Giglio, A passive microwave technique for estimating rainfall and vertical structure information from space. Part I: Algorithm description, *Journal of Applied Meteorology*, 33, 3–18, 1994.
21. D. Tsintikidis, J.L. Haferman, E.N. Anagnostou, W.F. Krajewski, and T.F. Smith, A neural network approach to estimating rainfall from spaceborne microwave data, *IEEE Transactions on Geoscience and Remote Sensing*, 35(5), 1079–1093, 1997.
22. T. Kawanishi, T. Sezai, Y. Ito, K. Imaoka, T. Takeshima, Y. Ishido, A. Shibata, M. Miura, H. Inahata, and R.W. Spencer, The Advanced Microwave Scanning Radiometer for the Earth observing system (AMSR-E), NASDA's contribution to the EOS for global energy and water cycle studies, *IEEE Transactions on Geoscience and Remote Sensing*, 41(2), 184–194, 2003.
23. T. Wilheit, C.D. Kummerow, and R. Ferraro, Rainfall algorithms for AMSR-E, *IEEE Transactions on Geoscience and Remote Sensing*, 41(2), 204–213, 2003.
24. J. Ikai and K. Nakamura, Comparison of rain rates over the ocean derived from TRMM microwave imager and precipitation radar, *Journal of Atmospheric and Oceanic Technology*, 20, 1709–1726, 2003.
25. C. Kummerow, W. Barnes, T. Kozu, J. Shiue, and J. Simpson, The tropical rainfall measuring mission (TRMM) sensor package, *Journal of Atmospheric and Oceanic Technology*, 15, 809–817, 1998.
26. C. Kummerow and L. Giglio, A passive microwave technique for estimating rainfall and vertical structure information from space. Part I: Algorithm description, *Journal of Applied Meteorology*, 33, 3–18, 1994.
27. T. Wilheit, C.D. Kummerow, and R. Ferraro, Rainfall algorithms for AMSR-E, *IEEE Transactions on Geoscience and Remote Sensing*, 41(2), 204–213, 2003.
28. T.J. Hewison and R. Saunders, Measurements of the AMSU-B antenna pattern, *IEEE Transactions on Geoscience and Remote Sensing*, 34(2), 405–412, 1996.
29. T. Mo, AMSU-A antenna pattern corrections, *IEEE Transactions on Geoscience and Remote Sensing*, 37(1), 103–112, 1999.
30. P.W. Rosenkranz, Improved rapid transmittance algorithm for microwave sounding channels, *Proceedings of the 1998 IEEE International Geoscience and Remote Sensing Symposium*, 2, 728–730, 1998.
31. B.H. Lambrigtsen, Calibration of the AIRS microwave instruments, *IEEE Transactions on Geoscience and Remote Sensing*, 41(2), 369–378, 2003.
32. H.H. Aumann, M.T. Chahine, C. Gautier, M.D. Goldberg, E. Kalnay, L.M. McMillin, H. Revercomb, P.W. Rosenkranz, W.L. Smith, D.H. Staelin, L.L. Strow, and J. Susskind, AIRS/AMSU/HSB on the aqua mission: design, science objectives, data products, and processing systems, *IEEE Transactions on Geoscience and Remote Sensing*, 41(2), 253–264, 2003.
33. C.L. Parkinson, Aqua: an Earth-observing satellite mission to examine water and other climate variables, *IEEE Transactions on Geoscience and Remote Sensing*, 41(2), 173–183, 2003.
34. W.J. Blackwell, *Retrieval of Cloud-Cleared Atmospheric Temperature Profiles from Hyperspectral Infrared and Microwave Observations*, Sc.D. thesis, Massachusetts Institute of Technology, Department of Electrical Engineering and Computer Science, June 2002.
35. W.J. Blackwell, Retrieval of atmospheric temperature and moisture profiles from hyperspectral sounding data using a projected principal components transform and a neural network, *Proceedings of the 2003 IEEE International Geoscience and Remote Sensing Symposium*, 3, 2078–2081, 2003.
36. C.R. Cabrera-Mercader, *Robust Compression of Multispectral Remote Sensing Data*, Ph.D. thesis, Massachusetts Institute of Technology, Department of Electrical Engineering and Computer Science, Boston, MA, 1999.
37. F.W. Chen, *Characterization of Clouds in Atmospheric Temperature Profile Retrievals*, M.E. thesis, Massachusetts Institute of Technology, Department of Electrical Engineering and Computer Science, Boston, MA, June 1998.

38. J. Lee, *Blind Noise Estimation and Compensation for Improved Characterization of Multivariate Processes*, Ph.D. thesis, Department of Electrical Engineering and Computer Science, Massachusetts Institute of Technology, March 2000.
39. J. Lee and D.H. Staelin, Iterative signal-order and noise estimation for multivariate data, *IEE Electronics Letters*, 37(2), 134–135, 2001.
40. A. Mueller, *Iterative Blind Seperation of Gaussian Data of Unknown Order*, M.E. thesis, Massachusetts Institute of Technology, Department of Electrical Engineering and Computer, June 2003.
41. I.T. Joliffe, *Principal Component Analysis*, Springer-Verlag, New York, 2002.
42. L. Wald, Some terms of reference in data fusion, *IEEE Transactions on Geoscience and Remote Sensing*, 37(3), 1190–1193, 1999.
43. D.L. Hall and J. Llinas, An introduction to multisensor data fusion, *Proceedings of the IEEE*, 85(1), 6–23, 1997.
44. C. Pohl and J.L. van Genderen, Multisensor image fusion in remote sensing: concepts, methods, and application, *International Journal of Remote Sensing*, 19(5), 823–854, 1998.
45. V.J.D. Tsai, Frequency-based fusion of multiresolution images, *Proceedings of the 2003 IEEE International Geoscience and Remote Sensing Symposium*, 6, 3665–3667, 2003.
46. P.W. Rosenkranz, Inversion of data from diffraction-limited multiwavelength remote sensors, 1. Linear case, *Radio Science*, 13(6), 1003–1010, 1978.
47. P.W. Rosenkranz, Inversion of data from diffraction-limited multiwavelength remote sensors, 2. Nonlinear dependence of observables on the geophysical parameters, *Radio Science*, 17(1), 245–256, 1982.
48. P.W. Rosenkranz, Inversion of data from diffraction-limited multiwavelength remote sensors, 3. Scanning multichannel microwave radiometer data, *Radio Science*, 17(1), 257–267, 1982.
49. R.P. Lippmann, An introduction to computing with neural nets, *IEEE Acoustics, Speech, and Signal Processing Magazine*, 4(2), 4–22, 1987.
50. S. Haykin, *Neural Networks: A Comprehensive Foundation*, Macmillan, New York, 1994.
51. M.T. Hagan and M.B. Menhaj, Training feedforward networks with the Marquardt algorithm, *IEEE Transactions on Neural Networks*, 5(6), 989–993, 1994.
52. D.W. Marquardt, An algorithm for least-squares estimation of nonlinear parameters, *Journal of the Society for Industrial and Applied Mathematics*, 11(2), 431–441, 1963.
53. D. Nguyen and B. Widrow, Improving the learning speed of 2-layer neural networks by choosing initial values of the adaptive weights, *Proceedings of the International Joint Conference on Neural Networks*, 3, 21–26, 1990.
54. *Neural Network Toolbox User's Guide*, The MathWorks, Inc., Natick, Massachusetts, 1998.
55. W.J. Blackwell et al., NPOESS aircraft sounder testbed-microwave (NAST-M): instrument description and initial flight results, *IEEE Transactions on Geoscience and Remote Sensing*, 39, 2444–2453, 2001.
56. A.J. Gasiewski, D.M. Jackson, J.R. Wang, P.E. Racette, and D.S. Zacharias, Airborne imaging of tropospheric emission at millimeter and submillimeter wavelengths, in *Proceedings IGARSS*, 2, 663–665, 1994.
57. D.H. Staelin and F.W. Chen, Precipitation observations near 54 and 183 GHz using the NOAA-15 satellite, *IEEE Transactions on Geoscience and Remote Sensing*, 38, 2322–2332, 2000.
58. P.A. Arkin and P. Xie, The global precipitation climatology project: first algorithm intercomparison project, *Bulletin of the American Meteorological Society*, 75(3), 401–420, 1994.
59. E.A. Smith et al., Results of WetNet PIP-2 project, *Journal of the Atmospheric Sciences*, 55(9), 1483–1536, 1998.
60. R.F. Adler et al., Intercomparison of global precipitation products: the third precipitation intercomparison project (PIP-3), *Bulletin of the American Meteorological Society*, 82(7), 1377–1396, 2001.
61. M.D. Conner and G.W. Petty, Validation and intercomparison of SSM/I rain-rate retrieval methods over the continental U.S., *Journal of Applied Meteorology*, 37(7), 679–700, 1998.
62. A. Hyvärinen, J. Karhunen, and E. Oja, *Independent Component Analysis*, John Wiley & Sons, New York, 2001.
63. F.R. Bach and M.I. Jordan, Kernel independent component analysis, *Journal of Machine Learning Research*, 3, 1–48, 2002.

64. J.C. Shiue, The advanced technology microwave sounder, a new atmospheric temperature and humidity sounder for operational polar-orbiting weather satellites, *Proceedings of the SPIE*, 4540, 159–165, 2001.
65. C. Muth, P.S. Lee, J.C. Shiue, and W.A. Webb, Advanced technology microwave sounder on NPOESS and NPP, *Proceedings of the 2004 IEEE International Geoscience Remote Sensing Symposium*, 4, 2454–2458, 2004.
66. F.W. Chen and D.H. Staelin, AIRS/AMSU/HSB precipitation estimates, *IEEE Transactions on Geoscience and Remote Sensing*, 41(2), 410–417, 2003.
67. F.W. Chen, Global Estimation of Precipitation using opaque Microwave Bands, PhD thesis, Massachusetts Institute of Technology, Massachusetts, 2004.

Index

A
Active sonar data, 189–187, 193, 197, 199–200
Adaptive filtering, 130–131, 141
Adaptive noise cancellation, 130–131, 149
Adaptive signal detection, 200 (In ref only)
Advanced Microwave Sounding Unit (AMSU), 206, 232
Affine transformation, 174
Akaike information criterion (AIC), 133
Artifact elimination, 175–176, 183
Atmospheric temperature profile, 234, 241, 246–247
Autoregressive (AR) model, 132, 134–136, 144, 161
Autoregressive integrated moving average (ARIMA) model, 161, 167
Autoregressive moving average (ARMA), 132, 143

B
Backward matrix, 193–194
Bandpass filtering, 38–39
Bedrosian theorem, 2–3, 9
Best linear unbiased estimate (BLUE), 156
Blind separation of convulutive mixture (BSCM) algorithm, 190–199
Blind signal (source) separation (BSS), 108, 188
 frequency domain, 28, 192
Bright temperature, 211–212, 221–222, 233–235, 239–240, 248–252, 258

C
Chen–Staelin algorithm, 245–253, 258
Classifiers
 infrasound event, 31–33
Climate data, 154–155, 168
Color noise, 134–137, 141, 144–146
Combining classifiers, 24–25
Common midpoint gather, 75
Component images, 10
 longer scales, 13–16
 midrange scales, 16
 shortest scales, 13–15
Confusion matrix (error matrix), 42, 44, 48, 50
Contextual information, 24
Convolutive mixtures, 187–200
Correct classification rate (CCR), 42
Cross correlation function (CCF), 158
Cross-validation, 172, 181
 sum of squares (CVSS), 179, 181
Cubic spline fit, 4

D
Data fusion, 25, 243, 258
Data smoothing, 161–163
Deconvolution, 57–58, 71, 103
Digital elevation model (DEM), 171
Digital seismic processing, 56–57
Dimensionality reduction, 25
Doppler effect, 187–189, 194–196

E
Empirical mode decomposition (EMD), 2
 normalized, (NEMD), 2–4, 7, 9–10, 18–19
Event migration, 72
Expectation-maximization (EM) algorithm, 109

F
Factor analysis, 101–120
 common factors in, 122–123
 maximum likelihood, 123–125
Feature extraction and selection, 24–25, 28, 39, 42

G
Gabor's approach, 8–9
Geophones, 54
Geophysical parameter retrieval, 204
 using neural networks, 207
Ground penetration radar (GPR), 128

H
High-order singular value decomposition (HOSVD), 74–198
Hilbert spectral analysis (HSA), 4
Hilbert transform, 1–20
 normalized (NHT), 1–20
Hughes phenomenon, 25
Huygen's construction, 56, 58–59, 61, 71
Hydrophones, 54
Hyperbolic frequency modulated (HFM) target echo, 195–198
Hyperspectral sounding data, 208–215

I
Image registration, 174–175
Image sharpening, 231–258

Imaging by seismic processing, 58–59
Independent component analysis (ICA), 23, 74–77, 108
 fast (FastICA), 78
 kernel, 261 (in ref only)
 partial (unimodal), 74
Independent factor analysis (IFA), 109
Indirect signal estimation, 138, 144
Infrared (IR) images, 11–16
Infrared cloud perturbation, 212–214
Infrared wavelengths, 204
Infrasound signals, 40–43
Instantaneous frequency (IF), 1, 7–10
Interference mitigation, 130–134
International Monitoring System (IMS), 31
Interpolation, 34, 178–179, 181–185
 Laplacian, 250
Intrinsic mode function (IMF), 2
Isosurface visualization, 17

J
Joint cumulants, 29 (in ref only)

K
Kalman filtering, 127–149
 conventional, 134–135, 138, 143–146
Keetch–Byram drought index (KBDI), 152–168
K-mean algorithm, 24

L
Laplacian filtering, 240–241
Least square estimation, 86, 107, 153
 linear, 207
 nonlinear, 207, 245
Levenberg–Marquardt method, 217

M
Marmousi data set, 109–110
Matched filter, output, 188–189, 198
Maximal diagonality (MD) method, 78
Microbaroms, 31, 38
Microwave radiometric data, 258
Microwave wavelengths, 204
Migration, 56, 61–62, 104
 prestack, 59
Mine (landmine) detection, 128, 140–144
MIT Geophysical Analysis Group, 57
Mixture of Gaussian (MoG), 108
Moisture profile, 203, 216
Multicomponent data, 2, 85
Multiple reflections, 56
Multi-segment Burg's algorithm, 144
Multi-way (3-mode) array data set, 83–85
Multi-way array processing, 85–98

Muting procedure, 111
Mutual information, 108

N
Nearest neighbor decision rule (NNDR), 24
Neural networks, 24–25, 32–51, 231–258
 multilayer feedforward (MLP), 204, 216–218
 parallel-bank (PBNNC), 33, 44
 power spectral density (PSD), 39
 radial basis function, 26, 33–38
No reflection (NR) area, 54–55
Nonlinear distorted waves, 1
Nonstationary process, 1, 190–191
Normal moveout (NMO), 103–104, 110–114
Normalized difference infrared index (NDII), 152
Normalized difference vegetation index (NDVI), 152
Normalized difference water index (NDWI), 152–153, 155–160, 162–168
Nuttall theorem, 3, 10–11

O
Ocean waves, 32
Opaque microwave channels, 235
Optimal feature number, 44–45, 47

P
Performance evaluation, 145–148, 253–258
Phase function, 4, 8
Pixel classification, 26
Precipitation-rate retrievals, 252, 254, 256–257
Prediction-error (PE) filter, 171–186
 matrix representation of, 176, 179
Principal component analysis (PCA), 28, 107–108, 118–120, 193, 241–242, 252, 260
 basic, 242
 constrained, 242, 254
 noise adjusted, 209, 213–214, 241
Principal component inverse (PCI), 187–188, 193
Projected principal component (PPC) transform, 204, 212, 214–215, 217–218, 225–226

Q
Quadratic polynomial, 161
Quadrupole resonance (QR), 128, 140–148

R
Radio frequency interference (RFI), 129, 141–142
Ray equations, 68–69
Ray tracing, 53–71
Receiver operating characteristics (ROC), 24, 44–45
 3-D, 33, 48

Index

Reflector and reflections, 55, 59, 71, 104–105
Regression analysis, 165–166
Retrieval algorithms, 204, 238, 258
Reverberations, 56–57
   cancellation of, 187–200

S

Satellite-based precipitation retrieval, 231–258
Scan-angle-dependent effects, 248
Seasonal metrics, 163–168
Seismic data decomposition, 74–98
   multi-dimensional, 74–98
Seismic images, 53–71
Seismic interpretation, 28, 59–60
Seismic migrations, 58–60
Seismic profiling, 101–120
Seismic signal processing, 102–105, 109–120
Seismic trace, 53, 55, 57, 62, 102
Self-organizing map, 24, 26
Serial correlation, 152–153, 156–161, 166–167
Shuttle radar topography mission (SRTM), 172–176, 179–182, 184–186
Signal enhancement, 57–58, 71–72
Signal model, 129
   harmonic, 129
   interference, 129–130, 132–133
   postmitigation, 131–133
Signal-generated noise, 57, 59, 71–72
Signal-to-reverberation ratio (SRR), 188
Simulated interferogram, 183, 185
Singular value decomposition (SVD), 74, 241
   higher order (HOSVD), 85–98
Slowness (or reciprocal velocity), 60, 66–71
Space borne atmospheric remote sensing, 205–208
Space-borne radiometer, 232–233
Stacking, 103–104, 110–120
Statistical pattern recognition, 23–28
Statistical signal processing, 23, 26–28, 188
Subspace method, 76–77
   matrix-based, 74

SVD, 79–83, 89–91, 93, 95
   three-dimensional, 97
   three-mode, 88
Subsurface structure, 55–56
Support vector machine, 24, 26
Surface wave processes, 12
SVD-ICA, 79
Synthetic aperture radar (SAR), 25
   inferometric (InSAR), 172, 183–185
   polarimetric (POLSAR), 26

T

"tanh" nonlinearity, 78
Time series analysis, 156–161, 166–167
Topographic synthetic aperture radar (TOPSAR), 172
Transform methods, 23, 25, 28
Traveltime (or eikonal), 61–64, 66–68, 103

U

Unmixing matrix, 193

V

Vegetation moisture dynamics, 151–168
Vegetation water content (VWC), 151
Velocity analysis, 104, 110–111
Vertical seismic profile (VSP), 60, 95
Vibrators, 54

W

Water surface waves, 16, 38
Wave number, 14
Wave velocity, 60–61, 67
Wavelet projection decomposition, 10
Weak signal detection, 129–149
Weak signal estimation, 128, 133–134
Weibull distribution, 187
Wigner–Ville distribution, 2

Z

Zenith opacity, 206, 235